SMALL UNMANNED AIRCRAFT

SMALL UNMANNED AIRCRAFT

Theory and Practice

RANDAL W. BEARD
TIMOTHY W. McLAIN

PRINCETON UNIVERSITY PRESS
Princeton and Oxford

Copyright © 2012 by Princeton University Press
Published by Princeton University Press, 41 William Street,
Princeton, New Jersey 08540
In the United Kingdom: Princeton University Press,
6 Oxford Street, Woodstock, Oxfordshire OX20 1TW
press.prenceton.edu

Jacket art: *Tempest and Farmhouse*, 06/02/2010.
University of Colorado Boulder. © Jack Elston.

All Rights Reserved

ISBN 978-0-691-14921-9

Library of Congress Cataloging-in-Publication Data

Beard, Randal W.
Small unmanned aircraft: theory and practice / Randal W. Beard, Timothy W. McLain.
   p. cm.
Includes bibliographical references and index.
ISBN 978-0-691-14921-9 (hardcover: alk. paper)   1. Drone aircraft—Control systems.   2. Drone aircraft—Automatic control.   3. Guidance systems (Flight)   I. McLain, Timothy W., 1963-II. Title.
UG1242.D7B43 2012
623.74'69–dc23
    2011025926

British Library Cataloging-in-Publication Data is available

This book has been composed in ITC Stone

Printed on acid-free paper. ∞

Typeset by S R Nova Pvt Ltd, Bangalore, India
Printed in the United States of America

10 9 8 7 6 5 4 3 2

To our families

Andrea, Laurann, Kaitlyn, Aubrey, Kelsey

Amber, Regan, Caitlin, Tab, Colby, Rorie

Contents

Preface — xi

1 Introduction — 1
1.1 System Architecture — 1
1.2 Design Models — 4
1.3 Design Project — 6

2 Coordinate Frames — 8
2.1 Rotation Matrices — 9
2.2 MAV Coordinate Frames — 12
2.3 Airspeed, Wind Speed, and Ground Speed — 18
2.4 The Wind Triangle — 20
2.5 Differentiation of a Vector — 24
2.6 Chapter Summary — 25
2.7 Design Project — 27

3 Kinematics and Dynamics — 28
3.1 State Variables — 28
3.2 Kinematics — 30
3.3 Rigid-body Dynamics — 31
3.4 Chapter Summary — 37
3.5 Design Project — 38

4 Forces and Moments — 39
4.1 Gravitational Forces — 39
4.2 Aerodynamic Forces and Moments — 40
4.3 Propulsion Forces and Moments — 52
4.4 Atmospheric Disturbances — 54
4.5 Chapter Summary — 57
4.6 Design Project — 58

5 Linear Design Models — 60
5.1 Summary of Nonlinear Equations of Motion — 60
5.2 Coordinated Turn — 64
5.3 Trim Conditions — 65
5.4 Transfer Function Models — 68

	5.5 Linear State-space Models	77
	5.6 Reduced-order Modes	87
	5.7 Chapter Summary	91
	5.8 Design Project	92
6	**Autopilot Design Using Successive Loop Closure**	**95**
	6.1 Successive Loop Closure	95
	6.2 Saturation Constraints and Performance	97
	6.3 Lateral-directional Autopilot	99
	6.4 Longitudinal Autopilot	105
	6.5 Digital Implementation of PID Loops	114
	6.6 Chapter Summary	117
	6.7 Design Project	118
7	**Sensors for MAVs**	**120**
	7.1 Accelerometers	120
	7.2 Rate Gyros	124
	7.3 Pressure Sensors	126
	7.4 Digital Compasses	131
	7.5 Global Positioning System	134
	7.6 Chapter Summary	141
	7.7 Design Project	141
8	**State Estimation**	**143**
	8.1 Benchmark Maneuver	143
	8.2 Low-pass Filters	144
	8.3 State Estimation by Inverting the Sensor Model	145
	8.4 Dynamic-observer Theory	149
	8.5 Derivation of the Continuous-discrete Kalman Filter	151
	8.6 Attitude Estimation	156
	8.7 GPS Smoothing	158
	8.8 Chapter Summary	161
	8.9 Design Project	162
9	**Design Models for Guidance**	**164**
	9.1 Autopilot Model	164
	9.2 Kinematic Model of Controlled Flight	165
	9.3 Kinematic Guidance Models	168
	9.4 Dynamic Guidance Model	170
	9.5 Chapter Summary	172
	9.6 Design Project	173

10 Straight-line and Orbit Following — 174
10.1 Straight-line Path Following — 175
10.2 Orbit Following — 181
10.3 Chapter Summary — 183
10.4 Design Project — 185

11 Path Manager — 187
11.1 Transitions Between Waypoints — 187
11.2 Dubins Paths — 194
11.3 Chapter Summary — 202
11.4 Design Project — 204

12 Path Planning — 206
12.1 Point-to-Point Algorithms — 207
12.2 Coverage Algorithms — 220
12.3 Chapter Summary — 223
12.4 Design Project — 224

13 Vision-guided Navigation — 226
13.1 Gimbal and Camera Frames and Projective Geometry — 226
13.2 Gimbal Pointing — 229
13.3 Geolocation — 231
13.4 Estimating Target Motion in the Image Plane — 234
13.5 Time to Collision — 238
13.6 Precision Landing — 240
13.7 Chapter Summary — 244
13.8 Design Project — 245

APPENDIX A: Nomenclature and Notation — 247

APPENDIX B: Quaternions — 254
B.1 Quaternion Rotations — 254
B.2 Aircraft Kinematic and Dynamic Equations — 255
B.3 Conversion Between Euler Angles and Quaternions — 259

APPENDIX C: Animations in Simulink — 260
C.1 Handle Graphics in Matlab — 260
C.2 Animation Example: Inverted Pendulum — 261

C.3 Animation Example: Spacecraft Using Lines 263
C.4 Animation Example: Spacecraft Using Vertices and Faces 268

APPENDIX D: Modeling in Simulink Using S-Functions 270
D.1 Example: Second-order Differential Equation 270

APPENDIX E: Airframe Parameters 275
E.1 Zagi Flying Wing 275
E.2 Aerosonde UAV 276

APPENDIX F: Trim and Linearization in Simulink 277
F.1 Using the Simulink `trim` Command 277
F.2 Numerical Computation of Trim 278
F.3 Using the Simulink `linmod` Command to Generate a State-space Model 282
F.4 Numerical Computation of State-space Model 284

APPENDIX G: Essentials from Probability Theory 286

APPENDIX H: Sensor Parameters 288
H.1 Rate Gyros 288
H.2 Accelerometers 288
H.3 Pressure Sensors 289
H.4 Digital Compass/Magnetometer 289
H.5 GPS 290

Bibliography 291
Index 299

Preface

Unmanned aircraft systems (UAS) are playing increasingly prominent roles in defense programs and defense strategy around the world. Technology advancements have enabled the development of both large unmanned aircraft (e.g., Global Hawk, Predator) and smaller, increasingly capable unmanned aircraft (e.g., Wasp, Nighthawk). As recent conflicts have demonstrated, there are numerous military applications for unmanned aircraft, including reconnaissance, surveillance, battle damage assessment, and communications relays.

Civil and commercial applications are not as well developed, although potential applications are extremely broad in scope, including environmental monitoring (e.g., pollution, weather, and scientific applications), forest fire monitoring, homeland security, border patrol, drug interdiction, aerial surveillance and mapping, traffic monitoring, precision agriculture, disaster relief, ad hoc communications networks, and rural search and rescue. For many of these applications to develop to maturity, the reliability of UAS needs to increase, their capabilities need to be extended further, their ease of use needs to be improved, and their cost must decrease. In addition to these technical and economic challenges, the regulatory challenge of integrating unmanned aircraft into national and international air space needs to be overcome.

The terminology *unmanned aircraft system* refers not only to the aircraft, but also to all of the supporting equipment used in the system, including sensors, microcontrollers, software, groundstation computers, user interfaces, and communications hardware. This text focuses on the aircraft and its guidance, navigation, and control subsystems. Unmanned aircraft (UA) can generally be divided into two categories: fixed-wing aircraft and rotorcraft. Both types of aircraft have distinctive characteristics that make autonomous behavior difficult to design. In this book we focus exclusively on fixed-wing aircraft, which can be roughly categorized by size. We use the term *small unmanned aircraft* to refer to the class of fixed-wing aircraft with a wing span between 5 and 10 feet. Small unmanned aircraft are usually gas powered and typically require a runway for take off and landing, although the Boeing ScanEagle, which uses a catapult for take off and a skyhook for recovery, is a notable exception. Small unmanned aircraft are typically designed to operate on the order of 10 to 12 hours, with payloads of approximately 10 to 50 pounds.

The term *miniature air vehicle*, which we will denote with the acronym MAV[1], will be used to refer to the class of fixed-wing aircraft with wingspans less than 5 feet. MAVs are typically battery powered, hand launched, and belly landed, and therefore do not require a runway for take off or landing. They are designed to operate from 20 minutes to several hours. Payloads range from ounces to several pounds. The small payload severely restricts the sensor suite that can be placed on MAVs, and also restricts the computer that can be put on board. These restrictions pose interesting challenges for designing autonomous modes of operation. While many of the concepts described in this book also apply to larger unmanned aircraft and smaller micro air vehicles, the primary focus of the book is on the challenges that are inherent in guiding and controlling limited-payload small and miniature aircraft systems.

This textbook was inspired by our desire to teach a course to our graduate students that prepared them to do work in the area of cooperative control for unmanned aircraft. Most of our students come from a background in electrical engineering, computer engineering, mechanical engineering, or computer science. Only a few of them have had courses in aerodynamics, and the electrical and computer engineering and computer science students generally have not had courses in kinematics, dynamics, or fluid mechanics. Most of our students, however, have had courses in signals and systems, feedback control, robotics, and computer vision.

There are a large number of textbooks that cover aircraft dynamics and control; however, most of them assume a background in aeronautics and assume that the student has not had exposure to feedback control. Therefore, textbooks like [1, 2, 3, 4, 5, 6] discuss aerodynamic forces without discussing basic ideas in fluid mechanics and aeronautics. On the other hand, they typically include a detailed introduction to feedback control concepts like root locus. The textbook [7] is much more in line with what we wanted to teach our students, but the focus of that text is on stability augmentation systems as opposed to autonomous operations. Autonomous operations require more than a simple autopilot; they require autonomous take off and landing, path planning, and path-following operations, integrated with higher-level decision-making processes. To our knowledge, another textbook is not currently available that covers aircraft dynamic models, low-level autopilot

---

[1] We acknowledge that MAV is more commonly used as an abbreviation for micro air vehicles, which have wingspans of 10 inches or smaller. Given that MAV is a single-syllable word that is easily pronounced in singular and plural forms, we could not resist adopting it to represent miniature air vehicles.

design, state estimation, and high-level path planning. Our hope is that this book fills that gap. Our target audience is electrical engineering, computer engineering, mechanical engineering, and computer science students who have had a course in introductory feedback control or robotics. We hope that the textbook is also interesting to aeronautics engineers seeking an introduction to autonomous systems.

In writing this book, our objective was to have 13 chapters, each of which could be covered in three one-hour lectures and thereby fit comfortably into a one-semester course. While some of the chapters are longer than we had originally hoped, our experience is that the material in this book can be covered in one semester. The additional material will allow the instructor some flexibility in the topics that are covered.

One of the unique features of the book is the associated design project. As we have taught this course, we have evolved from assigning pencil-and-paper homework assignments, to assigning computer simulation assignments. We have found that students are more engaged in the material and tend to understand it better when they implement the concepts in a computer simulation.

When we teach the course, we have our students develop a complete end-to-end flight simulator that includes realistic flight dynamics, sensor models, autopilot design, and path planning. By the end of the course, they have implemented each piece of the puzzle and therefore understand how each piece fits together. In addition, they understand the inner workings of a fairly sophisticated flight simulation package that can be used in their future research projects.

Our design exercises were originally developed for implementation in C/C++, requiring the students to be proficient programmers. This sometimes created an unnecessary burden on both the instructor and the students. Subsequently, we have altered the project exercises so that they are completed in Matlab/Simulink. We feel that this allows the students to focus more intently on the key concepts related to MAVs rather than on the details of programming. The appendices provide supplemental information that describes the main Matlab/Simulink tools that will be used in the development of the MAV simulation. The book also has an associated website that includes skeleton simulation files that will help the reader in the project assignments.

SMALL
UNMANNED
AIRCRAFT

1

Introduction

1.1 System Architecture

The objective of this book is to prepare the reader to do research in the exciting and rapidly developing field of autonomous navigation, guidance, and control of unmanned air vehicles. The focus is on the design of the software algorithms required for autonomous and semi-autonomous flight. To work in this area, researchers must be familiar with a wide range of topics, including coordinate transformations, aerodynamics, autopilot design, state estimation, path planning, and computer vision. The aim of this book is to cover these essential topics, focusing in particular on their application to small and miniature air vehicles, which we denote by the acronym MAV.

In the development of the topics, we have in mind the software architecture shown in figure 1.1. The block labeled *unmanned aircraft* in figure 1.1 is the six-degree-of-freedom (DOF) physical aircraft that responds to servo command inputs (elevator, aileron, rudder, and throttle) and wind and other disturbances. The mathematical models required to understand fixed-wing flight are complicated and are covered in chapters 2 to 5 and chapter 9. In particular, in chapter 2 we discuss coordinate frames and transformations between frames. A study of coordinate frames is required since most specifications for MAVs are given in the inertial frame (e.g., orbit a specific coordinate), whereas most of the sensor measurements are with respect to the body frame, and the actuators exert forces and torques in the body frame. In chapter 3 we develop the kinematic and dynamic equations of motion of a rigid body. In chapter 4 we describe the aerodynamic forces and moments that act on fixed-wing aircraft. Chapter 5 begins by combining the results of chapters 3 and 4 to obtain a six-DOF, 12-state, nonlinear dynamic model for a MAV. While incorporating the fidelity desired for simulation purposes, the six-DOF model is fairly complicated and cumbersome. The design and analysis of aircraft control approaches are more easily accomplished using lower-order linear models. Linear models that describe small deviations from trim are derived in chapter 5, including linear transfer function and state-space models.

The block labeled *autopilot* in figure 1.1 refers to the low-level control algorithms that maintain roll and pitch angles, airspeed, altitude, and course heading. Chapter 6 introduces the standard technique of

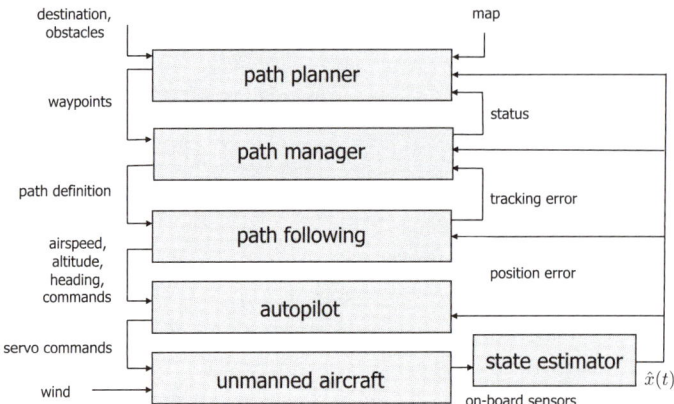

Figure 1.1 The system architecture that will be assumed throughout the book. The path planner produces straight-line or Dubins paths through an obstacle field. The path manager switches between orbit following and straight-line path following to maneuver along the waypoint paths. The path-following block produces commands to the low-level autopilot, which controls the airframe. Each of the blocks relies on estimates of the states produced by filtering the onboard sensors.

successive loop closure to design the autopilot control laws. Nested control loops are closed one at a time, with inner loops maintaining roll and pitch angles and outer loops maintaining airspeed, altitude, and course.

The autopilot and the higher level blocks rely on accurate state estimates obtained by dynamically filtering the onboard sensors, which include accelerometers, rate gyros, pressure sensors, magnetometers, and GPS receivers. A description of these sensors and their mathematical models is given in chapter 7. Because it is not possible to measure all the states of small unmanned aircraft using standard sensors, state estimation plays an important role. Descriptions of several state-estimation techniques that are effective for MAVs are given in chapter 8.

A complete model of the flight dynamics coupled with the autopilot and state estimation techniques represents a high dimensional, highly complex, nonlinear system of equations. The full model of the system is too complicated to facilitate the development of high level guidance algorithms. Therefore, chapter 9 develops low-order nonlinear equations that model the closed-loop behavior of the system. These models are used in subsequent chapters to develop guidance algorithms.

One of the primary challenges with MAVs is flight in windy conditions. Since airspeeds in the range of 20 to 40 mph are typical for MAVs, and since wind speeds at several hundred feet above ground level (AGL) almost always exceed 10 mph, MAVs must be able to maneuver effectively in wind. Traditional trajectory tracking methods used in robotics do not work well for MAVs. The primary difficulty with these methods is the requirement to be in a particular location at a particular time, which cannot properly take into account the variations in ground speed caused by the unknown and changing effects of the wind. Alternatively, path-following methods that simply maintain the vehicle on a desired path have proven to be effective in flight tests. Chapter 10 describes the algorithms and methods used to provide the capabilities of the *path following* block in figure 1.1. We will focus exclusively on straight-line paths and circular orbits and arcs. Other useful paths can be built up from these straight-line and circular path primitives.

The block labeled *path manager* in figure 1.1 is a finite-state machine that converts a sequence of waypoint configurations (positions and orientations) into sequences of straight-line paths and circular arcs that can be flown by the MAV. This makes it possible to simplify the path planning problem so that the *path planner* produces either a sequence of straight-line paths that maneuver the MAV through an obstacle field, or a Dubin's path that maneuvers through the obstacle field. Chapter 11 describes the *path manager*, while chapter 12 describes the *path planner*. For path planning we consider two classes of problems. The first class of problems is point-to-point algorithms, where the objective is to maneuver from a start position to an end position while avoiding a set of obstacles. The second class of problems is search algorithms, where the objective is to cover a region, potentially having no-go regions, with a sensor footprint.

Almost all applications involving MAVs require the use of an onboard electro-optical/infrared (EO/IR) video camera. The typical objective of the camera is to provide visual information to the end user. Since MAV payload capacities are limited, however, it makes sense to also use the video camera for navigation, guidance, and control. Effective use of camera information is currently an active research topic. In chapter 13 we discuss several potential uses of video cameras on MAVs, including geolocation and vision-based landing. Geolocation uses a sequence of images as well as the onboard sensors to estimate the world coordinates of objects on the ground. Vision-based landing uses video images captured by the MAV to guide it to a target identified in the image plane. We feel that an understanding of these problems will enable further investigations in vision-based guidance of MAVs.

In chapter 13, we use the software architecture shown in figure 1.2, where the *path planner* block has been replaced with the block *vision-based guidance*. However, the vision-based guidance laws interact with

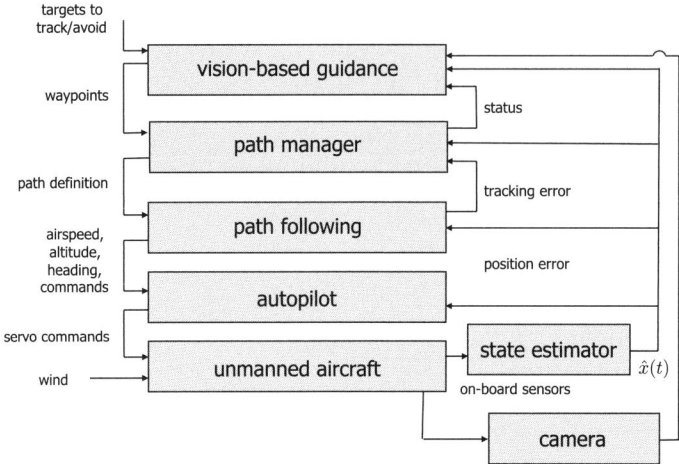

Figure 1.2 System architecture for vision-based navigation, guidance, and control. A video camera is added as an additional sensor and the path planner has been replaced with a vision-based guidance block.

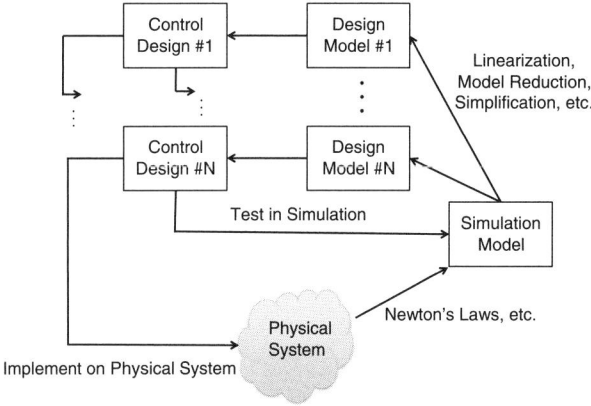

Figure 1.3 The design process. Using principles of physics, the physical system is modeled mathematically, resulting in the simulation model. The simulation model is simplified to create design models that are used for the control design. The control design is then tested and debugged in simulation and finally implemented on the physical system.

the architecture in the same manner as the path planner. The modularity of the architecture is one of its most appealing features.

1.2 Design Models

The design philosophy that we follow throughout the book is illustrated schematically in figure 1.3. The unmanned aircraft operating in its environment is depicted in figure 1.3 as the "Physical System," and includes the actuators (control flaps and propeller) and the sensors (IMU, GPS, camera, etc.). The first step in the design process is to model the physical system using nonlinear differential equations. While approximations and simplifications will be necessary at this step, the hope is to capture in mathematics all of the important characteristics of the physical system. In this book, the model of the physical system includes rigid body kinematics and dynamics (chapter 3), aerodynamic forces and moments (chapter 4), and the onboard sensors (chapter 7). The resulting model is called the "Simulation Model" in figure 1.3 and will be used for the high fidelity computer simulation of the physical system. However, we should note that the simulation model is only an approximation of the physical system, and simply because a design is effective on the simulation model, we should not assume that it will function properly on the physical system.

The simulation model is typically nonlinear and high order and is too mathematically complex to be useful for control design. Therefore, to facilitate design, the simulation model is simplified and usually linearized to create lower-order design models. For any physical system, there may be multiple design models that capture certain aspects of the design process. For MAVs, we will use a variety of different design models for both low-level control and also for high-level guidance. In chapter 5, we will decompose the aircraft motion into longitudinal (pitching and climbing) motion and lateral (rolling and heading) motion, and we will have different design models for each type of motion. The linear design models developed in chapter 5 will be used in chapter 6 to develop low-level autopilot loops that control the airspeed, altitude, and course angle of the vehicle. In chapter 8, we show how to estimate the states needed for the autopilot loops using sensors typically found on small and micro air vehicles.

The mathematical equations describing the physics of the system, the low-level autopilot, and the state estimation routines, when considered as a whole, are very complex and are not useful for designing the higher level guidance routines. Therefore, in chapter 9 we develop nonlinear design models that model the closed-loop behavior of the system, where the input is commanded airspeed, altitude, and course angle, and the outputs are the inertial position and orientation of the aircraft. The design models developed in chapter 9 are used in chapters 10 through 13 to develop guidance strategies for the MAV.

As shown in figure 1.3, the design models are used to design the guidance and control systems. The designs are then tested against the high fidelity simulation model, which sometimes requires that the design models be modified or enhanced if they have not captured the essential features of the system. After the designs have been thoroughly tested against the simulation model, they are implemented on the physical system and are again tested and debugged, sometimes requiring that the simulation model be modified to more closely match the physical system.

1.3 Design Project

In this textbook we have decided to replace traditional pencil-and-paper homework problems with a complete and rather extensive design project. The design project is an integral part of the book, and we believe that it will play a significant role in helping the reader to internalize the material that is presented.

The design project involves building a MAV flight simulator from the ground up. The flight simulator will be built using Matlab/Simulink, and we have specifically designed the assignments so that additional add-on packages are not required.[1] The website for the book contains a number of different Matlab and Simulink files that will assist you in developing the flight simulator. Our strategy is to provide you with the basic skeleton files that pass the right information between blocks, but to have you write the internal workings of each block. The project builds upon itself and requires the successful completion of each chapter before it is possible to move to the next chapter. To help you know when the design from each chapter is working, we have included graphs and pictures on the website that show the output of our simulator at each stage.

The project assignment in chapter 2 is to develop an animation of an aircraft and to ensure that you can properly rotate the body of the aircraft on the screen. A tutorial on animating graphics in Matlab is provided in appendix C. The assignment in chapter 3 is to drive the animation using a mathematical model of the rigid body equations of motion. In chapter 4 the force and moments acting on a fixed wing aircraft are added to the simulation. The assignment in chapter 5 is to use the Simulink commands `trim` and `linmod` to find the trim conditions of the aircraft and to derive linear transfer function and state-space models of the system. The assignment in chapter 6 adds an autopilot block that uses the real states to control the aircraft. In chapter 7, a model of the sensors is added to the simulator, and in chapter 8, state estimation

---

[1] We have also taught the course using the public domain flight simulator Aviones, which is available for download at Sourceforge.net. For those who do not have access to Matlab/Simulink and would prefer to develop the project in C/C++, we encourage the use of Aviones.

schemes are added to estimate the states needed for the autopilot using the available sensors. The result of the project assignment in chapter 8 is a closed-loop system that controls airspeed, altitude, and course angle using only available sensor information. The assignment in chapter 9 is to approximate the closed-loop behavior using simple design models and to tune the parameters of the design model so that it essentially matches the behavior of the closed-loop high-fidelity simulation. The assignment in chapter 10 is to develop simple guidance algorithms for following straight-lines and circular orbits in the presence of wind. In chapter 11, straight-line and orbit following are used to synthesize more complicated paths, with emphasis on following Dubins paths. The assignment in chapter 12 is to implement the RRT path planning scheme to plan Dubins paths through an obstacle field. The project assignment in chapter 13 is to point a camera at a moving ground target and to estimate the inertial position of the target using camera data and onboard sensors (geolocation).

2
Coordinate Frames

In studying unmanned aircraft systems, it is important to understand how different bodies are oriented relative to each other. Most obviously, we need to understand how the aircraft is oriented with respect to the earth. We may also want to know how a sensor (e.g., a camera) is oriented relative to the aircraft or how an antenna is oriented relative to a signal source on the ground. This chapter describes the various coordinate systems used to describe the position and orientation of the aircraft and its sensors, and the transformation between these coordinate systems. It is necessary to use several different coordinate systems for the following reasons:

- Newton's equations of motion are derived relative to a fixed, inertial reference frame. However, motion is most easily described in a body-fixed frame.

- Aerodynamic forces and torques act on the aircraft body and are most easily described in a body-fixed reference frame.

- On-board sensors like accelerometers and rate gyros measure information with respect to the body frame. Alternatively, GPS measures position, ground speed, and course angle with respect to the inertial frame.

- Most mission requirements, like loiter points and flight trajectories, are specified in the inertial frame. In addition, map information is also given in an inertial frame.

One coordinate frame is transformed into another through two basic operations: rotation and translation. Section 2.1 describes rotation matrices and their use in transforming between coordinate frames. Section 2.2 describes the specific coordinate frames used for miniature air vehicle systems. In section 2.3 we define airspeed, ground speed, and wind speed and the relationship between these quantities. This leads to the more detailed discussion of the wind triangle in section 2.4. In section 2.5 we derive an expression for differentiating a vector in a rotating and translating frame.

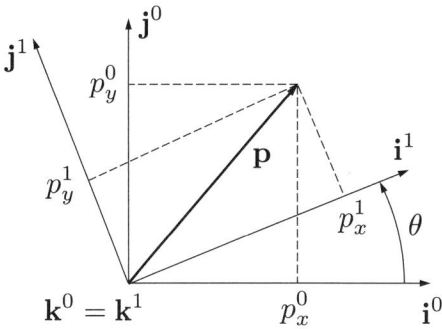

Figure 2.1 Rotation in 2D.

2.1 Rotation Matrices

We begin by considering the two coordinate frames shown in figure 2.1. The vector $\mathbf{p}$ can be expressed in both the $\mathcal{F}^0$ frame (specified by ($\mathbf{i}^0$, $\mathbf{j}^0$, $\mathbf{k}^0$)) and in the $\mathcal{F}^1$ frame (specified by ($\mathbf{i}^1$, $\mathbf{j}^1$, $\mathbf{k}^1$)). In the $\mathcal{F}^0$ frame we have

$$\mathbf{p} = p_x^0 \mathbf{i}^0 + p_y^0 \mathbf{j}^0 + p_z^0 \mathbf{k}^0.$$

Alternatively in the $\mathcal{F}^1$ frame we have

$$\mathbf{p} = p_x^1 \mathbf{i}^1 + p_y^1 \mathbf{j}^1 + p_z^1 \mathbf{k}^1.$$

The vector sets ($\mathbf{i}^0$, $\mathbf{j}^0$, $\mathbf{k}^0$) and ($\mathbf{i}^1$, $\mathbf{j}^1$, $\mathbf{k}^1$) are each mutually perpendicular sets of unit basis vectors.

Setting these two expressions equal to each other gives

$$p_x^1 \mathbf{i}^1 + p_y^1 \mathbf{j}^1 + p_z^1 \mathbf{k}^1 = p_x^0 \mathbf{i}^0 + p_y^0 \mathbf{j}^0 + p_z^0 \mathbf{k}^0.$$

Taking the dot product of both sides with $\mathbf{i}^1$, $\mathbf{j}^1$, and $\mathbf{k}^1$ respectively, and stacking the result into matrix form gives

$$\mathbf{p}^1 \triangleq \begin{pmatrix} p_x^1 \\ p_y^1 \\ p_z^1 \end{pmatrix} = \begin{pmatrix} \mathbf{i}^1 \cdot \mathbf{i}^0 & \mathbf{i}^1 \cdot \mathbf{j}^0 & \mathbf{i}^1 \cdot \mathbf{k}^0 \\ \mathbf{j}^1 \cdot \mathbf{i}^0 & \mathbf{j}^1 \cdot \mathbf{j}^0 & \mathbf{j}^1 \cdot \mathbf{k}^0 \\ \mathbf{k}^1 \cdot \mathbf{i}^0 & \mathbf{k}^1 \cdot \mathbf{j}^0 & \mathbf{k}^1 \cdot \mathbf{k}^0 \end{pmatrix} \begin{pmatrix} p_x^0 \\ p_y^0 \\ p_z^0 \end{pmatrix}.$$

From the geometry of figure 2.1 we get

$$\mathbf{p}^1 = \mathcal{R}_0^1 \mathbf{p}^0, \tag{2.1}$$

where

$$\mathcal{R}_0^1 \triangleq \begin{pmatrix} \cos\theta & \sin\theta & 0 \\ -\sin\theta & \cos\theta & 0 \\ 0 & 0 & 1 \end{pmatrix}.$$

The notation $\mathcal{R}_0^1$ is used to denote a rotation from coordinate frame $\mathcal{F}^0$ to coordinate frame $\mathcal{F}^1$.

Proceeding in a similar way, a right-handed rotation of the coordinate system about the y-axis gives

$$\mathcal{R}_0^1 \triangleq \begin{pmatrix} \cos\theta & 0 & -\sin\theta \\ 0 & 1 & 0 \\ \sin\theta & 0 & \cos\theta \end{pmatrix},$$

and a right-handed rotation of the coordinate system about the x-axis is

$$\mathcal{R}_0^1 \triangleq \begin{pmatrix} 1 & 0 & 0 \\ 0 & \cos\theta & \sin\theta \\ 0 & -\sin\theta & \cos\theta \end{pmatrix}.$$

As pointed out in [7], the negative sign on the sine term appears above the line with only ones and zeros.

The matrix $\mathcal{R}_0^1$ in the above equations is an example of a more general class of *orthonormal* rotation matrices that have the following properties:

P.1. $(\mathcal{R}_a^b)^{-1} = (\mathcal{R}_a^b)^\top = \mathcal{R}_b^a.$

P.2. $\mathcal{R}_b^c \mathcal{R}_a^b = \mathcal{R}_a^c.$

P.3. $\det(\mathcal{R}_a^b) = 1,$

where $\det(\cdot)$ is the determinant of a matrix.

In the derivation of equation (2.1), note that the vector $\mathbf{p}$ remains constant and the new coordinate frame $\mathcal{F}^1$ was obtained by rotating $\mathcal{F}^0$ through a *right-handed* rotation of angle θ. Alternatively, rotation matrices can be used to rotate a vector through a prescribed angle in a fixed reference frame. As an example, consider the *left-handed* rotation of a vector $\mathbf{p}$ in frame $\mathcal{F}^0$ about the $\mathbf{k}^0$-axis by the angle θ, as shown in figure 2.2.

Coordinate Frames

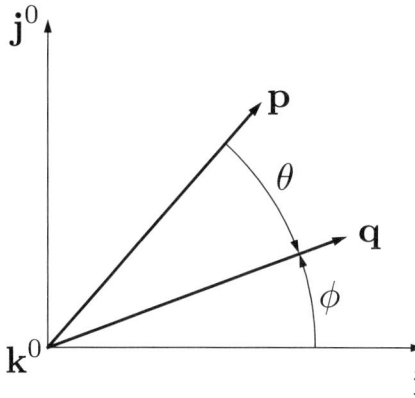

Figure 2.2 Rotation of **p** about the $\mathbf{k}^0$-axis.

Assuming **p** and **q** are confined to the $\mathbf{i}^0$-$\mathbf{j}^0$ plane, we can write the components of **p** and **q** as

$$\mathbf{p} = \begin{pmatrix} p\cos(\theta + \phi) \\ p\sin(\theta + \phi) \\ 0 \end{pmatrix}$$

$$= \begin{pmatrix} p\cos\theta\cos\phi - p\sin\theta\sin\phi \\ p\sin\theta\cos\phi + p\cos\theta\sin\phi \\ 0 \end{pmatrix} \quad (2.2)$$

and

$$\mathbf{q} = \begin{pmatrix} q\cos\phi \\ q\sin\phi \\ 0 \end{pmatrix}, \quad (2.3)$$

where $p \triangleq |\mathbf{p}| = q \triangleq |\mathbf{q}|$. Expressing equation (2.2) in terms of (2.3) gives

$$\mathbf{p} = \begin{pmatrix} \cos\theta & -\sin\theta & 0 \\ \sin\theta & \cos\theta & 0 \\ 0 & 0 & 1 \end{pmatrix} \mathbf{q}$$

$$= (\mathcal{R}_0^1)^\top \mathbf{q}$$

and

$$\mathbf{q} = \mathcal{R}_0^1 \mathbf{p}.$$

In this case, the rotation matrix $\mathcal{R}_0^1$ can be interpreted as a left-handed rotation of the vector **p** through the angle θ to a new vector **q** in the same reference frame. Notice that a right-handed rotation of a vector (in this case from **q** to **p**) can be obtained by using $(\mathcal{R}_0^1)^\top$. This

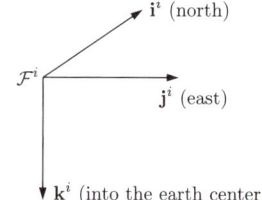

Figure 2.3 The inertial coordinate frame. The $\mathbf{i}^i$-axis points north, the $\mathbf{j}^i$-axis points east, and the $\mathbf{k}^i$-axis points into the earth.

interpretation contrasts with our original use of the rotation matrix to transform a fixed vector $\mathbf{p}$ from an expression in frame $\mathcal{F}^0$ to an expression in frame $\mathcal{F}^1$ where $\mathcal{F}^1$ has been obtained from $\mathcal{F}^0$ by a right-handed rotation.

2.2 MAV Coordinate Frames

To derive and understand the dynamic behavior of MAVs, several coordinate systems are of interest. In this section, we will define and describe the following coordinate frames: the inertial frame, the vehicle frame, the vehicle-1 frame, the vehicle-2 frame, the body frame, the stability frame, and the wind frame. The inertial and vehicle frames are related by a translation, while the remaining frames are related by rotations. The angles defining the relative orientations of the vehicle, vehicle-1, vehicle-2, and body frames are the roll, pitch, and yaw angles that describe the attitude of the aircraft. These angles are commonly known as Euler angles. The rotation angles that define the relative orientation of the body, stability, and wind coordinate frames are the angle of attack and sideslip angles. Throughout the book we assume a flat, non-rotating earth—a valid assumption for MAVs.

2.2.1 The inertial frame $\mathcal{F}^i$

The inertial coordinate system is an earth-fixed coordinate system with its origin at the defined home location. As shown in figure 2.3, the unit vector $\mathbf{i}^i$ is directed north, $\mathbf{j}^i$ is directed east, and $\mathbf{k}^i$ is directed toward the center of the earth, or down. This coordinate system is sometimes referred to as a north-east-down (NED) reference frame. It is common for north to be referred to as the inertial x direction, east to be referred to as the inertial y direction, and down to be referred to as the inertial z direction.

2.2.2 The vehicle frame $\mathcal{F}^v$

The origin of the vehicle frame is at the center of mass of the MAV. However, the axes of $\mathcal{F}^v$ are aligned with the axis of the inertial frame

Coordinate Frames 13

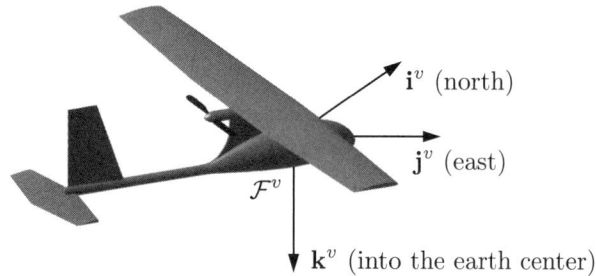

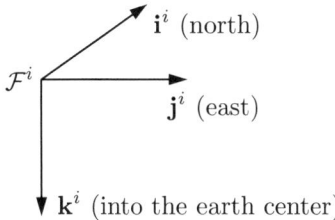

Figure 2.4 The vehicle coordinate frame. The $\mathbf{i}^v$-axis points north, the $\mathbf{j}^v$-axis points east, and the $\mathbf{k}^v$-axis points into the earth.

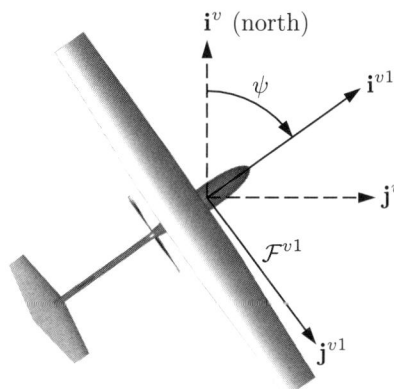

Figure 2.5 The vehicle-1 frame. The $\mathbf{i}^{v1}$-axis points out the nose of the aircraft, the $\mathbf{j}^{v1}$-axis points out the right wing, and the $\mathbf{k}^{v1}$-axis points into the earth.

$\mathcal{F}^i$. In other words, the unit vector $\mathbf{i}^v$ points north, $\mathbf{j}^v$ points east, and $\mathbf{k}^v$ points toward the center of the earth, as shown in figure 2.4.

2.2.3 The vehicle-1 frame $\mathcal{F}^{v1}$

The origin of the vehicle-1 frame is identical to the vehicle frame: the center of mass of the aircraft. However, $\mathcal{F}^{v1}$ is rotated in the positive right-handed direction about $\mathbf{k}^v$ by the heading (or yaw) angle ψ. In the absence of additional rotations, $\mathbf{i}^{v1}$ points out the nose of the airframe, $\mathbf{j}^{v1}$ points out the right wing, and $\mathbf{k}^{v1}$ is aligned with $\mathbf{k}^v$ and points into the earth. The vehicle-1 frame is shown in figure 2.5.

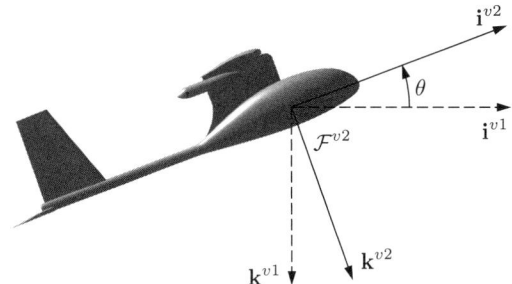

Figure 2.6 The vehicle-2 frame. The $\mathbf{i}^{v2}$-axis points out the nose of the aircraft, the $\mathbf{j}^{v2}$-axis points out the right wing, and the $\mathbf{k}^{v2}$-axis points out the belly.

The transformation from $\mathcal{F}^v$ to $\mathcal{F}^{v1}$ is given by

$$\mathbf{p}^{v1} = \mathcal{R}_v^{v1}(\psi)\mathbf{p}^v,$$

where

$$\mathcal{R}_v^{v1}(\psi) = \begin{pmatrix} \cos\psi & \sin\psi & 0 \\ -\sin\psi & \cos\psi & 0 \\ 0 & 0 & 1 \end{pmatrix}.$$

2.2.4 The vehicle-2 frame $\mathcal{F}^{v2}$

The origin of the vehicle-2 frame is again the center of mass of the aircraft and is obtained by rotating the vehicle-1 frame in a right-handed rotation about the $\mathbf{j}^{v1}$ axis by the pitch angle θ. The unit vector $\mathbf{i}^{v2}$ points out the nose of the aircraft, $\mathbf{j}^{v2}$ points out the right wing, and $\mathbf{k}^{v2}$ points out the belly, as shown in figure 2.6.

The transformation from $\mathcal{F}^{v1}$ to $\mathcal{F}^{v2}$ is given by

$$\mathbf{p}^{v2} = \mathcal{R}_{v1}^{v2}(\theta)\mathbf{p}^{v1},$$

where

$$\mathcal{R}_{v1}^{v2}(\theta) = \begin{pmatrix} \cos\theta & 0 & -\sin\theta \\ 0 & 1 & 0 \\ \sin\theta & 0 & \cos\theta \end{pmatrix}.$$

2.2.5 The body frame $\mathcal{F}^b$

The body frame is obtained by rotating the vehicle-2 frame in a right-handed rotation about $\mathbf{i}^{v2}$ by the roll angle ϕ. Therefore, the origin is the center of mass, $\mathbf{i}^b$ points out the nose of the airframe, $\mathbf{j}^b$ points out the right wing, and $\mathbf{k}^b$ points out the belly. The body frame is shown

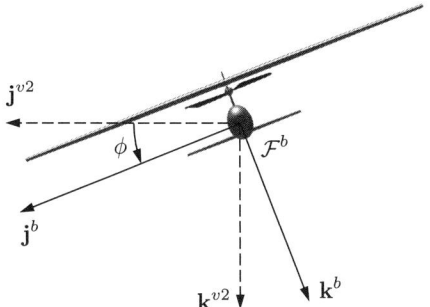

Figure 2.7 The body frame. The **i**^{b}-axis points out the nose of the airframe, the **j**^{b}-axis points out the right wing, and the **k**^{b}-axis points out the belly.

in figure 2.7. The directions indicated by the **i**^{b}, **j**^{b}, and **k**^{b} unit vectors are sometimes referred to as the body x, the body y, and the body z directions, respectively.

The transformation from $\mathcal{F}^{v2}$ to $\mathcal{F}^b$ is given by

$$\mathbf{p}^b = \mathcal{R}_{v2}^b(\phi)\mathbf{p}^{v2},$$

where

$$\mathcal{R}_{v2}^b(\phi) = \begin{pmatrix} 1 & 0 & 0 \\ 0 & \cos\phi & \sin\phi \\ 0 & -\sin\phi & \cos\phi \end{pmatrix}.$$

The transformation from the vehicle frame to the body frame is given by

$$\mathcal{R}_v^b(\phi, \theta, \psi) = \mathcal{R}_{v2}^b(\phi)\mathcal{R}_{v1}^{v2}(\theta)\mathcal{R}_v^{v1}(\psi) \tag{2.4}$$

$$= \begin{pmatrix} 1 & 0 & 0 \\ 0 & \cos\phi & \sin\phi \\ 0 & -\sin\phi & \cos\phi \end{pmatrix} \begin{pmatrix} \cos\theta & 0 & -\sin\theta \\ 0 & 1 & 0 \\ \sin\theta & 0 & \cos\theta \end{pmatrix} \begin{pmatrix} \cos\psi & \sin\psi & 0 \\ -\sin\psi & \cos\psi & 0 \\ 0 & 0 & 1 \end{pmatrix}$$

$$= \begin{pmatrix} c_\theta c_\psi & c_\theta s_\psi & -s_\theta \\ s_\phi s_\theta c_\psi - c_\phi s_\psi & s_\phi s_\theta s_\psi + c_\phi c_\psi & s_\phi c_\theta \\ c_\phi s_\theta c_\psi + s_\phi s_\psi & c_\phi s_\theta s_\psi - s_\phi c_\psi & c_\phi c_\theta \end{pmatrix}, \tag{2.5}$$

where $c_\phi \triangleq \cos\phi$ and $s_\phi \triangleq \sin\phi$. The angles ϕ, θ, and ψ are commonly referred to as Euler angles. Euler angles are commonly used because they provide an intuitive means for representing the orientation of a body in three dimensions. The rotation sequence ψ-θ-ϕ is commonly used for aircraft and is just one of several Euler angle systems in use [8].

The physical interpretation of Euler angles is clear and this contributes to their widespread use. Euler angle representations,

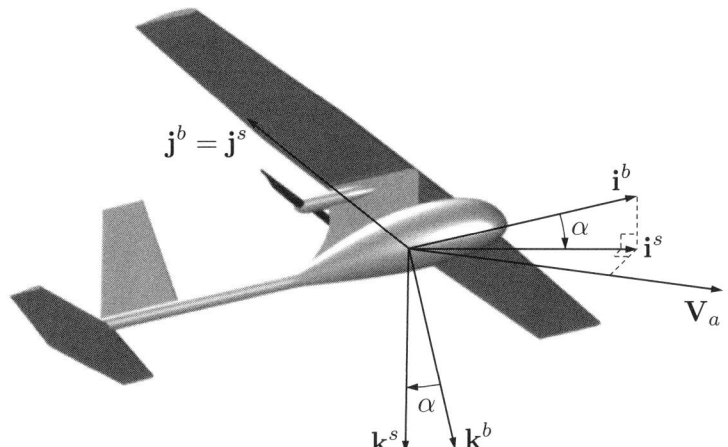

Figure 2.8 The stability frame. The i^s-axis points along the projection of the airspeed vector onto the i^b-k^b plane of the body frame, the j^s-axis is identical to the j^b-axis of the body frame, and the k^s-axis is constructed to make a right-handed coordinate system. Note that the angle of attack is defined as a *left*-handed rotation about the body j^b-axis.

however, have a mathematical singularity that can cause computational instabilities. For the ψ-θ-ϕ Euler angle sequence, there is a singularity when the pitch angle θ is ± 90 deg, in which case the yaw angle is not defined. This singularity is commonly referred to as gimbal lock. A common alternative to Euler angles is the quaternion. While the quaternion attitude representation lacks the intuitive appeal of Euler angles, they are free of mathematical singularities and are computationally more efficient. Quaternion attitude representations are discussed in appendix B.

2.2.6 The stability frame $\mathcal{F}^s$

Aerodynamic forces are generated as the airframe moves through the air surrounding it. We refer to the velocity of the aircraft relative to the surrounding air as the airspeed vector, denoted $\mathbf{V}_a$. The magnitude of the airspeed vector is simply referred to as the airspeed, V_a. To generate lift, the wings of the airframe must fly at a positive angle with respect to the airspeed vector. This angle is called the angle of attack and is denoted by α. As shown in figure 2.8, the angle of attack is defined as a *left*-handed rotation about j^b and is such that i^s aligns with the projection of $\mathbf{V}_a$ onto the plane spanned by i^b and k^b. The need for a left-handed rotation is caused by the definition of positive angle of attack, which is positive for a right-handed rotation from the stability frame i^s axis to the body frame i^b axis.

Coordinate Frames

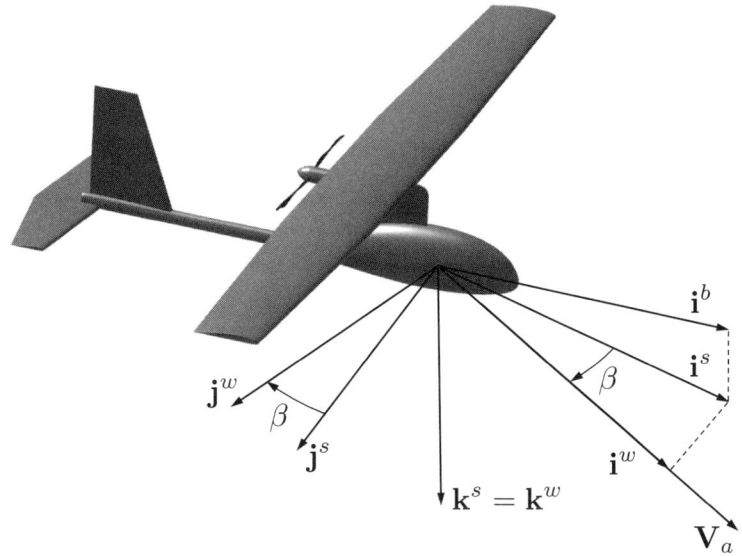

Figure 2.9 The wind frame. The $\mathbf{i}^w$-axis points along the airspeed vector.

Since α is given by a left-handed rotation, the transformation from $\mathcal{F}^b$ to $\mathcal{F}^s$ is given by

$$\mathbf{p}^s = \mathcal{R}^s_b(\alpha)\mathbf{p}^b,$$

where

$$\mathcal{R}^s_b(\alpha) = \begin{pmatrix} \cos\alpha & 0 & \sin\alpha \\ 0 & 1 & 0 \\ -\sin\alpha & 0 & \cos\alpha \end{pmatrix}.$$

2.2.7 The wind frame $\mathcal{F}^w$

The angle between the airspeed vector and the $\mathbf{i}^b$-$\mathbf{k}^b$ plane is called the side-slip angle and is denoted by β. As shown in figure 2.9, the wind frame is obtained by rotating the stability frame by a right-handed rotation of β about $\mathbf{k}^s$. The unit vector $\mathbf{i}^w$ is aligned with the airspeed vector $\mathbf{V}_a$.

The transformation from $\mathcal{F}^s$ to $\mathcal{F}^w$ is given by

$$\mathbf{p}^w = \mathcal{R}^w_s(\beta)\mathbf{p}^s,$$

where

$$\mathcal{R}^w_s(\beta) = \begin{pmatrix} \cos\beta & \sin\beta & 0 \\ -\sin\beta & \cos\beta & 0 \\ 0 & 0 & 1 \end{pmatrix}.$$

The total transformation from the body frame to the wind frame is given by

$$\mathcal{R}_b^w(\alpha, \beta) = \mathcal{R}_s^w(\beta)\mathcal{R}_b^s(\alpha)$$

$$= \begin{pmatrix} \cos\beta & \sin\beta & 0 \\ -\sin\beta & \cos\beta & 0 \\ 0 & 0 & 1 \end{pmatrix} \begin{pmatrix} \cos\alpha & 0 & \sin\alpha \\ 0 & 1 & 0 \\ -\sin\alpha & 0 & \cos\alpha \end{pmatrix}$$

$$= \begin{pmatrix} \cos\beta\cos\alpha & \sin\beta & \cos\beta\sin\alpha \\ -\sin\beta\cos\alpha & \cos\beta & -\sin\beta\sin\alpha \\ -\sin\alpha & 0 & \cos\alpha \end{pmatrix}.$$

Alternatively, the transformation from the wind frame to the body frame is

$$\mathcal{R}_w^b(\alpha, \beta) = (\mathcal{R}_b^w)^\top(\alpha, \beta) = \begin{pmatrix} \cos\beta\cos\alpha & -\sin\beta\cos\alpha & -\sin\alpha \\ \sin\beta & \cos\beta & 0 \\ \cos\beta\sin\alpha & -\sin\beta\sin\alpha & \cos\alpha \end{pmatrix}.$$

2.3 Airspeed, Wind Speed, and Ground Speed

When developing the dynamic equations of motion for a MAV, it is important to remember that the inertial forces experienced by the MAV are dependent on velocities and accelerations relative to a fixed (inertial) reference frame. The aerodynamic forces, however, depend on the velocity of the airframe relative to the surrounding air. When wind is not present, these velocities are the same. However, wind is almost always present with MAVs and we must carefully distinguish between airspeed, represented by the velocity with respect to the surrounding air $\mathbf{V}_a$, and the ground speed, represented by the velocity with respect to the inertial frame $\mathbf{V}_g$. These velocities are related by the expression

$$\mathbf{V}_a = \mathbf{V}_g - \mathbf{V}_w, \tag{2.6}$$

where $\mathbf{V}_w$ is the wind velocity relative to the inertial frame.

The MAV velocity $\mathbf{V}_g$ can be expressed in the body frame in terms of components along the $\mathbf{i}^b$, $\mathbf{j}^b$, and $\mathbf{k}^b$ axes

$$\mathbf{V}_g^b = \begin{pmatrix} u \\ v \\ w \end{pmatrix},$$

where $\mathbf{V}_g^b$ is the velocity of the MAV *with respect to the inertial frame,* as expressed in the body frame. Similarly, if we define the north, east, and down components of the wind as w_n, w_e, and w_d respectively, we can write an expression for the wind velocity in the body frame as

$$\mathbf{V}_w^b = \begin{pmatrix} u_w \\ v_w \\ w_w \end{pmatrix} = \mathcal{R}_v^b(\phi, \theta, \psi) \begin{pmatrix} w_n \\ w_e \\ w_d \end{pmatrix}.$$

Keeping in mind that the airspeed vector $\mathbf{V}_a$ is the velocity of the MAV with respect to the wind, it can be expressed in the wind frame as

$$\mathbf{V}_a^w = \begin{pmatrix} V_a \\ 0 \\ 0 \end{pmatrix}.$$

Defining u_r, v_r, and w_r as the body-frame components of the airspeed vector,[1] it can be written in the body frame as

$$\mathbf{V}_a^b = \begin{pmatrix} u_r \\ v_r \\ w_r \end{pmatrix} = \begin{pmatrix} u - u_w \\ v - v_w \\ w - w_w \end{pmatrix}.$$

When developing a MAV simulation, u_r, v_r, and w_r are used to calculate the aerodynamic forces and moments acting on the MAV. The body-frame velocity components u, v, and w are states of the MAV system and are readily available from the solution of the equations of motion. The wind velocity components u_w, v_w, and w_w typically come from a wind model as inputs to the equations of motion. Combining expressions, we can express the airspeed vector body-frame components in terms of the airspeed magnitude, angle of attack, and sideslip angle as

$$\mathbf{V}_a^b = \begin{pmatrix} u_r \\ v_r \\ w_r \end{pmatrix} = \mathcal{R}_w^b \begin{pmatrix} V_a \\ 0 \\ 0 \end{pmatrix}$$

$$= \begin{pmatrix} \cos\beta\cos\alpha & -\sin\beta\cos\alpha & -\sin\alpha \\ \sin\beta & \cos\beta & -\sin\beta\sin\alpha \\ \cos\beta\sin\alpha & 0 & \cos\alpha \end{pmatrix} \begin{pmatrix} V_a \\ 0 \\ 0 \end{pmatrix},$$

which implies that

$$\begin{pmatrix} u_r \\ v_r \\ w_r \end{pmatrix} = V_a \begin{pmatrix} \cos\alpha\cos\beta \\ \sin\beta \\ \sin\alpha\cos\beta \end{pmatrix}. \qquad (2.7)$$

---

[1] Some flight mechanics textbooks define u, v, and w as the body-frame components of the airspeed vector. We define u, v, and w as the body-frame components of the ground speed vector and u_r, v_r, and w_r as the body-frame components of the airspeed vector to clearly distinguish between the two.

Inverting this relationship gives

$$V_a = \sqrt{u_r^2 + v_r^2 + w_r^2}$$
$$\alpha = \tan^{-1}\left(\frac{w_r}{u_r}\right) \quad (2.8)$$
$$\beta = \sin^{-1}\left(\frac{v_r}{\sqrt{u_r^2 + v_r^2 + w_r^2}}\right).$$

Given that aerodynamic forces and moments are commonly expressed in terms of V_a, α, and β, these expressions are essential in formulating the equations of motion for a MAV.

2.4 The Wind Triangle

For MAVs, the wind speed is often in the range of 20 to 50 percent of the airspeed. The significant effect of wind on MAVs is important to understand, more so than for larger conventional aircraft, where the airspeed is typically much greater than the wind speed. Having introduced the concepts of reference frames, airframe velocity, wind velocity, and the airspeed vector, we can discuss some important definitions relating to the navigation of MAVs.

The direction of the ground speed vector relative to an inertial frame is specified using two angles. These angles are the course angle χ and the (inertial referenced) flight path angle γ. Figure 2.10 shows how these two angles are defined. The flight path angle γ is defined as the angle between the horizontal plane and the ground velocity vector $\mathbf{V}_g$, while the course χ is the angle between the projection of the ground velocity vector onto the horizontal plane and true north.

The relationship between the groundspeed vector, the airspeed vector, and the wind vector, which is given by equation (2.6) is called the wind triangle. A more detailed depiction of the wind triangle is given in the horizontal plane in figure 2.11 and in the vertical plane in figure 2.12. Figure 2.11 shows an air vehicle following a ground track represented by the dashed line. The north direction is indicated by the $\mathbf{i}^i$ vector, and the direction that the vehicle is pointed is shown by the $\mathbf{i}^b$ vector, which is fixed in the direction of the body x-axis. For level flight, the heading (yaw) angle ψ, is the angle between $\mathbf{i}^i$ and $\mathbf{i}^b$ and defines the direction that the vehicle is pointed. The direction the vehicle is traveling with respect to the surrounding air mass is given by the airspeed vector $\mathbf{V}_a$. In steady, level flight, $\mathbf{V}_a$ is commonly aligned with $\mathbf{i}^b$, meaning the sideslip angle β is zero.

Coordinate Frames

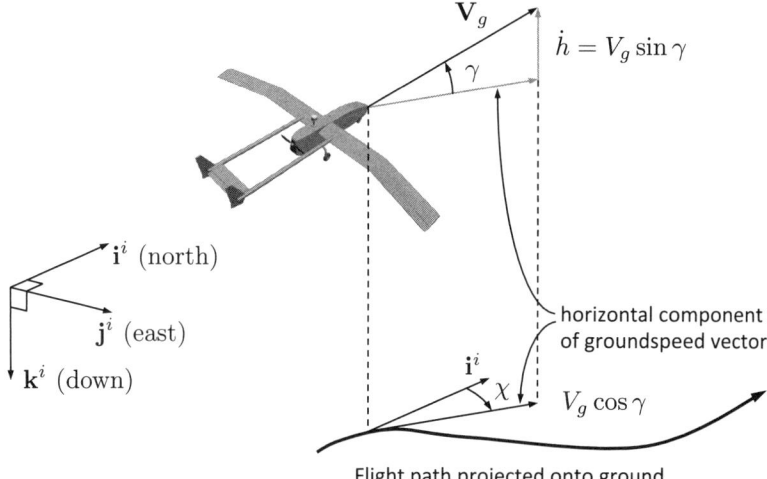

Figure 2.10 The flight path angle γ and the course angle χ.

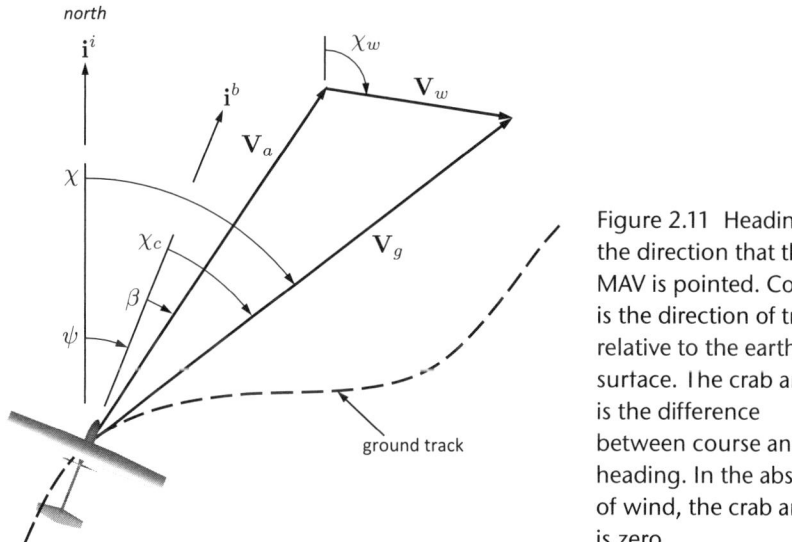

Figure 2.11 Heading is the direction that the MAV is pointed. Course is the direction of travel relative to the earth's surface. The crab angle is the difference between course and heading. In the absence of wind, the crab angle is zero.

The direction the vehicle is traveling with respect to the ground is shown by the velocity vector $\mathbf{V}_g$. The angle between the inertial north and the inertial velocity vector projected onto the local north-east plane is called the course angle χ. If there is a constant ambient wind, the aircraft will need to crab into the wind in order to follow a ground track that is not aligned with the wind. The *crab angle* χ_c is defined as the difference between the course and the heading angles

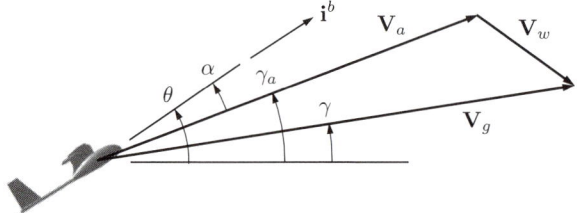

Figure 2.12 The wind triangle projected onto the vertical plane.

as follows:

$$\chi_c \triangleq \chi - \psi.$$

Figure 2.12 depicts the vertical component of the wind triangle. When there is a down component of wind, we define the angle from the inertial north-east plane to $\mathbf{V}_a$ as the *air-mass-referenced flight-path angle* and denote it by γ_a. The relationship between the air-mass-referenced flight-path angle, the angle of attack, and the pitch angle is given by

$$\gamma_a = \theta - \alpha.$$

In the absence of wind $\gamma_a = \gamma$.

The ground speed vector in the inertial frame can be expressed as

$$\mathbf{V}_g^i = \begin{pmatrix} \cos\chi & -\sin\chi & 0 \\ \sin\chi & \cos\chi & 0 \\ 0 & 0 & 1 \end{pmatrix} \begin{pmatrix} \cos\gamma & 0 & \sin\gamma \\ 0 & 1 & 0 \\ -\sin\gamma & 0 & \cos\gamma \end{pmatrix} \begin{pmatrix} V_g \\ 0 \\ 0 \end{pmatrix} = V_g \begin{pmatrix} \cos\chi\cos\gamma \\ \sin\chi\cos\gamma \\ -\sin\gamma \end{pmatrix},$$

where $V_g = \|\mathbf{V}_g\|$. Similarly, the airspeed vector in the inertial frame can be expressed as

$$\mathbf{V}_a^i = V_a \begin{pmatrix} \cos\psi\cos\gamma_a \\ \sin\psi\cos\gamma_a \\ -\sin\gamma_a \end{pmatrix},$$

where $V_a = \|\mathbf{V}_a\|$. Therefore, the wind triangle can be expressed in inertial coordinates as

$$V_g \begin{pmatrix} \cos\chi\cos\gamma \\ \sin\chi\cos\gamma \\ -\sin\gamma \end{pmatrix} - \begin{pmatrix} w_n \\ w_e \\ w_d \end{pmatrix} = V_a \begin{pmatrix} \cos\psi\cos\gamma_a \\ \sin\psi\cos\gamma_a \\ -\sin\gamma_a \end{pmatrix}. \qquad (2.9)$$

Equation (2.9) allows us to derive relationships between V_g, V_a, χ, ψ, γ and γ_a. Specifically, we will consider the case where χ, γ, the wind components (w_n, w_e, w_d), and either V_g or V_a are known. Taking the squared norm of both sides of equation (2.9) results in the expression

$$V_g^2 - 2V_g \begin{pmatrix} \cos\chi\cos\gamma \\ \sin\chi\cos\gamma \\ -\sin\gamma \end{pmatrix}^\top \begin{pmatrix} w_n \\ w_e \\ w_d \end{pmatrix} + V_w^2 - V_a^2 = 0 \qquad (2.10)$$

where $V_w = \|\mathbf{V}_w\| = \sqrt{w_n^2 + w_e^2 + w_d^2}$ is the wind speed. Given χ, γ, and the components of wind, equation (2.10) can be solved for V_a given V_g, or V_g given V_a, depending on the need. When solving the quadratic equation for V_g, the positive root is taken since V_g must be positive.

With both V_a and V_g known, the third row of equation (2.9) can be solved for γ_a to obtain

$$\gamma_a = \sin^{-1}\left(\frac{V_g \sin\gamma + w_d}{V_a}\right). \qquad (2.11)$$

To derive an expression for ψ, multiply both sides of equation (2.9) by $(-\sin\chi, \cos\chi, 0)$ to get the expression

$$0 = V_a \cos\gamma_a \left(-\sin\chi\cos\psi + \cos\chi\sin\psi\right) + \begin{pmatrix} w_n \\ w_e \end{pmatrix}^\top \begin{pmatrix} -\sin\chi \\ \cos\chi \end{pmatrix}.$$

Solving for ψ gives

$$\psi = \chi - \sin^{-1}\left(\frac{1}{V_a \cos\gamma_a} \begin{pmatrix} w_n \\ w_e \end{pmatrix}^\top \begin{pmatrix} -\sin\chi \\ \cos\chi \end{pmatrix}\right). \qquad (2.12)$$

Using equations (2.10) through (2.12), we can calculate ψ and γ provided we have knowledge of the wind components and either V_g or V_a. Similar expressions allowing us to determine χ and γ from ψ and γ_a can be derived from equation (2.9).

Because wind typically has a significant impact on the flight behavior of small unmanned aircraft, we have tried to carefully account for it throughout the text. If wind effects are negligible, however, some important simplifications result. For example, when $V_w = 0$, we also have that $V_a = V_g$, $u = u_r$, $v = v_r$, $w = w_r$, $\psi = \chi$ (assuming also that $\beta = 0$), and $\gamma = \gamma_a$.

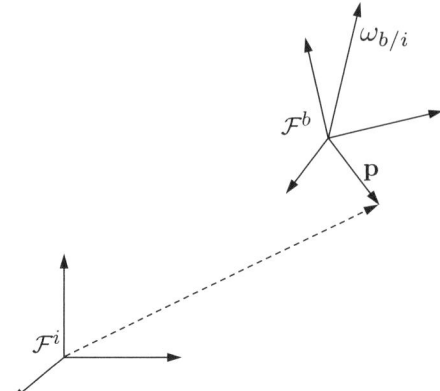

Figure 2.13 A vector in a rotating reference frame.

2.5 Differentiation of a Vector

In the process of deriving equations of motion for a MAV, it is necessary to compute derivatives of vectors in reference frames that are moving with respect to one another. Suppose that we are given two coordinate frames, $\mathcal{F}^i$ and $\mathcal{F}^b$, as shown in figure 2.13. For example, $\mathcal{F}^i$ might represent the inertial frame and $\mathcal{F}^b$ might represent the body frame of a MAV. Suppose that the vector $\mathbf{p}$ is moving in $\mathcal{F}^b$ and that $\mathcal{F}^b$ is rotating (but not translating) with respect to $\mathcal{F}^i$. Our objective is to find the time derivative of $\mathbf{p}$ as seen from frame $\mathcal{F}^i$. To do this, denote the angular velocity of frame $\mathcal{F}^b$ in $\mathcal{F}^i$ as $\boldsymbol{\omega}_{b/i}$ and express the vector $\mathbf{p}$ in terms of its vector components as

$$\mathbf{p} = p_x \mathbf{i}^b + p_y \mathbf{j}^b + p_z \mathbf{k}^b. \tag{2.13}$$

The time derivative of $\mathbf{p}$ with respect to frame $\mathcal{F}^i$ can be found by differentiating equation (2.13) as

$$\frac{d}{dt_i}\mathbf{p} = \dot{p}_x \mathbf{i}^b + \dot{p}_y \mathbf{j}^b + \dot{p}_z \mathbf{k}^b + p_x \frac{d}{dt_i}\mathbf{i}^b + p_y \frac{d}{dt_i}\mathbf{j}^b + p_z \frac{d}{dt_i}\mathbf{k}^b, \tag{2.14}$$

where d/dt_i represents time differentiation with respect to the inertial frame. The first three terms on the right-hand side of equation (2.14) represent the change in $\mathbf{p}$ as viewed by an observer in the rotating $\mathcal{F}^b$ frame. Thus, the differentiation is carried out in the moving frame. We denote this local derivative term by

$$\frac{d}{dt_b}\mathbf{p} = \dot{p}_x \mathbf{i}^b + \dot{p}_y \mathbf{j}^b + \dot{p}_z \mathbf{k}^b. \tag{2.15}$$

The next three terms on the right-hand side of equation (2.14) represent the change in $\mathbf{p}$ due to the rotation of frame $\mathcal{F}^b$ relative to $\mathcal{F}^i$. Given that

Coordinate Frames

$\mathbf{i}^b$, $\mathbf{j}^b$, and $\mathbf{k}^b$ are fixed in the $\mathcal{F}^b$ frame, their derivatives can be calculated as shown in [9] as

$$\dot{\mathbf{i}}^b = \boldsymbol{\omega}_{b/i} \times \mathbf{i}^b$$
$$\dot{\mathbf{j}}^b = \boldsymbol{\omega}_{b/i} \times \mathbf{j}^b$$
$$\dot{\mathbf{k}}^b = \boldsymbol{\omega}_{b/i} \times \mathbf{k}^b.$$

We can rewrite the last three terms of equation (2.14) as

$$p_x \dot{\mathbf{i}}^b + p_y \dot{\mathbf{j}}^b + p_z \dot{\mathbf{k}}^b = p_x(\boldsymbol{\omega}_{b/i} \times \mathbf{i}^b) + p_y(\boldsymbol{\omega}_{b/i} \times \mathbf{j}^b) + p_z(\boldsymbol{\omega}_{b/i} \times \mathbf{k}^b)$$
$$= \boldsymbol{\omega}_{b/i} \times \mathbf{p}. \qquad (2.16)$$

Combining results from equations (2.14), (2.15), and (2.16), we obtain the desired relation

$$\frac{d}{dt_i}\mathbf{p} = \frac{d}{dt_b}\mathbf{p} + \boldsymbol{\omega}_{b/i} \times \mathbf{p}, \qquad (2.17)$$

which expresses the derivative of the vector $\mathbf{p}$ in frame $\mathcal{F}^i$ in terms of its change as observed in frame $\mathcal{F}^b$ and the relative rotation of the two frames. We will use this relation as we derive equations of motion for the MAV in chapter 3.

2.6 Chapter Summary

In this chapter, we have introduced the coordinate frames important to describing the orientation of MAVs. We have described how rotation matrices can be used to transform coordinates in one frame of reference to coordinates in another frame of reference. We have introduced the 3-2-1 Euler angles (ψ, θ, and ϕ) as a means to rotate from the inertial coordinate frame to the body frame. We have also introduced the angle of attack α and the sideslip angle β to describe the relative orientation of the body frame, the stability frame, and the wind frame. An understanding of these orientations is essential to the derivation of equations of motion and the modeling of aerodynamic forces involved in MAV flight. We have introduced the wind triangle and have made the relationships between airspeed, ground speed, wind speed, heading, course, flight-path angle, and air-mass-referenced flight-path angle explicit. We have also derived an expression for the differentiation of a vector in a rotating reference frame.

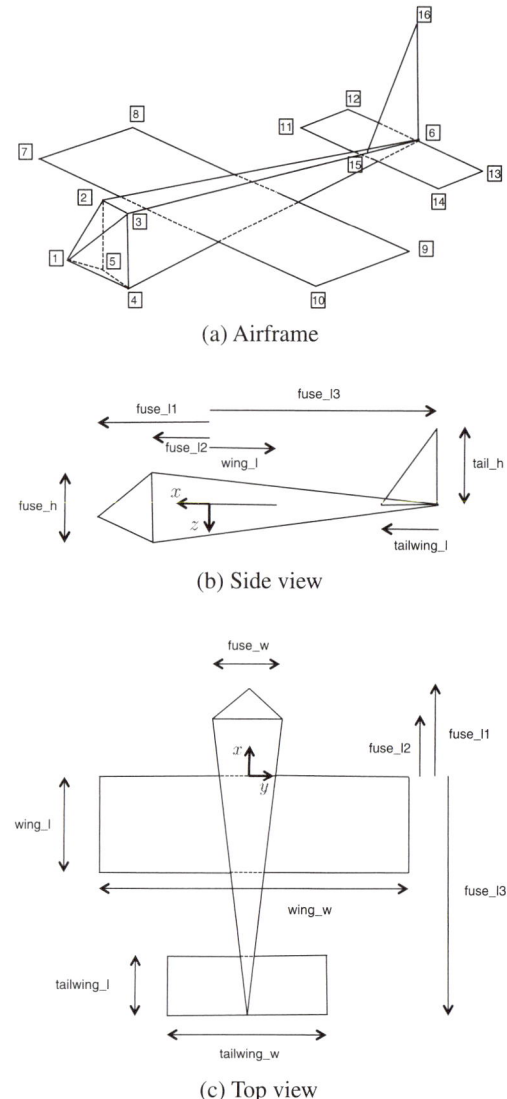

Figure 2.14 Specifications for animation of aircraft for the design project.

Notes and References

There are many references on coordinate frames and rotations matrices. A particularly good overview of rotation matrices is [10]. Overviews of attitude representations are included in [8, 11]. The definition of the different aircraft frames can be found in [4, 1, 7, 12]. A particularly good explanation is given in [13]. Vector differentiation is discussed in most textbooks on mechanics, including [14, 15, 16, 9].

2.7 Design Project

The objective of this assignment is to create a 3-D graphic of a MAV that is correctly rotated and translated to the desired configuration. Creating animations in Simulink is described in appendix C and example files are contained on the textbook website.

2.1. Read appendix C and study carefully the spacecraft animation using vertices and faces given at the textbook website.

2.2. Create an animation drawing of the aircraft shown in figure 2.14.

2.3. Using a Simulink model like the one given on the website, verify that the aircraft is correctly rotated and translated in the animation.

2.4. In the animation file, switch the order of rotation and translation so that the aircraft is first translated and then rotated, and observe the effect.

3

Kinematics and Dynamics

The first step in developing navigation, guidance, and control strategies for MAVs is to develop appropriate dynamic models. Deriving the nonlinear equations of motion for a MAV is the focus of chapters 3 and 4. In chapter 5, we linearize the equations of motion to create transfer-function and state-space models appropriate for control design.

In this chapter, we derive the expressions for the kinematics and the dynamics of a rigid body. We will apply Newton's laws: for example, $\mathbf{f} = m\dot{\mathbf{v}}$ in the case of the linear motion. In this chapter, we will focus on defining the relations between positions and velocities (the kinematics) and relations between forces and moments and the momentum (dynamics). In chapter 4, we will concentrate on the definition of the forces and moments involved, particularly the aerodynamic forces and moments. In chapter 5, we will combine these relations to form the complete nonlinear equations of motion. While the expressions derived in this chapter are general to any rigid body, we will use notation and coordinate frames that are typical in the aeronautics literature. In particular, in section 3.1 we define the notation that will be used for MAV state variables. In section 3.2 we derive the kinematics, and in section 3.3 we derive the dynamics.

3.1 State Variables

In developing the equations of motion for a MAV, twelve state variables will be introduced. There are three position states and three velocity states associated with the translational motion of the MAV. Similarly, there are three angular position and three angular velocity states associated with the rotational motion. The state variables are listed in table 3.1.

The state variables are shown schematically in figure 3.1. The northeast-down positions of the MAV (p_n, p_e, p_d) are defined relative to the inertial frame. We will sometimes use $h = -p_d$ to denote the altitude. The linear velocities (u, v, w) and the angular velocities (p, q, r) of the MAV are defined with respect to the body frame. The

TABLE 3.1
State variables for MAV equations of motion

Name	Description
p_n	Inertial north position of the MAV along $\mathbf{i}^i$ in $\mathcal{F}^i$
p_e	Inertial east position of the MAV along $\mathbf{j}^i$ in $\mathcal{F}^i$
p_d	Inertial down position (negative of altitude) of the MAV measured along $\mathbf{k}^i$ in $\mathcal{F}^i$
u	Body frame velocity measured along $\mathbf{i}^b$ in $\mathcal{F}^b$
v	Body frame velocity measured along $\mathbf{j}^b$ in $\mathcal{F}^b$
w	Body frame velocity measured along $\mathbf{k}^b$ in $\mathcal{F}^b$
ϕ	Roll angle defined with respect to $\mathcal{F}^{v2}$
θ	Pitch angle defined with respect to $\mathcal{F}^{v1}$
ψ	Heading (yaw) angle defined with respect to $\mathcal{F}^v$
p	Roll rate measured along $\mathbf{i}^b$ in $\mathcal{F}^b$
q	Pitch rate measured along $\mathbf{j}^b$ in $\mathcal{F}^b$
r	Yaw rate measured along $\mathbf{k}^b$ in $\mathcal{F}^b$

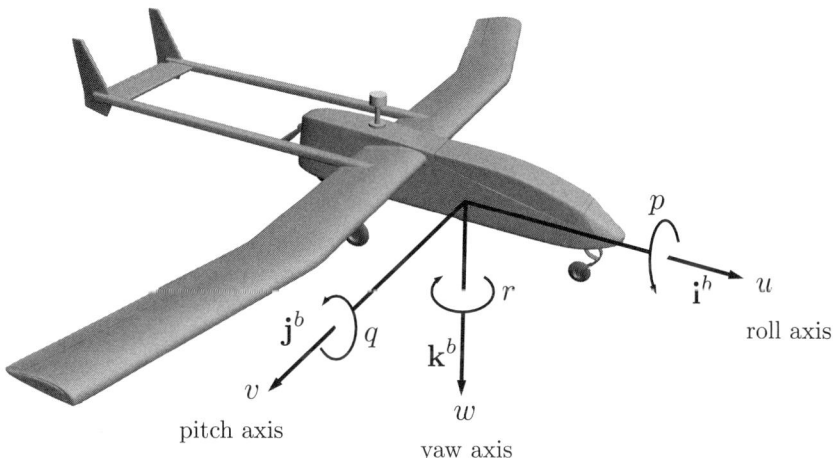

Figure 3.1 Definition of axes of motion.

Euler angles—roll ϕ, pitch θ, and heading (yaw) ψ—are defined with respect to the vehicle-2 frame, the vehicle-1 frame, and the vehicle frame, respectively. Because the Euler angles are defined relative to intermediate frames of reference, we cannot say that the angular rates (p, q, r) are simply the time derivatives of the attitude angles (ϕ, θ, ψ).

As we will show in the following section, $p = \dot{\phi}$, $q = \dot{\theta}$, and $r = \dot{\psi}$ only at the instant that $\phi = \theta = 0$. Generally, the angular rates p, q, and r are functions of the time derivatives of the attitude angles, $\dot{\phi}$, $\dot{\theta}$, and $\dot{\psi}$ and the angles ϕ and θ. The remainder of this chapter is devoted to formulating the equations of motion corresponding to each of the states listed in table 3.1.

3.2 Kinematics

The translational velocity of the MAV is commonly expressed in terms of the velocity components along each of the axes in a body-fixed coordinate frame. The components u, v, and w correspond to the inertial velocity of the vehicle projected onto the $\mathbf{i}^b$, $\mathbf{j}^b$, and $\mathbf{k}^b$ axes, respectively. On the other hand, the translational position of the MAV is usually measured and expressed in an inertial reference frame. Relating the translational velocity and position requires differentiation and a rotational transformation

$$\frac{d}{dt}\begin{pmatrix} p_n \\ p_e \\ p_d \end{pmatrix} = \mathcal{R}_b^v \begin{pmatrix} u \\ v \\ w \end{pmatrix} = (\mathcal{R}_v^b)^\top \begin{pmatrix} u \\ v \\ w \end{pmatrix},$$

which using equation (2.5) gives

$$\begin{pmatrix} \dot{p}_n \\ \dot{p}_e \\ \dot{p}_d \end{pmatrix} = \begin{pmatrix} c_\theta c_\psi & s_\phi s_\theta c_\psi - c_\phi s_\psi & c_\phi s_\theta c_\psi + s_\phi s_\psi \\ c_\theta s_\psi & s_\phi s_\theta s_\psi + c_\phi c_\psi & c_\phi s_\theta s_\psi - s_\phi c_\psi \\ -s_\theta & s_\phi c_\theta & c_\phi c_\theta \end{pmatrix} \begin{pmatrix} u \\ v \\ w \end{pmatrix}, \quad (3.1)$$

where we have used the shorthand notation $c_x \triangleq \cos x$ and $s_x \triangleq \sin x$. This is a kinematic relation in that it relates the derivative of position to velocity: forces or accelerations are not considered.

The relationship between angular positions ϕ, θ, and ψ and the angular rates p, q, and r is also complicated by the fact that these quantities are defined in different coordinate frames. The angular rates are defined in the body frame $\mathcal{F}^b$. The angular positions (Euler angles) are defined in three different coordinate frames: the roll angle ϕ is a rotation from $\mathcal{F}^{v2}$ to $\mathcal{F}^b$ about the $\mathbf{i}^{v2} = \mathbf{i}^b$ axis; the pitch angle θ is a rotation from $\mathcal{F}^{v1}$ to $\mathcal{F}^{v2}$ about the $\mathbf{j}^{v1} = \mathbf{j}^{v2}$ axis; and the yaw angle ψ is a rotation from $\mathcal{F}^{v}$ to $\mathcal{F}^{v1}$ about the $\mathbf{k}^{v} = \mathbf{k}^{v1}$ axis.

The body-frame angular rates can be expressed in terms of the derivatives of the Euler angles, provided that the proper rotational transformations are carried out as follows:

$$\begin{pmatrix} p \\ q \\ r \end{pmatrix} = \begin{pmatrix} \dot{\phi} \\ 0 \\ 0 \end{pmatrix} + \mathcal{R}_{v2}^{b}(\phi) \begin{pmatrix} 0 \\ \dot{\theta} \\ 0 \end{pmatrix} + \mathcal{R}_{v2}^{b}(\phi)\mathcal{R}_{v1}^{v2}(\theta) \begin{pmatrix} 0 \\ 0 \\ \dot{\psi} \end{pmatrix}$$

$$= \begin{pmatrix} \dot{\phi} \\ 0 \\ 0 \end{pmatrix} + \begin{pmatrix} 1 & 0 & 0 \\ 0 & \cos\phi & \sin\phi \\ 0 & -\sin\phi & \cos\phi \end{pmatrix} \begin{pmatrix} 0 \\ \dot{\theta} \\ 0 \end{pmatrix}$$

$$+ \begin{pmatrix} 1 & 0 & 0 \\ 0 & \cos\phi & \sin\phi \\ 0 & -\sin\phi & \cos\phi \end{pmatrix} \begin{pmatrix} \cos\theta & 0 & -\sin\theta \\ 0 & 1 & 0 \\ \sin\theta & 0 & \cos\theta \end{pmatrix} \begin{pmatrix} 0 \\ 0 \\ \dot{\psi} \end{pmatrix}$$

$$= \begin{pmatrix} 1 & 0 & -\sin\theta \\ 0 & \cos\phi & \sin\phi\cos\theta \\ 0 & -\sin\phi & \cos\phi\cos\theta \end{pmatrix} \begin{pmatrix} \dot{\phi} \\ \dot{\theta} \\ \dot{\psi} \end{pmatrix}. \qquad (3.2)$$

Inverting this expression yields

$$\begin{pmatrix} \dot{\phi} \\ \dot{\theta} \\ \dot{\psi} \end{pmatrix} = \begin{pmatrix} 1 & \sin\phi\tan\theta & \cos\phi\tan\theta \\ 0 & \cos\phi & -\sin\phi \\ 0 & \sin\phi\sec\theta & \cos\phi\sec\theta \end{pmatrix} \begin{pmatrix} p \\ q \\ r \end{pmatrix}, \qquad (3.3)$$

which expresses the derivatives of the three angular position states in terms of the angular positions ϕ and θ and the body rates p, q, and r.

3.3 Rigid-body Dynamics

To derive the dynamic equations of motion for the MAV, we will apply Newton's second law—first to the translational degrees of freedom and then to the rotational degrees of freedom. Newton's laws hold in inertial reference frames, meaning the motion of the body of interest must be referenced to a fixed (i.e., inertial) frame of reference, which in our case is the ground. We will assume a flat earth model, which is appropriate for small and miniature air vehicles. Even though motion is referenced to a fixed frame, it can be *expressed* using vector components associated with other frames, such as the body frame.

We do this with the MAV velocity vector $\mathbf{V}_g$, which for convenience is most commonly expressed in the body frame as $\mathbf{V}_g^b = (u, v, w)^\top$. $\mathbf{V}_g^b$ is the velocity of the MAV with respect to the ground as expressed in the body frame.

3.3.1 Translational Motion

Newton's second law applied to a body undergoing translational motion can be stated as

$$\mathsf{m}\frac{d\mathbf{V}_g}{dt_i} = \mathbf{f}, \tag{3.4}$$

where m is the mass of the MAV,[1] $\frac{d}{dt_i}$ is the time derivative in the inertial frame, and $\mathbf{f}$ is the sum of all external forces acting on the MAV. The external forces include gravity, aerodynamic forces, and propulsion forces.

The derivative of velocity taken in the inertial frame can be written in terms of the derivative in the body frame and the angular velocity according to equation (2.17) as

$$\frac{d\mathbf{V}_g}{dt_i} = \frac{d\mathbf{V}_g}{dt_b} + \boldsymbol{\omega}_{b/i} \times \mathbf{V}_g, \tag{3.5}$$

where $\boldsymbol{\omega}_{b/i}$ is the angular velocity of the MAV with respect to the inertial frame. Combining (3.4) and (3.5) results in an alternative representation of Newton's second law with differentiation carried out in the body frame:

$$\mathsf{m}\left(\frac{d\mathbf{V}_g}{dt_b} + \boldsymbol{\omega}_{b/i} \times \mathbf{V}_g\right) = \mathbf{f}.$$

In the case of a maneuvering aircraft, we can most easily apply Newton's second law by expressing the forces and velocities in the body frame as

$$\mathsf{m}\left(\frac{d\mathbf{V}_g^b}{dt_b} + \boldsymbol{\omega}_{b/i}^b \times \mathbf{V}_g^b\right) = \mathbf{f}^b, \tag{3.6}$$

---

[1] Mass is denoted with a sans serif font m to distinguish it from m, which will be introduced as the sum of moments about the body-fixed $\mathbf{j}^b$ axis.

Kinematics and Dynamics

where $\mathbf{V}_g^b = (u, v, w)^\top$ and $\boldsymbol{\omega}_{b/i}^b = (p, q, r)^\top$. The vector $\mathbf{f}^b$ represents the sum of the externally applied forces and is defined in terms of its body-frame components as $\mathbf{f}^b \triangleq (f_x, f_y, f_z)^\top$.

The expression $\frac{d\mathbf{V}_g^b}{dt_b}$ is the rate of change of the velocity expressed in the body frame, as viewed by an observer on the moving body. Since u, v, and w are the instantaneous projections of $\mathbf{V}_g^b$ onto the $\mathbf{i}^b$, $\mathbf{j}^b$, and $\mathbf{k}^b$ axes, it follows that

$$\frac{d\mathbf{V}_g^b}{dt_b} = \begin{pmatrix} \dot{u} \\ \dot{v} \\ \dot{w} \end{pmatrix}.$$

Expanding the cross product in equation (3.6) and rearranging terms, we get

$$\begin{pmatrix} \dot{u} \\ \dot{v} \\ \dot{w} \end{pmatrix} = \begin{pmatrix} rv - qw \\ pw - ru \\ qu - pv \end{pmatrix} + \frac{1}{m} \begin{pmatrix} f_x \\ f_y \\ f_z \end{pmatrix}. \tag{3.7}$$

3.3.2 Rotational Motion

For rotational motion, Newton's second law states that

$$\frac{d\mathbf{h}}{dt_i} = \mathbf{m},$$

where $\mathbf{h}$ is the angular momentum in vector form and $\mathbf{m}$ is the sum of all externally applied moments. This expression is true provided that moments are summed about the center of mass of the MAV. The derivative of angular momentum taken in the inertial frame can be expanded using equation (2.17) as

$$\frac{d\mathbf{h}}{dt_i} = \frac{d\mathbf{h}}{dt_b} + \boldsymbol{\omega}_{b/i} \times \mathbf{h} = \mathbf{m}.$$

As with translational motion, it is most convenient to express this equation in the body frame, giving

$$\frac{d\mathbf{h}^b}{dt_b} + \boldsymbol{\omega}_{b/i}^b \times \mathbf{h}^b = \mathbf{m}^b. \tag{3.8}$$

For a rigid body, angular momentum is defined as the product of the *inertia matrix* $\mathbf{J}$ and the angular velocity vector: $\mathbf{h}^b \triangleq \mathbf{J}\boldsymbol{\omega}_{b/i}^b$

where **J** is given by

$$\mathbf{J} = \begin{pmatrix} \int (y^2 + z^2)\,dm & -\int xy\,dm & -\int xz\,dm \\ -\int xy\,dm & \int (x^2 + z^2)\,dm & -\int yz\,dm \\ -\int xz\,dm & -\int yz\,dm & \int (x^2 + y^2)\,dm \end{pmatrix}$$

$$\triangleq \begin{pmatrix} J_x & -J_{xy} & -J_{xz} \\ -J_{xy} & J_y & -J_{yz} \\ -J_{xz} & -J_{yz} & J_z \end{pmatrix}. \tag{3.9}$$

The diagonal terms of **J** are called the *moments of inertia*, while the off-diagonal terms are called the *products of inertia*. The moments of inertia are measures of the aircraft's tendency to oppose acceleration about a specific axis of rotation. For example, J_x can be conceptually thought of as taking the product of the mass of each element composing the aircraft (dm) and the square of the distance of the mass element from the body x axis ($y^2 + z^2$) and adding them up. The larger J_x is in value, the more the aircraft opposes angular acceleration about the x axis. This line of thinking, of course, applies to the moments of inertia J_y and J_z as well. In practice, the inertia matrix is not calculated using equation (3.9). Instead, it is numerically calculated from mass properties using CAD models or it is measured experimentally using equipment such as a bifilar pendulum [17, 18].

Because the integrals in equation (3.9) are calculated with respect to the $\mathbf{i}^b$, $\mathbf{j}^b$, and $\mathbf{k}^b$ axes fixed in the (rigid) body, **J** is constant when viewed from the body frame, hence $\frac{d\mathbf{J}}{dt_b} = 0$. Taking derivatives and substituting into equation (3.8), we get

$$\mathbf{J}\frac{d\boldsymbol{\omega}_{b/i}^b}{dt_b} + \boldsymbol{\omega}_{b/i}^b \times \left(\mathbf{J}\boldsymbol{\omega}_{b/i}^b\right) = \mathbf{m}^b. \tag{3.10}$$

The expression $\frac{d\boldsymbol{\omega}_{b/i}^b}{dt_b}$ is the rate of change of the angular velocity expressed in the body frame, as viewed by an observer on the moving body. Since p, q, and r are the instantaneous projections of $\boldsymbol{\omega}_{b/i}^b$ onto the $\mathbf{i}^b$, $\mathbf{j}^b$, and $\mathbf{k}^b$ axes, it follows that

$$\dot{\boldsymbol{\omega}}_{b/i}^b = \frac{d\boldsymbol{\omega}_{b/i}^b}{dt_b} = \begin{pmatrix} \dot{p} \\ \dot{q} \\ \dot{r} \end{pmatrix}.$$

Rearranging equation (3.10), we get

$$\dot{\boldsymbol{\omega}}_{b/i}^b = \mathbf{J}^{-1}\left[-\boldsymbol{\omega}_{b/i}^b \times \left(\mathbf{J}\boldsymbol{\omega}_{b/i}^b\right) + \mathbf{m}^b\right]. \tag{3.11}$$

Aircraft are often symmetric about the plane spanned by $\mathbf{i}^b$ and $\mathbf{k}^b$. In that case $J_{xy} = J_{yz} = 0$, which implies that

$$\mathbf{J} = \begin{pmatrix} J_x & 0 & -J_{xz} \\ 0 & J_y & 0 \\ -J_{xz} & 0 & J_z \end{pmatrix}.$$

Under this symmetry assumption, the inverse of $\mathbf{J}$ is given by

$$\mathbf{J}^{-1} = \frac{\mathrm{adj}(\mathbf{J})}{\det(\mathbf{J})}$$

$$= \frac{\begin{pmatrix} J_y J_z & 0 & J_y J_{xz} \\ 0 & J_x J_z - J_{xz}^2 & 0 \\ J_{xz} J_y & 0 & J_x J_y \end{pmatrix}}{J_x J_y J_z - J_{xz}^2 J_y}$$

$$= \begin{pmatrix} \frac{J_z}{\Gamma} & 0 & \frac{J_{xz}}{\Gamma} \\ 0 & \frac{1}{J_y} & 0 \\ \frac{J_{xz}}{\Gamma} & 0 & \frac{J_x}{\Gamma} \end{pmatrix},$$

where $\Gamma \triangleq J_x J_z - J_{xz}^2$.

Defining the components of the externally applied moment about the $\mathbf{i}^b$, $\mathbf{j}^b$, and $\mathbf{k}^b$ axes as $\mathbf{m}^b \triangleq (l, m, n)^\top$, we can write equation (3.11) in component form as

$$\begin{pmatrix} \dot{p} \\ \dot{q} \\ \dot{r} \end{pmatrix} = \begin{pmatrix} \frac{J_z}{\Gamma} & 0 & \frac{J_{xz}}{\Gamma} \\ 0 & \frac{1}{J_y} & 0 \\ \frac{J_{xz}}{\Gamma} & 0 & \frac{J_x}{\Gamma} \end{pmatrix} \left[\begin{pmatrix} 0 & r & -q \\ -r & 0 & p \\ q & -p & 0 \end{pmatrix} \begin{pmatrix} J_x & 0 & -J_{xz} \\ 0 & J_y & 0 \\ -J_{xz} & 0 & J_z \end{pmatrix} \begin{pmatrix} p \\ q \\ r \end{pmatrix} + \begin{pmatrix} l \\ m \\ n \end{pmatrix} \right]$$

$$= \begin{pmatrix} \frac{J_z}{\Gamma} & 0 & \frac{J_{xz}}{\Gamma} \\ 0 & \frac{1}{J_y} & 0 \\ \frac{J_{xz}}{\Gamma} & 0 & \frac{J_x}{\Gamma} \end{pmatrix} \left[\begin{pmatrix} J_{xz} pq + (J_y - J_z) qr \\ J_{xz}(r^2 - p^2) + (J_z - J_x) pr \\ (J_x - J_y) pq - J_{xz} qr \end{pmatrix} + \begin{pmatrix} l \\ m \\ n \end{pmatrix} \right]$$

$$= \begin{pmatrix} \Gamma_1 pq - \Gamma_2 qr + \Gamma_3 l + \Gamma_4 n \\ \Gamma_5 pr - \Gamma_6 (p^2 - r^2) + \frac{1}{J_y} m \\ \Gamma_7 pq - \Gamma_1 qr + \Gamma_4 l + \Gamma_8 n \end{pmatrix}, \qquad (3.12)$$

where

$$\Gamma_1 = \frac{J_{xz}(J_x - J_y + J_z)}{\Gamma}$$

$$\Gamma_2 = \frac{J_z(J_z - J_y) + J_{xz}^2}{\Gamma}$$

$$\Gamma_3 = \frac{J_z}{\Gamma}$$

$$\Gamma_4 = \frac{J_{xz}}{\Gamma} \qquad (3.13)$$

$$\Gamma_5 = \frac{J_z - J_x}{J_y}$$

$$\Gamma_6 = \frac{J_{xz}}{J_y}$$

$$\Gamma_7 = \frac{(J_x - J_y)J_x + J_{xz}^2}{\Gamma}$$

$$\Gamma_8 = \frac{J_x}{\Gamma}.$$

The six-degree-of-freedom, 12-state model for the MAV kinematics and dynamics are given by equations (3.1), (3.3), (3.7), and (3.12), and are summarized as follows:

$$\begin{pmatrix} \dot{p}_n \\ \dot{p}_e \\ \dot{p}_d \end{pmatrix} = \begin{pmatrix} c_\theta c_\psi & s_\phi s_\theta c_\psi - c_\phi s_\psi & c_\phi s_\theta c_\psi + s_\phi s_\psi \\ c_\theta s_\psi & s_\phi s_\theta s_\psi + c_\phi c_\psi & c_\phi s_\theta s_\psi - s_\phi c_\psi \\ -s_\theta & s_\phi c_\theta & c_\phi c_\theta \end{pmatrix} \begin{pmatrix} u \\ v \\ w \end{pmatrix} \qquad (3.14)$$

$$\begin{pmatrix} \dot{u} \\ \dot{v} \\ \dot{w} \end{pmatrix} = \begin{pmatrix} rv - qw \\ pw - ru \\ qu - pv \end{pmatrix} + \frac{1}{m}\begin{pmatrix} f_x \\ f_y \\ f_z \end{pmatrix}, \qquad (3.15)$$

$$\begin{pmatrix} \dot{\phi} \\ \dot{\theta} \\ \dot{\psi} \end{pmatrix} = \begin{pmatrix} 1 & \sin\phi \tan\theta & \cos\phi \tan\theta \\ 0 & \cos\phi & -\sin\phi \\ 0 & \frac{\sin\phi}{\cos\theta} & \frac{\cos\phi}{\cos\theta} \end{pmatrix} \begin{pmatrix} p \\ q \\ r \end{pmatrix} \qquad (3.16)$$

$$\begin{pmatrix} \dot{p} \\ \dot{q} \\ \dot{r} \end{pmatrix} = \begin{pmatrix} \Gamma_1 pq - \Gamma_2 qr \\ \Gamma_5 pr - \Gamma_6(p^2 - r^2) \\ \Gamma_7 pq - \Gamma_1 qr \end{pmatrix} + \begin{pmatrix} \Gamma_3 l + \Gamma_4 n \\ \frac{1}{J_y} m \\ \Gamma_4 l + \Gamma_8 n \end{pmatrix}. \qquad (3.17)$$

Equations (3.14)–(3.17) represent the dynamics of the MAV. They are not complete in that the externally applied forces and moments are not yet defined. Models for forces and moments due to gravity, aerodynamics, and propulsion will be derived in chapter 4. In appendix B, an alternative formulation to these equations that uses quaternions to represent the MAV attitude is given.

3.4 Chapter Summary

In this chapter, we have derived a six-degree-of-freedom, 12-state dynamic model for a MAV from first principles. This model will be the basis for analysis, simulation, and control design that will be discussed in forthcoming chapters.

Notes and References

The material in this chapter is standard, and similar discussions can be found in textbooks on mechanics [14, 15, 19], space dynamics [20, 21], flight dynamics [1, 2, 5, 7, 12, 22] and robotics [10, 23].

Equations (3.14) and (3.15) are expressed in terms of inertially referenced velocities u, v, and w. Alternatively, they can be expressed in terms of velocities referenced to the air-mass surrounding the aircraft u_r, v_r, and w_r as

$$\begin{pmatrix} \dot{p}_n \\ \dot{p}_e \\ \dot{p}_d \end{pmatrix} = \mathcal{R}_b^v(\phi, \theta, \psi) \begin{pmatrix} u_r \\ v_r \\ w_r \end{pmatrix} + \begin{pmatrix} w_n \\ w_e \\ w_d \end{pmatrix} \tag{3.18}$$

$$\begin{pmatrix} \dot{u}_r \\ \dot{v}_r \\ \dot{w}_r \end{pmatrix} = \begin{pmatrix} rv_r - qw_r \\ pw_r - ru_r \\ qu_r - pv_r \end{pmatrix} + \frac{1}{m} \begin{pmatrix} f_x \\ f_y \\ f_z \end{pmatrix} - \mathcal{R}_v^b(\phi, \theta, \psi) \begin{pmatrix} \dot{w}_n \\ \dot{w}_e \\ \dot{w}_d \end{pmatrix}, \tag{3.19}$$

where

$$\mathcal{R}_b^v(\phi, \theta, \psi) = (\mathcal{R}_v^b)^\top(\phi, \theta, \psi) = \begin{pmatrix} c_\theta c_\psi & s_\phi s_\theta c_\psi - c_\phi s_\psi & c_\phi s_\theta c_\psi + s_\phi s_\psi \\ c_\theta s_\psi & s_\phi s_\theta s_\psi + c_\phi c_\psi & c_\phi s_\theta s_\psi - s_\phi c_\psi \\ -s_\theta & s_\phi c_\theta & c_\phi c_\theta \end{pmatrix}.$$

The choice of which equations to use to express the aircraft kinematics is a matter of personal preference. In equations (3.14) and (3.15), the velocity states u, v, and w represent the aircraft motion with respect to the ground (inertial frame). In equations (3.18) and (3.19), the velocity states u_r, v_r, and w_r represent the aircraft motion with respect to the air mass surrounding the aircraft. To correctly represent the motion of the aircraft in the inertial frame using u_r, v_r, and w_r as states, the effect of wind speed and wind acceleration must be taken into account.

3.5 Design Project

3.1. Read appendix D on building s-functions in Simulink, and also the Matlab documentation on s-functions.

3.2. Implement the MAV equations of motion given in equations (3.14) through (3.17) using a Simulink s-function. Assume that the inputs to the block are the forces and moments applied to the MAV in the body frame. Block parameters should include the mass, the moments and products of inertia, and the initial conditions for each state. Use the parameters given in appendix E. Simulink templates are provided on the website.

3.3. Connect the equations of motion to the animation block developed in the previous chapter. Verify that the equations of motion are correct by individually setting the forces and moments along each axis to a nonzero value and convincing yourself that the motion is appropriate.

3.4. Since J_{xz} is non-zero, there is gyroscopic coupling between roll and yaw. To test your simulation, set J_{xz} to zero and place nonzero moments on l and n and verify that there is no coupling between the roll and yaw axes. Verify that when J_{xz} is not zero, there is coupling between the roll and yaw axes.

4

Forces and Moments

The objective of this chapter is to describe the forces and moments that act on a MAV. Following [5], we will assume that the forces and moments are primarily due to three sources, namely, gravity, aerodynamics, and propulsion. Letting $\mathbf{f}_g$ be the force due to gravity, $(\mathbf{f}_a, \mathbf{m}_a)$ be the forces and moments due to aerodynamics, and $(\mathbf{f}_p, \mathbf{m}_p)$ be the forces and moments due to propulsion, we have

$$\mathbf{f} = \mathbf{f}_g + \mathbf{f}_a + \mathbf{f}_p$$
$$\mathbf{m} = \mathbf{m}_a + \mathbf{m}_p,$$

where $\mathbf{f}$ is the total force acting on the airframe and $\mathbf{m}$ is the total moment acting on the airframe.

In this chapter, we derive expressions for each of the forces and moments. Gravitational forces are discussed in section 4.1, aerodynamic forces and torques are described in section 4.2, and the forces and torques due to propulsion are described in section 4.3. Atmospheric disturbances, described in section 4.4, are modeled as changes in the wind speed and enter the equations of motion through the aerodynamic forces and torques.

4.1 Gravitational Forces

The effect of the earth's gravitational field on a MAV can be modeled as a force proportional to the mass acting at the center of mass. This force acts in the $\mathbf{k}^i$ direction and is proportional to the mass of the MAV by the gravitational constant g. In the vehicle frame $\mathcal{F}^v$, the gravity force acting on the center of mass is given by

$$\mathbf{f}_g^v = \begin{pmatrix} 0 \\ 0 \\ mg \end{pmatrix}.$$

When applying Newton's second law in chapter 3, we summed forces along the axes in the body frame. Therefore, we must transform the

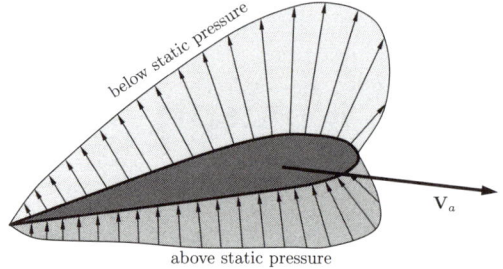

Figure 4.1 Pressure distribution around an airfoil.

gravitational force into its body-frame components to give

$$\mathbf{f}_g^b = \mathcal{R}_v^b \begin{pmatrix} 0 \\ 0 \\ mg \end{pmatrix}$$

$$= \begin{pmatrix} -mg \sin \theta \\ mg \cos \theta \sin \phi \\ mg \cos \theta \cos \phi \end{pmatrix}.$$

Since the gravity force acts through the center of mass of the MAV, there are no moments produced by gravity.

4.2 Aerodynamic Forces and Moments

As a MAV passes through the air, a pressure distribution is generated around the MAV body, as depicted in figure 4.1. The strength and distribution of the pressure acting on the MAV is a function of the airspeed, the air density, and the shape and attitude of the MAV. Accordingly, the dynamic pressure is given by $\frac{1}{2}\rho V_a^2$, where ρ is the air density and V_a is the speed of the MAV through the surrounding air mass.

Instead of attempting to characterize the pressure distribution around the wing, the common approach is to capture the effect of the pressure with a combination of forces and a moment. For example, if we consider the longitudinal ($\mathbf{i}^b$-$\mathbf{k}^b$) plane, the effect of the pressure acting on the MAV body can be modeled using a lift force, a drag force, and a moment. As shown in figure 4.2, the lift and drag forces are applied at the quarter-chord point, also known as the aerodynamic center.

Forces and Moments

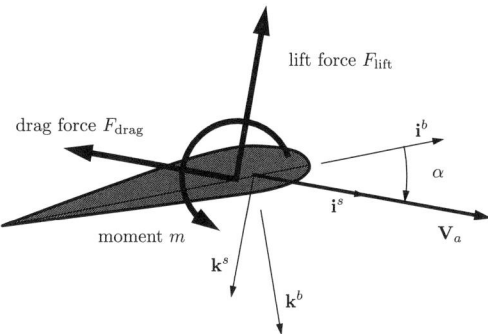

Figure 4.2 Effect of pressure distribution can be modeled using a lift force, a drag force, and a moment.

The lift, drag, and moment are commonly expressed as

$$F_{\text{lift}} = \frac{1}{2}\rho V_a^2 S C_L$$

$$F_{\text{drag}} = \frac{1}{2}\rho V_a^2 S C_D \qquad (4.1)$$

$$m = \frac{1}{2}\rho V_a^2 S c C_m,$$

where C_L, C_D, and C_m are nondimensional aerodynamic coefficients, S is the planform area of the MAV wing, and c is the mean chord of the MAV wing. For airfoils generally, the lift, drag, and pitching moment coefficients are significantly influenced by the airfoil shape, Reynolds number, Mach number, and the angle of attack. For the range of airspeeds flown by small and miniature aircraft, the Reynolds number and Mach number effects are approximately constant. We will consider the effects of the angles α and β; the angular rates p, q, and r; and the deflection of control surfaces on the aerodynamic coefficients.

It is common to decompose the aerodynamic forces and moments into two groups: longitudinal and lateral. The longitudinal forces and moments act in the $\mathbf{i}^b$-$\mathbf{k}^b$ plane, also called the pitch plane. They include the forces in the $\mathbf{i}^b$ and $\mathbf{k}^b$ directions (caused by lift and drag) and the moment about the $\mathbf{j}^b$ axis. The lateral forces and moments include the force in the $\mathbf{j}^b$ direction and the moments about the $\mathbf{i}^b$ and $\mathbf{k}^b$ axes.

4.2.1 Control Surfaces

Before giving detailed equations that describe the aerodynamic forces and moments due to the lifting surfaces, we need to define the control surfaces that are used to maneuver the aircraft. The control surfaces

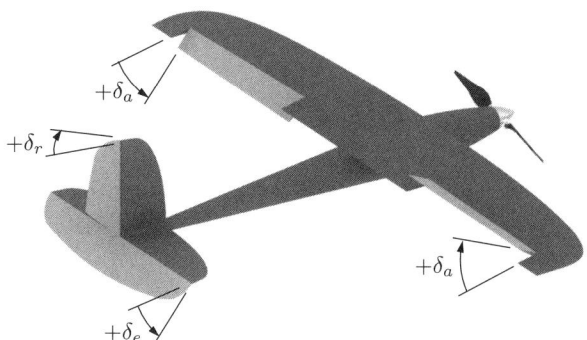

Figure 4.3 Control surfaces for a standard aircraft configuration. The ailerons are used to control the roll angle ϕ. The elevators are used to control the pitch angle θ. The rudder directly effects the yaw angle ψ.

are used to modify the aerodynamic forces and moments. For standard aircraft configurations, the control surfaces include the elevator, the aileron, and the rudder. Other surfaces, including spoilers, flaps, and canards, will not be discussed in this book but are modeled similarly.

Figure 4.3 shows the standard configuration, where the aileron deflection is denoted by δ_a, the elevator deflection is denoted by δ_e, and the rudder deflection is denoted by δ_r. The positive direction of a control surface deflection can be determined by applying the right-hand rule to the hinge axis of the control surface. For example, the hinge axis of the elevator is aligned with the body $\mathbf{j}^b$ axis. Applying the right-hand rule about the $\mathbf{j}^b$ axis implies that a positive deflection for the elevator is trailing edge down. Similarly, positive deflection for the rudder is trailing edge left. Finally, positive aileron deflection is trailing edge down on each aileron. The aileron deflection δ_a can be thought of as a composite deflection defined by

$$\delta_a = \frac{1}{2}(\delta_{a\text{-left}} - \delta_{a\text{-right}}).$$

Therefore a positive δ_a is produced when the left aileron is trailing edge down and the right aileron is trailing edge up.

For small aircraft, there are two other standard configurations. The first is the v-tail configuration as shown in figure 4.4. The control surfaces for a v-tail are called ruddervators. The angular deflection of the right ruddervator is denoted as δ_{rr}, and the angular deflection of the left ruddervator is denoted as δ_{rl}. Driving the ruddervators differentially has the same effect as a rudder, producing a torque about $\mathbf{k}^b$. Driving the ruddervators together has the same effect as an elevator, producing

Forces and Moments

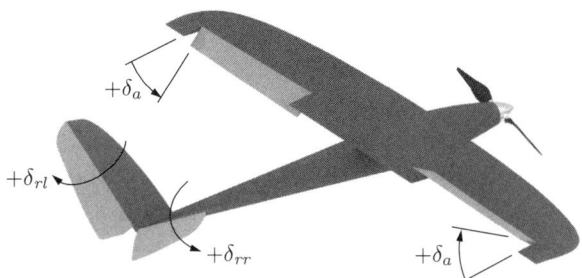

Figure 4.4 Ruddervators are used to control a v-tail aircraft. The ruddervators replace the rudder and the elevator. Driving the ruddervators together has the same effect as an elevator, and driving them differentially has the same effect as a rudder.

torque about $\mathbf{j}^b$. Mathematically, we can convert between ruddervators and rudder-elevator signals as

$$\begin{pmatrix} \delta_e \\ \delta_r \end{pmatrix} = \begin{pmatrix} 1 & 1 \\ -1 & 1 \end{pmatrix} \begin{pmatrix} \delta_{rr} \\ \delta_{rl} \end{pmatrix}.$$

Using this relation, the mathematical model for forces and torques for v-tail aircraft can be expressed in terms of standard rudder-elevator notation.

The other standard configuration for small aircraft is the flying wing depicted in figure 4.5. The control surfaces for a flying wing are called elevons. The angular deflection of the right elevon is denoted as δ_{er}, and the angular deflection of the left elevon is denoted as δ_{el}. Driving the elevons differentially has the same effect as ailerons, producing a torque about $\mathbf{i}^b$, while driving the elevons together has the same effect as an elevator, causing a torque about $\mathbf{j}^b$. Mathematically, we can convert between elevons and aileron-elevator signals with

$$\begin{pmatrix} \delta_e \\ \delta_a \end{pmatrix} = \begin{pmatrix} 1 & 1 \\ -1 & 1 \end{pmatrix} \begin{pmatrix} \delta_{er} \\ \delta_{el} \end{pmatrix}.$$

Therefore, the mathematical model for forces and torques for flying-wing aircraft can be expressed in terms of standard aileron-elevator notation.

4.2.2 Longitudinal Aerodynamics

The longitudinal aerodynamic forces and moments cause motion in the body $\mathbf{i}^b$-$\mathbf{k}^b$ plane, also known as the pitch plane. They are the aerodynamic forces and moment with which we are perhaps most

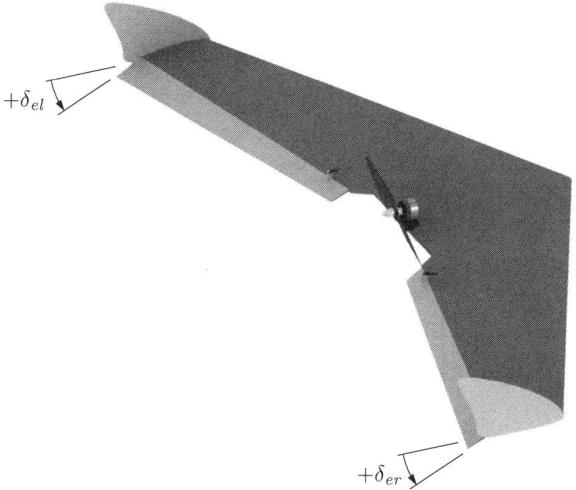

Figure 4.5 Elevons are used to control a flying-wing aircraft. The elevons replace the aileron and the elevator. Driving the elevons together has the same effect as an elevator, and driving them differentially has the same effect as ailerons.

familiar: lift, drag, and pitching moment. By definition, the lift and drag forces are aligned with the axes of the stability frame, as shown in figure 4.2. When represented as a vector, the pitching moment also aligns with the $\mathbf{j}^s$ axis of the stability frame. The lift and drag forces and the pitching moment are heavily influenced by the angle of attack. The pitch rate q and the elevator deflection δ_e also influence the longitudinal forces and moment. Based on this, we can rewrite the equations for lift, drag, and pitching moment to express this functional dependence on α, q and δ_e as

$$F_{\text{lift}} = \frac{1}{2} \rho V_a^2 S C_L(\alpha, q, \delta_e)$$

$$F_{\text{drag}} = \frac{1}{2} \rho V_a^2 S C_D(\alpha, q, \delta_e)$$

$$m = \frac{1}{2} \rho V_a^2 S c C_m(\alpha, q, \delta_e).$$

In general, these force and moment equations are nonlinear. For small angles of attack, however, the flow over the wing will remain laminar and attached. Under these conditions, the lift, drag, and pitching moment can be modeled with acceptable accuracy using linear approximations. Working with the lift equation as an example, a first-order

Taylor series approximation of the lift force can be written as

$$F_{\text{lift}} = \frac{1}{2}\rho V_a^2 S \left[C_{L_0} + \frac{\partial C_L}{\partial \alpha}\alpha + \frac{\partial C_L}{\partial q}q + \frac{\partial C_L}{\partial \delta_e}\delta_e \right]. \tag{4.2}$$

The coefficient C_{L_0} is the value of the C_L when $\alpha = q = \delta_e = 0$. It is common to nondimensionalize the partial derivatives of this linear approximation. Since C_L and the angles α and δ_e (expressed in radians) are dimensionless, the only partial requiring nondimensionalization is $\delta C_L / \delta q$. Since the units of q are rad/s, a standard factor to use is $c/(2V_a)$. We can then rewrite equation (4.2) as

$$F_{\text{lift}} = \frac{1}{2}\rho V_a^2 S \left[C_{L_0} + C_{L_\alpha}\alpha + C_{L_q}\frac{c}{2V_a}q + C_{L_{\delta_e}}\delta_e \right], \tag{4.3}$$

where the coefficients C_{L_0}, $C_{L_\alpha} \triangleq \frac{\partial C_L}{\partial \alpha}$, $C_{L_q} \triangleq \frac{\partial C_L}{\partial \frac{qc}{2V_a}}$, and $C_{L_{\delta_e}} \triangleq \frac{\partial C_L}{\partial \delta_e}$ are dimensionless quantities. C_{L_α} and C_{L_q} are commonly referred to as stability derivatives, while $C_{L_{\delta_e}}$ is an example of a control derivative. The label "derivative" comes from the fact that the coefficients originated as partial derivatives in the Taylor series approximation. In a similar manner, we express linear approximations for the aerodynamic drag force and pitching moment as

$$F_{\text{drag}} = \frac{1}{2}\rho V_a^2 S \left[C_{D_0} + C_{D_\alpha}\alpha + C_{D_q}\frac{c}{2V_a}q + C_{D_{\delta_e}}\delta_e \right] \tag{4.4}$$

$$m = \frac{1}{2}\rho V_a^2 S c \left[C_{m_0} + C_{m_\alpha}\alpha + C_{m_q}\frac{c}{2V_a}q + C_{m_{\delta_e}}\delta_e \right]. \tag{4.5}$$

Equations (4.3), (4.4), and (4.5) are commonly used as the basis for the longitudinal aerodynamic model. Under typical, low-angle-of-attack flight conditions, they are a sufficiently accurate representation of the forces and moments produced. The flow over the aircraft body is laminar and attached and the flow field over the aircraft is termed quasi-steady, meaning it only changes slowly with respect to time. The shape of the flow field is predictable and changes in response to changes in the angle of attack, pitch rate, and elevator deflection. The quasi-steady behavior of the flow field results in longitudinal aerodynamic forces and torques that are predictable and fairly straightforward to model, as shown above.

In contrast to the quasi-steady aerodynamics typically experienced by aircraft, unsteady aerodynamics are challenging to model and predict. Unsteady aerodynamics are characterized by nonlinear,

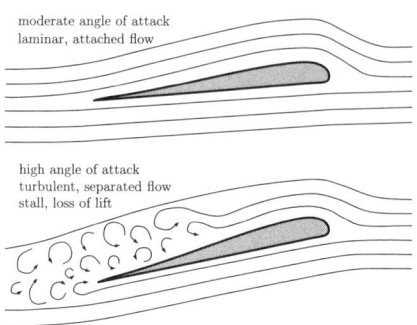

Figure 4.6 The upper drawing depicts a wing under normal flow conditions. The flow is laminar and follows the surface of the wing. The lower drawing shows a wing under stall conditions due to a high angle of attack. In this case, the flow separates from the top surface of the wing, leading to turbulent flow and a significant drop in the lift force produced by the wing.

three-dimensional, time-varying, separated flows that significantly affect the forces and moments experienced by the aircraft. Two unsteady flow scenarios of possible interest to MAV designers are high-angle-of-attack, high-angular-rate aircraft maneuvers, such as those performed by fighter aircraft, and flapping-wing flight. In fact, the efficiency and maneuverability demonstrated by insects and birds is in part because of their ability to exploit unsteady aerodynamic flow effects.

Perhaps the most important unsteady flow phenomena for MAV designers and users to understand is stall, which occurs when the angle of attack increases to the point that the flow separates from the wing, resulting in a drastic loss of lift. Under stall conditions, equations (4.3), (4.4), and (4.5) produce dangerously optimistic estimates of the aerodynamic forces on the aircraft. This wing stall phenomenon is depicted in figure 4.6. At low or moderate angles of attack, the flow over the wing is laminar and stays attached to the wing as it flows over it. It is this attached flow over the wing that produces the desired lift. When the angle of attack exceeds the critical stall angle, the flow begins to separate from the top surface of the wing causing turbulent flow and an abrupt drop in the lift produced by the wing, which can lead to catastrophic results for the aircraft. The key weakness of the linear aerodynamic model of equations (4.3) to (4.5) is that it fails to predict this abrupt drop in lift force with increasing angle of attack. Instead, it erroneously predicts that the lift force continues to increase as the angle of attack increases to physically unrealistic flight conditions. Given that the MAV dynamic model presented here could be used to design control laws for real aircraft and to and simulate their performance, it is important that the effects of wing stall be incorporated into the longitudinal aerodynamic model.

To incorporate wing stall into our longitudinal aerodynamic model, we modify equations (4.3) and (4.4) so that the lift and drag forces are nonlinear in angle of attack. This will allow us to more accurately model lift and drag over wider ranges of α. Lift and drag can be modeled more

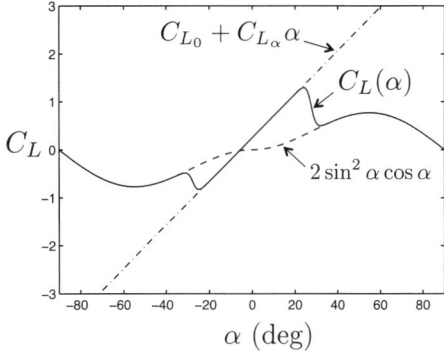

Figure 4.7 The lift coefficient as a function of α (solid) can be approximated by blending a linear function of alpha (dot-dashed), with the lift coefficient of a flat plate (dashed).

generally as

$$F_{\text{lift}} = \frac{1}{2}\rho V_a^2 S \left[C_L(\alpha) + C_{L_q}\frac{c}{2V_a}q + C_{L_{\delta_e}}\delta_e \right] \quad (4.6)$$

$$F_{\text{drag}} = \frac{1}{2}\rho V_a^2 S \left[C_D(\alpha) + C_{D_q}\frac{c}{2V_a}q + C_{D_{\delta_e}}\delta_e \right], \quad (4.7)$$

where C_L and C_D are now expressed nonlinear functions of α. For angles of attack that are beyond the onset of stall conditions, the wing acts roughly like a flat plate, whose lift coefficient can be modeled as [22]

$$C_{L,\text{flat plate}} = 2\,\text{sign}(\alpha)\sin^2\alpha\cos\alpha. \quad (4.8)$$

To obtain an accurate model of lift versus angle of attack for a specific wing design over a large range of angles of attack requires either wind tunnel testing or a detailed computational study. While for many simulation purposes it may not be necessary to have a high-fidelity lift model specific to the aircraft under consideration, it is desirable to have a lift model that incorporates the effects of stall. A lift model that incorporates the common linear lift behavior and the effects of stall is given by

$$C_L(\alpha) = (1-\sigma(\alpha))[C_{L_0} + C_{L_\alpha}\alpha] + \sigma(\alpha)\left[2\,\text{sign}(\alpha)\sin^2\alpha\cos\alpha\right], \quad (4.9)$$

where

$$\sigma(\alpha) = \frac{1 + e^{-M(\alpha-\alpha_0)} + e^{M(\alpha+\alpha_0)}}{(1+e^{-M(\alpha-\alpha_0)})(1+e^{M(\alpha+\alpha_0)})}, \quad (4.10)$$

and M and α_0 are positive constants. The sigmoid function in equation (4.10) is a blending function with cutoff at $\pm\alpha_0$ and transition rate M. Figure 4.7 shows the lift coefficient in equation (4.9) as a

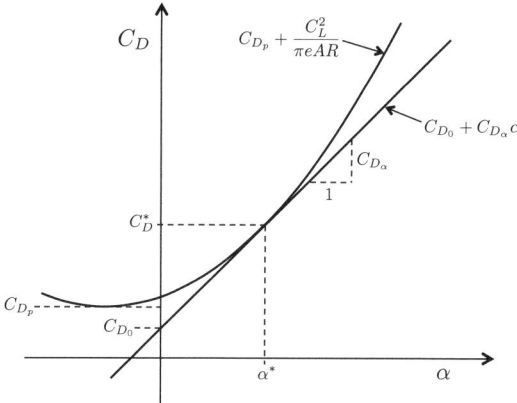

Figure 4.8 The drag coefficient as a function of angle of attack. Linear and quadratic models are represented.

blended function of the linear term $C_{L_0} + C_{L_\alpha}\alpha$ and the flat plate term in equation (4.8). For small aircraft, the linear lift coefficient can be reasonably approximated as

$$C_{L_\alpha} = \frac{\pi AR}{1 + \sqrt{1 + (AR/2)^2}},$$

where $AR \triangleq b^2/S$ is the wing aspect ratio, b is the wingspan, and S is the wing area.

The drag coefficient C_D is also a nonlinear function of the angle of attack. There are two contributions to the drag coefficient, namely induced drag and parasitic drag [22]. The parasitic drag, generated by the shear stress of air moving over the wing and other effects, is roughly constant and is denoted by C_{D_p}.[1] For small angles of attack, the induced drag is proportional to the square of the lift force. Combining the parasitic drag and the induced drag, we have

$$C_D(\alpha) = C_{D_p} + \frac{(C_{L_0} + C_{L_\alpha}\alpha)^2}{\pi e AR}. \tag{4.11}$$

The parameter e is the Oswald efficiency factor, which ranges between 0.8 and 1.0 [12].

Figure 4.8 shows typical plots of drag coefficient versus angle of attack for quadratic and linear models. The quadratic model correctly models

---

[1] The parasitic drag is commonly denoted in the aerodynamics literature as C_{D_0}. To avoid confusion with the constant term of equation (4.4), we will call it C_{D_p}. See [12, pp. 179-180] for a detailed explanation.

the drag force as an even function with respect to α. The drag force is always opposite the forward velocity of the aircraft, independent of the sign of the angle of attack. The linear model incorrectly predicts that the drag force becomes negative (pushing the aircraft forward) when the angle of attack becomes sufficiently negative. The figure clarifies the difference between the parasitic drag, C_{D_p}, also known as the zero-lift drag coefficient, and C_{D_0}, the drag coefficient predicted by the linear model at zero angle of attack. The parameters α^* and C_D^* are the angle of attack and the corresponding drag coefficient at a nominal operating condition $\alpha = \alpha^*$ about which C_D is linearized. While the quadratic model provides a more accurate representation of the influence of the angle of attack over a wider range of α, the linear model is sometimes used because of its simplicity and its fidelity under typical flight conditions.

The lift and drag forces expressed in equations (4.6) and (4.7) are expressed in the stability frame. To express lift and drag in the body frame requires a rotation by the angle of attack:

$$\begin{pmatrix} f_x \\ f_z \end{pmatrix} = \begin{pmatrix} \cos\alpha & -\sin\alpha \\ \sin\alpha & \cos\alpha \end{pmatrix} \begin{pmatrix} -F_{\text{drag}} \\ -F_{\text{lift}} \end{pmatrix}$$

$$= \frac{1}{2}\rho V_a^2 S \begin{pmatrix} [-C_D(\alpha)\cos\alpha + C_L(\alpha)\sin\alpha] \\ + [-C_{D_q}\cos\alpha + C_{L_q}\sin\alpha]\frac{c}{2V_a}q \\ + [-C_{D_{\delta_e}}\cos\alpha + C_{L_{\delta_e}}\sin\alpha]\delta_e \\ --- \\ [-C_D(\alpha)\sin\alpha - C_L(\alpha)\cos\alpha] \\ + [-C_{D_q}\sin\alpha - C_{L_q}\cos\alpha]\frac{c}{2V_a}q \\ + [-C_{D_{\delta_e}}\sin\alpha - C_{L_{\delta_e}}\cos\alpha]\delta_e \end{pmatrix}.$$

The functions $C_L(\alpha)$ and $C_D(\alpha)$ used in the force model above can be nonlinear functions expressed in equations (4.9) and (4.11), which are valid over a wide range of angles of attack. Alternatively, if simpler models are desired, the linear coefficient models given by

$$C_L(\alpha) = C_{L_0} + C_{L_\alpha}\alpha \tag{4.12}$$

$$C_D(\alpha) = C_{D_0} + C_{D_\alpha}\alpha \tag{4.13}$$

can be used.

The pitching moment of the aircraft is generally a nonlinear function of angle of attack and must be determined by wind tunnel or flight experiments for the specific aircraft of interest. For the purposes of simulation, we will use the linear model

$$C_m(\alpha) = C_{m_0} + C_{m_\alpha}\alpha,$$

where $C_{m_\alpha} < 0$ implies that the airframe is inherently pitch stable.

4.2.3 Lateral Aerodynamics

The lateral aerodynamic force and moments cause translational motion in the lateral direction along the $\mathbf{j}^b$ axis as well as rotational motions in roll and yaw that will result in directional changes in the flight path of the MAV. The lateral aerodynamics are most significantly influenced by the sideslip angle β. They are also influenced by the roll rate p, the yaw rate r, the deflection of the aileron δ_a, and the deflection of the rudder δ_r. Denoting the lateral force as f_y and the roll and yaw moments as l and n, respectively, we have

$$f_y = \frac{1}{2}\rho V_a^2 S C_Y(\beta, p, r, \delta_a, \delta_r)$$

$$l = \frac{1}{2}\rho V_a^2 S b C_l(\beta, p, r, \delta_a, \delta_r)$$

$$n = \frac{1}{2}\rho V_a^2 S b C_n(\beta, p, r, \delta_a, \delta_r),$$

where C_Y, C_l, and C_n are nondimensional aerodynamic coefficients, and b is the wingspan of the aircraft. As with the longitudinal aerodynamic forces and moments, the coefficients C_Y, C_l, and C_n are nonlinear in their constitutive parameters, in this case β, p, r, δ_a, and δ_r. These nonlinear relationships, however, are difficult to characterize. Further, linear aerodynamic models yield acceptable accuracy in most applications and provide valuable insights into the dynamic stability of the aircraft. We will follow the approach used in section 4.2.2 to produce the linear longitudinal aerodynamic models: first-order Taylor series approximation and nondimensionalization of the aerodynamic coefficients. Using this approach, linear relationships for lateral force, roll moment, and yaw moment are given by

$$f_y = \frac{1}{2}\rho V_a^2 S \left[C_{Y_0} + C_{Y_\beta}\beta + C_{Y_p}\frac{b}{2V_a}p + C_{Y_r}\frac{b}{2V_a}r + C_{Y_{\delta_a}}\delta_a + C_{Y_{\delta_r}}\delta_r \right] \tag{4.14}$$

$$l = \frac{1}{2}\rho V_a^2 S b \left[C_{l_0} + C_{l_\beta}\beta + C_{l_p}\frac{b}{2V_a}p + C_{l_r}\frac{b}{2V_a}r + C_{l_{\delta_a}}\delta_a + C_{l_{\delta_r}}\delta_r \right] \tag{4.15}$$

$$n = \frac{1}{2}\rho V_a^2 Sb \left[C_{n_0} + C_{n_\beta}\beta + C_{n_p}\frac{b}{2V_a}p + C_{n_r}\frac{b}{2V_a}r + C_{n_{\delta_a}}\delta_a + C_{n_{\delta_r}}\delta_r \right].$$

(4.16)

These forces and moments are aligned with the body axes of the aircraft and do not require a rotational transformation to be implemented in the equations of motion. The coefficient C_{Y_0} is the value of the lateral force coefficient C_Y when $\beta = p = r = \delta_a = \delta_a = 0$. For aircraft that are symmetrical about the $\mathbf{i}^b$-$\mathbf{k}^b$ plane, C_{Y_0} is typically zero. The coefficients C_{l_0} and C_{n_0} are defined similarly and are also typically zero for symmetric aircraft.

4.2.4 Aerodynamic Coefficients

The aerodynamic coefficients C_{m_α}, C_{l_β}, C_{n_β}, C_{m_q}, C_{l_p}, and C_{n_r} are referred to as *stability derivatives* because their values determine the static and dynamic stability of the MAV. Static stability deals with the direction of aerodynamic moments as the MAV is perturbed away from its nominal flight condition. If the moments tend to restore the MAV to its nominal flight condition, the MAV is said to be statically stable. Most aircraft are designed to be statically stable. The coefficients C_{m_α}, C_{l_β}, and C_{n_β} determine the static stability of the MAV. They represent the change in the moment coefficients with respect to changes in the direction of the relative airspeed, as represented by α and β.

C_{m_α} is referred to as the longitudinal static stability derivative. For the MAV to be statically stable, C_{m_α} must be less than zero. In this case, an increase in α due to an updraft would cause the MAV to nose down in order to maintain the nominal angle of attack.

C_{l_β} is called the roll static stability derivative and is typically associated with dihedral in the wings. For static stability in roll, C_{l_β} must be negative. A negative value for C_{l_β} will result in rolling moments that roll the MAV away from the direction of sideslip, thereby driving the sideslip angle β to zero.

C_{n_β} is referred to as the yaw static stability derivative and is sometimes called the weathercock stability derivative. If an aircraft is statically stable in yaw, it will naturally point into the wind like a weathervane (or weathercock). The value of C_{n_β} is heavily influenced by the design of the tail of the aircraft. The larger the tail and the further the tail is aft of the center of mass of the aircraft, the larger C_{n_β} will be. For the MAV to be stable in yaw, C_{n_β} must be positive. This simply implies that for a positive sideslip angle, a positive yawing moment will be induced. This yawing moment will yaw the MAV into the direction of the relative airspeed, driving the sideslip angle to zero.

Dynamic stability deals with the dynamic behavior of the airframe in response to disturbances. If a disturbance is applied to the MAV, the MAV is said to be dynamically stable if the response of the MAV damps out over time. If we use a second-order mass-spring-damper analogy to analyze the MAV, the stability derivatives C_{m_α}, C_{l_β}, and C_{n_β} behave like torsional springs, while the derivatives C_{m_q}, C_{l_p}, and C_{n_r} behave like torsional dampers. The moments of inertia of the MAV body provide the mass. As we will see in chapter 5, when we linearize the dynamic equations of motion for the MAV, the signs of the stability derivatives must be consistent in order to ensure that the characteristic roots of the MAV dynamics lie in the left half of the complex plane.

C_{m_q} is referred to as the pitch damping derivative, C_{l_p} is called the roll damping derivative, and C_{n_r} is referred to as the yaw damping derivative. Each of these damping derivatives is usually negative, meaning that a moment is produced that opposes the direction of motion, thus damping the motion.

The aerodynamic coefficients $C_{m_{\delta_e}}$, $C_{l_{\delta_a}}$, and $C_{n_{\delta_r}}$ are associated with the deflection of control surfaces and are referred to as the *primary control derivatives*. They are primary because the moments produced are the intended result of the specific control surface deflection. For example, the intended result of an elevator deflection δ_e is a pitching moment m. $C_{l_{\delta_r}}$ and $C_{n_{\delta_a}}$ are called *cross-control derivatives*. They define the off-axis moments that occur when the control surfaces are deflected. Control derivatives can be thought of as gains. The larger the value of the control derivative, the larger the magnitude of the moment produced for a given deflection of the control surface.

The sign convention described in section 4.2.1 implies that a positive elevator deflection results in a nose-down pitching moment (negative about $\mathbf{j}^b$), positive aileron deflection causes a right-wing-down rolling moment (positive about $\mathbf{i}^b$), and positive rudder deflection causes a nose-left yawing moment (negative about $\mathbf{k}^b$). We will define the signs of the primary control derivatives so that positive deflections cause positive moments. For this to be the case, $C_{m_{\delta_e}}$ will be negative, $C_{l_{\delta_a}}$ will be positive, and $C_{n_{\delta_r}}$ will be negative.

4.3 Propulsion Forces and Moments

4.3.1 Propeller Thrust

A simple model for the thrust generated by a propeller can be developed by applying Bernoulli's principle to calculate the pressure ahead of and behind the propeller and then applying the pressure difference to the propeller area. This approach will yield a model that is correct for a perfectly efficient propeller. While overly optimistic in its thrust predictions, this model will provide a reasonable starting point for a MAV simulation.

Using Bernoulli's equation, the total pressure upstream of the propeller can be written as

$$P_{\text{upstream}} = P_0 + \frac{1}{2}\rho V_a^2,$$

where P_0 is the static pressure and ρ is the air density. The pressure downstream of the propeller can be expressed as

$$P_{\text{downstream}} = P_0 + \frac{1}{2}\rho V_{\text{exit}}^2,$$

where V_{exit} is the speed of the air as it leaves the propeller. Ignoring the transients in the motor, there is a linear relationship between the pulse-width-modulation command δ_t and the angular velocity of the propeller. The propeller in turn creates an exit air speed of

$$V_{\text{exit}} = k_{\text{motor}}\delta_t.$$

If S_{prop} is the area swept out by the propeller, then the thrust produced by the motor is given by

$$F_{x_p} = S_{\text{prop}} C_{\text{prop}} (P_{\text{downstream}} - P_{\text{upstream}})$$
$$= \frac{1}{2}\rho S_{\text{prop}} C_{\text{prop}} [(k_{\text{motor}}\delta_t)^2 - V_a^2].$$

Therefore,

$$\mathbf{f}_p = \frac{1}{2}\rho S_{\text{prop}} C_{\text{prop}} \begin{pmatrix} (k_{\text{motor}}\delta_t)^2 - V_a^2 \\ 0 \\ 0 \end{pmatrix}.$$

Most MAVs are designed so that the thrust acts directly along the $\mathbf{i}^b$ body-axis of the aircraft. Therefore, the thrust does not produce any moments about the center of mass of the MAV.

4.3.2 Propeller Torque

As the MAV propeller spins, it applies force to the air that passes through the propeller, increasing the momentum of the air while generating a thrust force on the MAV. Equal and opposite forces are applied by the air on the propeller. The net effect of these forces is a torque about the propeller axis of rotation applied to the MAV. The torque applied by the motor to the propeller (and then to the air) results in an equal and opposite torque applied by the propeller to the motor that is fixed to the MAV body. This torque is opposite to the direction of the propeller rotation and proportional to the square of the propeller angular velocity as expressed by

$$T_p = -k_{T_p}(k_\Omega \delta_t)^2,$$

where $\Omega = k_\Omega \delta_t$ is the propeller speed and k_{T_p} is a constant determined by experiment. The moments due to the propulsion system are therefore

$$\mathbf{m}_p = \begin{pmatrix} -k_{T_p}(k_\Omega \delta_t)^2 \\ 0 \\ 0 \end{pmatrix}.$$

The effects of this propeller torque are usually relatively minor. If unaccounted for, the propeller torque will cause a slow rolling motion in the direction opposite the propeller rotation. It is easily corrected by applying a small aileron deflection, which generates a rolling moment to counteract the propeller torque.

4.4 Atmospheric Disturbances

In this section, we will discuss atmospheric disturbances, such as wind, and describe how these disturbances enter into the dynamics of the aircraft. In chapter 2, we defined $\mathbf{V}_g$ as the velocity of the airframe relative to the ground, $\mathbf{V}_a$ as the velocity of the airframe relative to the surrounding air mass, and $\mathbf{V}_w$ as the velocity of the air mass relative to the ground, or in other words, the wind velocity. As shown in equation (2.6), the relationship between ground velocity, air velocity, and wind velocity is given by

$$\mathbf{V}_g = \mathbf{V}_a + \mathbf{V}_w. \tag{4.17}$$

For simulation purposes, we will assume that the total wind vector can be represented as

$$\mathbf{V}_w = \mathbf{V}_{w_s} + \mathbf{V}_{w_g},$$

where $\mathbf{V}_{w_s}$ is a constant vector that represents a steady ambient wind, and $\mathbf{V}_{w_g}$ is a stochastic process that represents wind gusts and other atmospheric disturbances. The ambient (steady) wind is typically expressed in the *inertial* frame as

$$\mathbf{V}_{w_s}^i = \begin{pmatrix} w_{n_s} \\ w_{e_s} \\ w_{d_s} \end{pmatrix},$$

where w_{n_s} is the speed of the steady wind in the north direction, w_{e_s} is the speed of the steady wind in the east direction, and w_{d_s} is the speed of the steady wind in the down direction. The stochastic (gust) component of the wind is typically expressed in the aircraft *body* frame because the atmospheric effects experienced by the aircraft in the direction of its forward motion occur at a higher frequency than do those in the lateral and down directions. The gust portion of the wind can be written in terms of its body-frame components as

$$\mathbf{V}^b_{w_g} = \begin{pmatrix} u_{w_g} \\ v_{w_g} \\ w_{w_g} \end{pmatrix}.$$

Experimental results indicate that a good model for the non-steady gust portion of the wind model is obtained by passing white noise through a linear time-invariant filter given by the von Karmen turbulence spectrum in [22]. Unfortunately, the von Karmen spectrum does not result in a rational transfer function. A suitable approximation of the von Karmen model is given by the Dryden transfer functions

$$H_u(s) = \sigma_u \sqrt{\frac{2V_a}{L_u}} \frac{1}{s + \frac{V_a}{L_u}}$$

$$H_v(s) = \sigma_v \sqrt{\frac{3V_a}{L_v}} \frac{\left(s + \frac{V_a}{\sqrt{3}L_v}\right)}{\left(s + \frac{V_a}{L_v}\right)^2}$$

$$H_w(s) = \sigma_w \sqrt{\frac{3V_a}{L_w}} \frac{\left(s + \frac{V_a}{\sqrt{3}L_w}\right)}{\left(s + \frac{V_a}{L_w}\right)^2},$$

where σ_u, σ_v, and σ_w are the intensities of the turbulence along the vehicle frame axes; and L_u, L_v, and L_w are spatial wavelengths; and V_a is the airspeed of the vehicle. The Dryden models are typically implemented assuming a constant nominal airspeed V_{a_0}. The parameters for the Dryden gust model are defined in MIL-F-8785C. Suitable parameters for low and medium altitudes and light and moderate turbulence were presented in [24] and are shown in table 4.1.

Figure 4.9 shows how the steady wind and atmospheric disturbance components enter into the equations of motion. White noise is passed through the Dryden filters to produce the gust components expressed in the vehicle frame. The steady components of the wind are rotated

TABLE 4.1
Dryden gust model parameters [24]

gust description	altitude (m)	$L_u = L_v$ (m)	L_w (m)	$\sigma_u = \sigma_v$ (m/s)	σ_w (m/s)
low altitude, light turbulence	50	200	50	1.06	0.7
low altitude, moderate turbulence	50	200	50	2.12	1.4
medium altitude, light turbulence	600	533	533	1.5	1.5
medium altitude, moderate turbulence	600	533	533	3.0	3.0

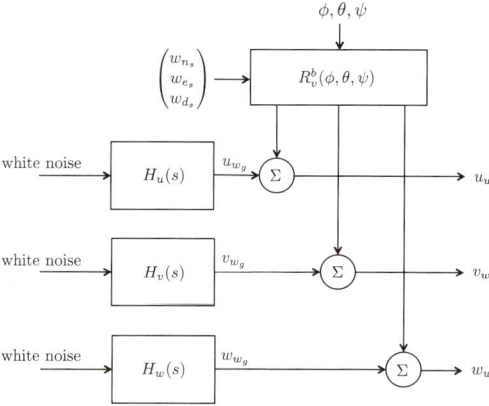

Figure 4.9 The wind is modeled as a constant wind field plus turbulence. The turbulence is generated by filtering white noise with a Dryden model.

from the inertial frame into the body frame and added to the gust components to produce the total wind in the body frame. The combination of steady and gust terms can be expressed mathematically as

$$\mathbf{V}_w^b = \begin{pmatrix} u_w \\ v_w \\ w_w \end{pmatrix} = \mathcal{R}_v^b(\phi, \theta, \psi) \begin{pmatrix} w_{n_s} \\ w_{e_s} \\ w_{d_s} \end{pmatrix} + \begin{pmatrix} u_{w_g} \\ v_{w_g} \\ w_{w_g} \end{pmatrix},$$

where $\mathcal{R}_v^b$ is the rotation matrix from the vehicle to the body frame given in equation (2.5). From the components of the wind velocity $\mathbf{V}_w^b$ and the ground velocity $\mathbf{V}_g^b$, we can calculate the body-frame components of the airspeed vector as

$$\mathbf{V}_a^b = \begin{pmatrix} u_r \\ v_r \\ w_r \end{pmatrix} = \begin{pmatrix} u - u_w \\ v - v_w \\ w - w_w \end{pmatrix}.$$

From the body-frame components of the airspeed vector, we can calculate the airspeed magnitude, the angle of attack, and the sideslip angle according to equation (2.8) as

$$V_a = \sqrt{u_r^2 + v_r^2 + w_r^2}$$

$$\alpha = \tan^{-1}\left(\frac{w_r}{u_r}\right)$$

$$\beta = \sin^{-1}\left(\frac{v_r}{\sqrt{u_r^2 + v_r^2 + w_r^2}}\right).$$

These expressions for V_a, α, and β are used to calculate the aerodynamic forces and moments acting on the vehicle. The key point to understand is that wind and atmospheric disturbances affect the airspeed, the angle of attack, and the sideslip angle. It is through these parameters that wind and atmospheric effects enter the calculation of the aerodynamic forces and moments and thereby influence the motion of the aircraft.

4.5 Chapter Summary

The total forces on the MAV can be summarized as follows:

$$\begin{pmatrix} f_x \\ f_y \\ f_z \end{pmatrix} = \begin{pmatrix} -mg\sin\theta \\ mg\cos\theta\sin\phi \\ mg\cos\theta\cos\phi \end{pmatrix}$$

$$+ \frac{1}{2}\rho V_a^2 S \begin{pmatrix} C_X(\alpha) + C_{X_q}(\alpha)\frac{c}{2V_a}q + C_{X_{\delta_e}}(\alpha)\delta_e \\ C_{Y_0} + C_{Y_\beta}\beta + C_{Y_p}\frac{b}{2V_a}p + C_{Y_r}\frac{b}{2V_a}r + C_{Y_{\delta_a}}\delta_a + C_{Y_{\delta_r}}\delta_r \\ C_Z(\alpha) + C_{Z_q}(\alpha)\frac{c}{2V_a}q + C_{Z_{\delta_e}}(\alpha)\delta_e \end{pmatrix}$$

$$+ \frac{1}{2}\rho S_{\text{prop}} C_{\text{prop}} \begin{pmatrix} (k_{\text{motor}}\delta_t)^2 - V_a^2 \\ 0 \\ 0 \end{pmatrix}, \qquad (4.18)$$

where

$$C_X(\alpha) \triangleq -C_D(\alpha)\cos\alpha + C_L(\alpha)\sin\alpha$$
$$C_{X_q}(\alpha) \triangleq -C_{D_q}\cos\alpha + C_{L_q}\sin\alpha$$
$$C_{X_{\delta_e}}(\alpha) \triangleq -C_{D_{\delta_e}}\cos\alpha + C_{L_{\delta_e}}\sin\alpha \quad (4.19)$$
$$C_Z(\alpha) \triangleq -C_D(\alpha)\sin\alpha - C_L(\alpha)\cos\alpha$$
$$C_{Z_q}(\alpha) \triangleq -C_{D_q}\sin\alpha - C_{L_q}\cos\alpha$$
$$C_{Z_{\delta_e}}(\alpha) \triangleq -C_{D_{\delta_e}}\sin\alpha - C_{L_{\delta_e}}\cos\alpha,$$

and where $C_L(\alpha)$ is given by equation (4.9) and $C_D(\alpha)$ is given by equation (4.11). The subscripts X and Z denote that the forces act in the X and Z directions in the body frame, which correspond to the directions of the $\mathbf{i}^b$ and the $\mathbf{k}^b$ vectors.

The total torques on the MAV can be summarized as follows:

$$\begin{pmatrix} l \\ m \\ n \end{pmatrix} = \frac{1}{2}\rho V_a^2 S \begin{pmatrix} b\left[C_{l_0} + C_{l_\beta}\beta + C_{l_p}\frac{b}{2V_a}p + C_{l_r}\frac{b}{2V_a}r + C_{l_{\delta_a}}\delta_a + C_{l_{\delta_r}}\delta_r\right] \\ c\left[C_{m_0} + C_{m_\alpha}\alpha + C_{m_q}\frac{c}{2V_a}q + C_{m_{\delta_e}}\delta_e\right] \\ b\left[C_{n_0} + C_{n_\beta}\beta + C_{n_p}\frac{b}{2V_a}p + C_{n_r}\frac{b}{2V_a}r + C_{n_{\delta_a}}\delta_a + C_{n_{\delta_r}}\delta_r\right] \end{pmatrix}$$
$$+ \begin{pmatrix} -k_{T_p}(k_\Omega\delta_t)^2 \\ 0 \\ 0 \end{pmatrix}. \quad (4.20)$$

Notes and References

The material in this chapter can be found in most textbooks on flight dynamics, including [12, 22, 1, 2, 5, 7, 25]. Our discussion on lift, drag, and moment coefficients is drawn primarily from [22]. Decomposing the wind vector into a constant and a random term follows [22]. Our discussion of aircraft aerodynamics and dynamics focused on effects of primary significance. A more thorough coverage of flight mechanics, including topics such as ground effect and gyroscopic effects, can be found in [25].

4.6 Design Project

4.1. Download the simulation files from the website. Modify the block forces_moments.m that implements the gravity, aerodynamic, and propulsion forces and torques described in this chapter. Use the parameters given in appendix E.

4.2. Modify the gust block to output wind gusts along the body axes. Modify `forces_moments.m` so that the outputs are the forces and moments resolved in the body frame, the airspeed V_a, the angle of attack α, the sideslip angle β, and the wind vectors resolved in the inertial frame $(w_n, w_e, w_d)^\top$.

4.3. Verify your simulation by setting the control surface deflections to different values. Observe the response of the MAV. Does it behave as you think it should?

5

Linear Design Models

As chapters 3 and 4 have shown, the equations of motion for a MAV are a fairly complicated set of 12 nonlinear, coupled, first-order, ordinary differential equations, which we will present in their entirety in section 5.1. Because of their complexity, designing controllers based on them is difficult and requires more straightforward approaches. In this chapter, we will linearize and decouple the equations of motion to produce reduced-order transfer function and state-space models more suitable for control system design. Low-level autopilot control loops for unmanned aircraft will be designed based on these linear design models, which capture the approximate dynamic behavior of the system under specific conditions. The objective of this chapter is to derive the linear design models that will be used in chapter 6 to design the autopilot.

The dynamics for fixed-wing aircraft can be approximately decomposed into longitudinal motion, which includes airspeed, pitch angle, and altitude, and into lateral motion, which includes roll and heading angles. While there is coupling between longitudinal and lateral motion, for most airframes the dynamic coupling is sufficiently small that its unwanted effects can be mitigated by control algorithms designed for disturbance rejection. In this chapter we will follow the standard convention and decompose the dynamics into lateral and longitudinal motion. Many of the linear models presented in this chapter are derived with respect to an equilibrium condition. In flight dynamics, force and moment equilibrium is called *trim*, which is discussed in section 5.3. Transfer functions for both the lateral and longitudinal dynamics are derived in section 5.4. State-space models are derived in section 5.5.

5.1 Summary of Nonlinear Equations of Motion

A variety of models for the aerodynamic forces and moments appear in the literature ranging from linear, uncoupled models to highly nonlinear models with significant cross coupling. In this section, we summarize the six-degree-of-freedom, 12-state equations of motion with the quasi-linear aerodynamic and propulsion models developed in chapter 4. We characterize them as quasi-linear because the lift and drag terms are nonlinear in the angle of attack, and the propeller thrust is nonlinear in the throttle command. For completeness, we will also present the linear models for lift and drag that are commonly used. Incorporating the aerodynamic and propulsion models described in chapter 4 into equations (3.14)–(3.17), we get the following equations of motion:

$$\dot{p}_n = (\cos\theta\cos\psi)u + (\sin\phi\sin\theta\cos\psi - \cos\phi\sin\psi)v$$
$$+ (\cos\phi\sin\theta\cos\psi + \sin\phi\sin\psi)w \tag{5.1}$$

$$\dot{p}_e = (\cos\theta\sin\psi)u + (\sin\phi\sin\theta\sin\psi + \cos\phi\cos\psi)v$$
$$+ (\cos\phi\sin\theta\sin\psi - \sin\phi\cos\psi)w \tag{5.2}$$

$$\dot{h} = u\sin\theta - v\sin\phi\cos\theta - w\cos\phi\cos\theta \tag{5.3}$$

$$\dot{u} = rv - qw - g\sin\theta$$
$$+ \frac{\rho V_a^2 S}{2m}\left[C_X(\alpha) + C_{X_q}(\alpha)\frac{cq}{2V_a} + C_{X_{\delta_e}}(\alpha)\delta_e\right]$$
$$+ \frac{\rho S_{\text{prop}} C_{\text{prop}}}{2m}\left[(k_{\text{motor}}\delta_t)^2 - V_a^2\right] \tag{5.4}$$

$$\dot{v} = pw - ru + g\cos\theta\sin\phi + \frac{\rho V_a^2 S}{2m}$$
$$\times \left[C_{Y_0} + C_{Y_\beta}\beta + C_{Y_p}\frac{bp}{2V_a} + C_{Y_r}\frac{br}{2V_a} + C_{Y_{\delta_a}}\delta_a + C_{Y_{\delta_r}}\delta_r\right] \tag{5.5}$$

$$\dot{w} = qu - pv + g\cos\theta\cos\phi$$
$$+ \frac{\rho V_a^2 S}{2m}\left[C_Z(\alpha) + C_{Z_q}(\alpha)\frac{cq}{2V_a} + C_{Z_{\delta_e}}(\alpha)\delta_e\right] \tag{5.6}$$

$$\dot{\phi} = p + q\sin\phi\tan\theta + r\cos\phi\tan\theta \tag{5.7}$$

$$\dot{\theta} = q\cos\phi - r\sin\phi \tag{5.8}$$

$$\dot{\psi} = q\sin\phi\sec\theta + r\cos\phi\sec\theta \tag{5.9}$$

$$\dot{p} = \Gamma_1 pq - \Gamma_2 qr + \frac{1}{2}\rho V_a^2 Sb$$
$$\times \left[C_{p_0} + C_{p_\beta}\beta + C_{p_p}\frac{bp}{2V_a} + C_{p_r}\frac{br}{2V_a} + C_{p_{\delta_a}}\delta_a + C_{p_{\delta_r}}\delta_r\right] \tag{5.10}$$

$$\dot{q} = \Gamma_5 pr - \Gamma_6(p^2 - r^2) + \frac{\rho V_a^2 Sc}{2J_y}$$
$$\times \left[C_{m_0} + C_{m_\alpha}\alpha + C_{m_q}\frac{cq}{2V_a} + C_{m_{\delta_e}}\delta_e \right] \quad (5.11)$$

$$\dot{r} = \Gamma_7 pq - \Gamma_1 qr + \frac{1}{2}\rho V_a^2 Sb$$
$$\times \left[C_{r_0} + C_{r_\beta}\beta + C_{r_p}\frac{bp}{2V_a} + C_{r_r}\frac{br}{2V_a} + C_{r_{\delta_a}}\delta_a + C_{r_{\delta_r}}\delta_r \right], \quad (5.12)$$

where $h = -p_d$ is the altitude and

$$C_{p_0} = \Gamma_3 C_{l_0} + \Gamma_4 C_{n_0}$$
$$C_{p_\beta} = \Gamma_3 C_{l_\beta} + \Gamma_4 C_{n_\beta}$$
$$C_{p_p} = \Gamma_3 C_{l_p} + \Gamma_4 C_{n_p}$$
$$C_{p_r} = \Gamma_3 C_{l_r} + \Gamma_4 C_{n_r}$$
$$C_{p_{\delta_a}} = \Gamma_3 C_{l_{\delta_a}} + \Gamma_4 C_{n_{\delta_a}}$$
$$C_{p_{\delta_r}} = \Gamma_3 C_{l_{\delta_r}} + \Gamma_4 C_{n_{\delta_r}}$$
$$C_{r_0} = \Gamma_4 C_{l_0} + \Gamma_8 C_{n_0}$$
$$C_{r_\beta} = \Gamma_4 C_{l_\beta} + \Gamma_8 C_{n_\beta}$$
$$C_{r_p} = \Gamma_4 C_{l_p} + \Gamma_8 C_{n_p}$$
$$C_{r_r} = \Gamma_4 C_{l_r} + \Gamma_8 C_{n_r}$$
$$C_{r_{\delta_a}} = \Gamma_4 C_{l_{\delta_a}} + \Gamma_8 C_{n_{\delta_a}}$$
$$C_{r_{\delta_r}} = \Gamma_4 C_{l_{\delta_r}} + \Gamma_8 C_{n_{\delta_r}}.$$

The inertia parameters specified by $\Gamma_1, \Gamma_2, \ldots, \Gamma_8$ are defined in equation (3.13). As shown in chapter 4, the aerodynamic force coefficients in the X and Z directions are nonlinear functions of the angle of attack. For completeness, we restate them here as

$$C_X(\alpha) \triangleq -C_D(\alpha)\cos\alpha + C_L(\alpha)\sin\alpha$$
$$C_{X_q}(\alpha) \triangleq -C_{D_q}\cos\alpha + C_{L_q}\sin\alpha$$
$$C_{X_{\delta_e}}(\alpha) \triangleq -C_{D_{\delta_e}}\cos\alpha + C_{L_{\delta_e}}\sin\alpha$$
$$C_Z(\alpha) \triangleq -C_D(\alpha)\sin\alpha - C_L(\alpha)\cos\alpha$$
$$C_{Z_q}(\alpha) \triangleq -C_{D_q}\sin\alpha - C_{L_q}\cos\alpha$$
$$C_{Z_{\delta_e}}(\alpha) \triangleq -C_{D_{\delta_e}}\sin\alpha - C_{L_{\delta_e}}\cos\alpha.$$

If we incorporate the effects of stall into the lift coefficient, we can model it as

$$C_L(\alpha) = (1 - \sigma(\alpha))[C_{L_0} + C_{L_\alpha}\alpha] + \sigma(\alpha)[2\,\text{sign}(\alpha)\sin^2\alpha\cos\alpha],$$

where

$$\sigma(\alpha) = \frac{1 + e^{-M(\alpha-\alpha_0)} + e^{M(\alpha+\alpha_0)}}{(1 + e^{-M(\alpha-\alpha_0)})(1 + e^{M(\alpha+\alpha_0)})},$$

and M and α_0 are positive constants.

Further, it is common to model drag as a nonlinear quadratic function of the lift as

$$C_D(\alpha) = C_{D_p} + \frac{(C_{L_0} + C_{L_\alpha}\alpha)^2}{\pi e AR},$$

where e is the Oswald efficiency factor and AR is the aspect ratio of the wing.

If we are interested in modeling MAV flight under low-angle-of-attack conditions, simpler, linear models for the lift and drag coefficients can be used, such as

$$C_L(\alpha) = C_{L_0} + C_{L_\alpha}\alpha$$
$$C_D(\alpha) = C_{D_0} + C_{D_\alpha}\alpha.$$

The equations provided in this section completely describe the dynamic behavior of a MAV in response to inputs from the throttle and the aerodynamic control surfaces (ailerons, elevator, and rudder). These equations are the basis for much of what we do in the remainder of the book and are the core of the MAV simulation environment developed as part of the project exercises at the end of each chapter.

An alternative form of these equations, utilizing quaternions to represent the MAV attitude, is given in appendix B. The quaternion-based equations are free of the gimbal-lock singularity and are more computationally efficient than the Euler-angle-based equations of motion. For this reason, the quaternion form of the equations of motion are often used as the basis for high-fidelity simulations. The quaternion representation of attitude is difficult to interpret physically. For this reason, the Euler-angle representation of attitude is preferred for the reduced-order, linear models that will be developed in this chapter. Furthermore, the gimbal-lock singularity is far removed from the flight conditions that will be considered subsequently, and thus will not cause issues with the models to be developed.

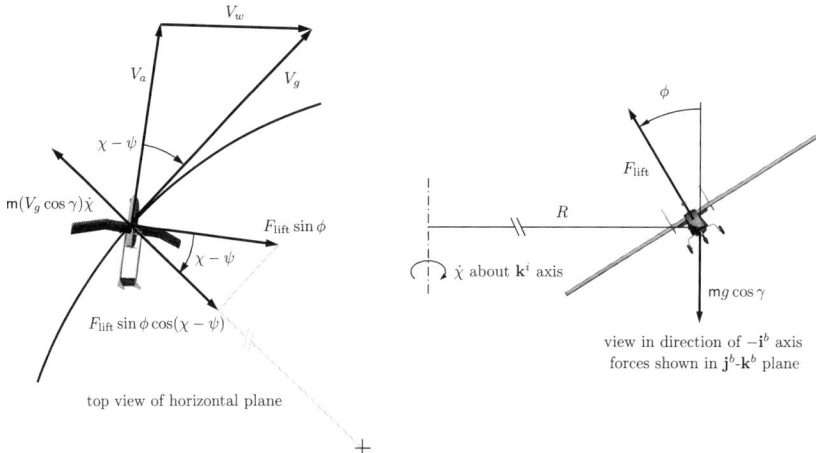

Figure 5.1 Free-body diagrams indicating forces on a MAV in a climbing coordinated turn.

5.2 Coordinated Turn

Referring to equation (5.9), we can see that heading rate is related to the pitch rate, yaw rate, pitch, and roll states of the aircraft. Each of these states is governed by an ordinary differential equation. Physically, we know that heading rate is related to the roll or bank angle of the aircraft, and we seek a simplified relationship to help us develop linear transfer function relationships in coming sections of this chapter. The coordinated-turn condition provides this relationship. The coordinated turn is a sought-after flight condition in manned flight for reasons of passenger comfort. During a coordinated turn, there is no lateral acceleration in the body frame of the aircraft. The aircraft "carves" the turn rather than skidding laterally. From an analysis perspective, the assumption of a coordinated turn allows us to develop a simplified expression that relates course (or heading) rate and bank angle, as shown by Phillips [25]. During a coordinated turn, the bank angle ϕ is set so that there is no net side force acting on the MAV. As shown in the free-body diagram of figure 5.1, the centrifugal force acting on the MAV is equal and opposite to the horizontal component of the lift force acting in the radial direction. Summing forces in the horizontal direction gives

$$\begin{aligned} F_{\text{lift}} \sin \phi \cos(\chi - \psi) &= \mathrm{m}\frac{v^2}{R} \\ &= \mathrm{m} v \omega \\ &= \mathrm{m}(V_g \cos \gamma)\dot{\chi}, \end{aligned} \qquad (5.13)$$

where F_{lift} is the lift force, γ is the flight path angle, V_g is the ground speed, and χ is the course angle.

The centrifugal force is calculated using the angular rate $\dot{\chi}$ about the inertial frame $\mathbf{k}^i$ axis and the horizontal component of the airspeed, $V_a \cos \gamma$. Similarly, the vertical component of the lift force is equal and opposite to the projection of the gravitational force onto the $\mathbf{j}^b$-$\mathbf{k}^b$ plane as shown in figure 5.1. Summing vertical force components gives

$$F_{\text{lift}} \cos \phi = mg \cos \gamma. \tag{5.14}$$

Dividing equation (5.13) by equation (5.14) and solving for $\dot{\chi}$ gives

$$\dot{\chi} = \frac{g}{V_g} \tan \phi \cos(\chi - \psi), \tag{5.15}$$

which is the equation for a coordinated turn. Given that the turning radius is given by $R = V_g \cos \gamma / \dot{\chi}$, we get

$$R = \frac{V_g^2 \cos \gamma}{g \tan \phi \cos(\chi - \psi)}. \tag{5.16}$$

In the absence of wind or sideslip, we have that $V_a = V_g$ and $\psi = \chi$, which leads to the following expressions for the coordinated turn

$$\dot{\chi} = \frac{g}{V_g} \tan \phi = \dot{\psi} = \frac{g}{V_a} \tan \phi.$$

These expressions for the coordinated turn will be used at several points in the text as a means for deriving simplified expressions for the turn dynamics of a MAV. Further discussion on coordinated turns can be found in [25, 26, 27, 130], and we will have more to say about coordinated turns in section 9.2, where we will show that

$$\dot{\psi} = \frac{g}{V_a} \tan \phi$$

also holds true in the presence of wind.

5.3 Trim Conditions

Given a nonlinear system described by the differential equations

$$\dot{x} = f(x, u),$$

where $f : \mathbb{R}^n \times \mathbb{R}^m \to \mathbb{R}^n$, x is the state of the system, and u is the input, the system is said to be in equilibrium at the state x^* and input u^* if

$$f(x^*, u^*) = 0.$$

When a MAV is in constant-altitude, wings-level steady flight, a subset of its states are in equilibrium. In particular, the altitude $h = -p_d$; the body frame velocities u, v, w; the Euler angles ϕ, θ, ψ; and the angular rates p, q, and r are all constant. In the aerodynamics literature, an aircraft in equilibrium is said to be in trim. In general, trim conditions may include states that are not constant. For example, in steady-climb, wings-level flight, $\dot{h}$ is constant and h grows linearly. Also, in a constant turn $\dot{\psi}$ is constant and ψ has linear growth. Therefore, in general, the conditions for trim are given by

$$\dot{x}^* = f(x^*, u^*).$$

In the process of performing trim calculations for the aircraft, we will treat wind as an unknown disturbance. Since its effect on the MAV is unknown, we will find trim assuming that the wind speed is zero: i.e., $V_a = V_g$, $\psi = \chi$, and $\gamma = \gamma_a$.

The objective is to compute trim states and inputs when the aircraft simultaneously satisfies the following three conditions:

- It is traveling at a constant speed V_a^*,
- It is climbing at a constant flight path angle of γ^*,
- It is in a constant orbit of radius R^*.

The three parameters V_a^*, γ^*, and R^*, are inputs to the trim calculations. We will assume that $R^* \geq R_{\min}$, where $R_{\min}$ is the minimum turning radius of the aircraft. The most common scenario where trim values are needed are in wings-level, constant-altitude flight. In that case we have $\gamma^* = 0$ and $R^* = \infty$. Another common scenario is a constant altitude orbit with radius R^*. In that case $\gamma^* = 0$.

For fixed-wing aircraft, the states are given by

$$x \triangleq (p_n, p_e, p_d, u, v, w, \phi, \theta, \psi, p, q, r)^\top \qquad (5.17)$$

and the inputs are given by

$$u \triangleq (\delta_e, \delta_t, \delta_a, \delta_r)^\top, \qquad (5.18)$$

and $f(x, u)$ is specified by the right-hand side of equations (5.1)–(5.12). Note however, that the right-hand side of equations (5.1)–(5.12) are independent of p_n, p_e, p_d. Therefore, trimmed flight is independent of position. In addition, since only $\dot{p}_n$ and $\dot{p}_e$ are dependent on ψ, trimmed flight is also independent of the heading ψ.

In a constant-climb orbit, the speed of the aircraft is not changing, which implies that $\dot{u}^* = \dot{v}^* = \dot{w}^* = 0$. Similarly, since the roll and pitch angles will be constant, we have that $\dot{\phi}^* = \dot{\theta}^* = \dot{p}^* = \dot{q}^* = 0$. The turn rate is constant and is given by

$$\dot{\psi}^* = \frac{V_a^*}{R^*} \cos \gamma^*, \qquad (5.19)$$

which implies that $\dot{r}^* = 0$. Finally, the climb rate is constant, and is given by

$$\dot{h}^* = V_a^* \sin \gamma^*. \qquad (5.20)$$

Therefore, given the parameters V_a^*, γ^*, and R^*, it is possible to specify $\dot{x}^*$, as

$$\dot{x}^* = \begin{pmatrix} \dot{p}_n^* \\ \dot{p}_e^* \\ \dot{h}^* \\ \dot{u}^* \\ \dot{v}^* \\ \dot{w}^* \\ \dot{\phi}^* \\ \dot{\theta}^* \\ \dot{\psi}^* \\ \dot{p}^* \\ \dot{q}^* \\ \dot{r}^* \end{pmatrix} = \begin{pmatrix} \text{[don't care]} \\ \text{[don't care]} \\ V_a^* \sin \gamma^* \\ 0 \\ 0 \\ 0 \\ 0 \\ 0 \\ \frac{V_a^*}{R^*} \cos \gamma^* \\ 0 \\ 0 \\ 0 \end{pmatrix}. \qquad (5.21)$$

The problem of finding x^* (with the exception of p_n^*, p_e^*, h^* and ψ^*) and u^* such that $\dot{x}^* = f(x^*, u^*)$, reduces to solving a nonlinear algebraic system of equations. There are numerous numerical techniques for solving this system of equations. Appendix F describes two methods for solving this system of equations. The first is to use the Simulink `trim` command. If Simulink is not available, however, then appendix F also describes the process required to write a dedicated trim routine.

5.4 Transfer Function Models

Transfer function models for the lateral dynamics are derived in section 5.4.1. These describe the motion of the aircraft in the horizontal plane. Transfer function models for the longitudinal dynamics that describe the motion of the aircraft in the vertical plane are derived in section 5.4.1.

5.4.1 Lateral Transfer Functions

For the lateral dynamics, the variables of interest are the roll angle ϕ, the roll rate p, the heading angle ψ, and the yaw rate r. The control surfaces used to influence the lateral dynamics are the ailerons δ_a, and the rudder δ_r. The ailerons are primarily used to influence the roll rate p, while the rudder is primarily used to control the yaw ψ of the aircraft.

Roll Angle

Our first task is to derive a transfer function from the the ailerons δ_a to the roll angle ϕ. From equation (5.7), we have

$$\dot{\phi} = p + q \sin\phi \tan\theta + r \cos\phi \tan\theta.$$

Since in most flight conditions, θ will be small, the primary influence on $\dot{\phi}$ is the roll rate p. Defining

$$d_{\phi_1} \triangleq q \sin\phi \tan\theta + r\cos\phi \tan\theta$$

and considering d_{ϕ_1} as a disturbance gives

$$\dot{\phi} = p + d_{\phi_1}. \tag{5.22}$$

Differentiating equation (5.22) and using equation (5.10) we get

$$\ddot{\phi} = \dot{p} + \dot{d}_{\phi_1}$$
$$= \Gamma_1 pq - \Gamma_2 qr$$
$$+ \frac{1}{2}\rho V_a^2 Sb \left[C_{p_0} + C_{p_\beta}\beta + C_{p_p}\frac{bp}{2V_a} + C_{p_r}\frac{br}{2V_a} + C_{p_{\delta_a}}\delta_a + C_{p_{\delta_r}}\delta_r \right] + \dot{d}_{\phi_1}$$
$$= \Gamma_1 pq - \Gamma_2 qr + \frac{1}{2}\rho V_a^2 Sb$$
$$\times \left[C_{p_0} + C_{p_\beta}\beta + C_{p_p}\frac{b}{2V_a}(\dot{\phi} - d_{\phi_1}) + C_{p_r}\frac{br}{2V_a} + C_{p_{\delta_a}}\delta_a + C_{p_{\delta_r}}\delta_r \right] + \dot{d}_{\phi_1}$$

Linear Design Models

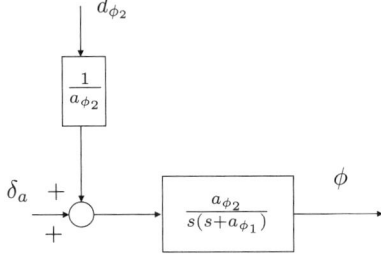

Figure 5.2 Block diagram for roll dynamics. The inputs are the ailerons δ_a and the disturbance d_{ϕ_2}.

$$= \left(\frac{1}{2}\rho V_a^2 S b C_{p_p} \frac{b}{2V_a}\right) \dot{\phi} + \left(\frac{1}{2}\rho V_a^2 S b C_{p_{\delta_a}}\right) \delta_a$$

$$+ \left\{ \Gamma_1 pq - \Gamma_2 qr + \frac{1}{2}\rho V_a^2 S b \left[C_{p_0} + C_{p_\beta}\beta - C_{p_p}\frac{b}{2V_a}(d_{\phi_1}) + C_{p_r}\frac{br}{2V_a}\right.\right.$$

$$\left.\left. + C_{p_{\delta_r}}\delta_r\right] + \dot{d}_{\phi_1} \right\}$$

$$= -a_{\phi_1}\dot{\phi} + a_{\phi_2}\delta_a + d_{\phi_2},$$

where

$$a_{\phi_1} \triangleq -\frac{1}{2}\rho V_a^2 S b C_{p_p} \frac{b}{2V_a} \tag{5.23}$$

$$a_{\phi_2} \triangleq \frac{1}{2}\rho V_a^2 S b C_{p_{\delta_a}} \tag{5.24}$$

$$d_{\phi_2} \triangleq \Gamma_1 pq - \Gamma_2 qr + \frac{1}{2}\rho V_a^2 S b$$

$$\times \left[C_{p_0} + C_{p_\beta}\beta - C_{p_p}\frac{b}{2V_a}(d_{\phi_1}) + C_{p_r}\frac{br}{2V_a} + C_{p_{\delta_r}}\delta_r\right] + \dot{d}_{\phi_1}, \tag{5.25}$$

where d_{ϕ_2} is considered a disturbance on the system.

In the Laplace domain, we have

$$\phi(s) = \left(\frac{a_{\phi_2}}{s(s + a_{\phi_1})}\right)\left(\delta_a(s) + \frac{1}{a_{\phi_2}}d_{\phi_2}(s)\right). \tag{5.26}$$

A block diagram is shown in figure 5.2, where the inputs to the block diagram are the ailerons δ_a and the disturbance d_{ϕ_2}.

Course and Heading

We can also derive a transfer function from the roll angle ϕ to the course angle χ. In a coordinated turn with zero wind, we have that

$$\dot{\chi} = \frac{g}{V_g} \tan \phi.$$

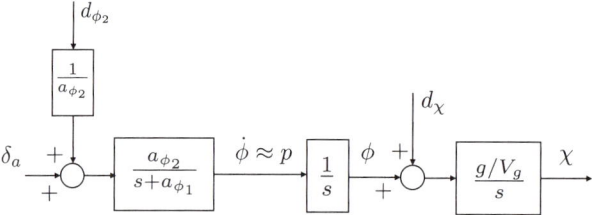

Figure 5.3 Block diagram for lateral dynamics. The roll rate p is shown explicitly because it can be obtained directly from the rate gyros and will be used as a feedback signal in chapter 6.

This equation can be rewritten as

$$\dot{\chi} = \frac{g}{V_g}\phi + \frac{g}{V_g}(\tan\phi - \phi)$$
$$= \frac{g}{V_g}\phi + \frac{g}{V_g}d_\chi,$$

where

$$d_\chi = \tan\phi - \phi$$

is a disturbance. In the Laplace domain, we have

$$\chi(s) = \frac{g/V_g}{s}(\phi(s) + d_\chi(s)). \tag{5.27}$$

This leads to the block diagram for the lateral dynamics controlled by the aileron shown in figure 5.3. To implement this transfer function, we need a value for the ground speed V_g. Since we have assumed zero wind, and we can further assume that the aircraft will track its commanded airspeed, we can use the commanded airspeed as the value for V_g. In chapter 6, we will design control laws to control the flight path of the aircraft relative to the ground. This, combined with the fact that course measurements are readily available from GPS, has lead us to express the transfer function in equation (5.27) in terms of the course angle χ. This transfer function could alternatively be expressed as

$$\psi(s) = \frac{g/V_a}{s}(\phi(s) + d_\chi(s)).$$

Sideslip

A second component of the lateral dynamics is the yaw behavior in response to rudder inputs. In the absence of wind, $v = V_a \sin\beta$. For

Linear Design Models

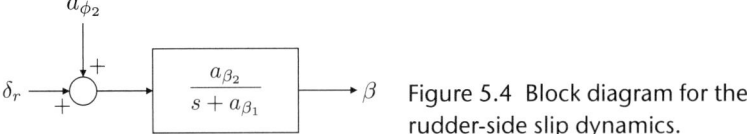

Figure 5.4 Block diagram for the rudder-side slip dynamics.

a constant airspeed, this results in $\dot{v} = (V_a \cos \beta)\dot{\beta}$. Therefore, from equation (5.5), we have

$$(V_a \cos \beta)\dot{\beta} = pw - ru + g \cos \theta \sin \phi + \frac{\rho V_a^2 S}{2m}$$

$$\times \left[C_{Y_0} + C_{Y_\beta}\beta + C_{Y_p}\frac{bp}{2V_a} + C_{Y_r}\frac{br}{2V_a} + C_{Y_{\delta_a}}\delta_a + C_{Y_{\delta_r}}\delta_r \right].$$

With the reasonable assumption that β is small, this leads to $\cos \beta \approx 1$ and

$$\dot{\beta} = -a_{\beta_1}\beta + a_{\beta_2}\delta_r + d_\beta,$$

where

$$a_{\beta_1} = -\frac{\rho V_a S}{2m} C_{Y_\beta}$$

$$a_{\beta_2} = \frac{\rho V_a S}{2m} C_{Y_{\delta_r}}$$

$$d_\beta = \frac{1}{V_a}(pw - ru + g \cos \theta \sin \phi)$$

$$+ \frac{\rho V_a S}{2m}\left[C_{Y_0} + C_{Y_p}\frac{bp}{2V_a} + C_{Y_r}\frac{br}{2V_a} + C_{Y_{\delta_a}}\delta_a \right].$$

In the Laplace domain, we have

$$\beta(s) = \frac{a_{\beta_2}}{s + a_{\beta_1}}(\delta_r(s) + d_\beta(s)). \tag{5.28}$$

This transfer function is depicted in block diagram form in figure 5.4.

Longitudinal Transfer Functions

In this section we will derive transfer function models for the longitudinal dynamics. The variables of interest are the pitch angle θ, the pitch rate q, the altitude $h = -p_d$, and the airspeed V_a. The control signals used to influence the longitudinal dynamics are the elevator δ_e and the throttle δ_t. The elevator will be used to directly influence the pitch angle θ. As we will show below, the pitch angle can be used to manipulate both the altitude h and the airspeed V_a. The airspeed can be used to manipulate the altitude, and the throttle is used to influence the airspeed. The transfer functions derived in this section will be used in chapter 6 to design an altitude control strategy.

Pitch Angle

We begin by deriving a simplified relationship between the elevator δ_e and the pitch angle θ. From equation (5.8), we have

$$\dot{\theta} = q \cos \phi - r \sin \phi$$

$$= q + q(\cos \phi - 1) - r \sin \phi$$

$$\triangleq q + d_{\theta_1},$$

where $d_{\theta_1} \triangleq q(\cos \phi - 1) - r \sin \phi$ and where d_{θ_1} is small for small roll angles ϕ. Differentiating, we get

$$\ddot{\theta} = \dot{q} + \dot{d}_{\theta_1}.$$

Using equation (5.11) and the relationship $\theta = \alpha + \gamma_a$, where $\gamma_a = \gamma$ is the flight path angle, we get

$$\ddot{\theta} = \Gamma_6(r^2 - p^2) + \Gamma_5 pr + \frac{\rho V_a^2 c S}{2 J_y} \left[C_{m_0} + C_{m_\alpha} \alpha + C_{m_q} \frac{cq}{2V_a} + C_{m_{\delta_e}} \delta_e \right] + \dot{d}_{\theta_1}$$

$$= \Gamma_6(r^2 - p^2) + \Gamma_5 pr + \frac{\rho V_a^2 c S}{2 J_y}$$

$$\times \left[C_{m_0} + C_{m_\alpha}(\theta - \gamma) + C_{m_q} \frac{c}{2V_a}(\dot{\theta} - d_{\theta 1}) + C_{m_{\delta_e}} \delta_e \right] + \dot{d}_{\theta_1}$$

$$= \left(\frac{\rho V_a^2 c S}{2 J_y} C_{m_q} \frac{c}{2V_a} \right) \dot{\theta} + \left(\frac{\rho V_a^2 c S}{2 J_y} C_{m_\alpha} \right) \theta + \left(\frac{\rho V_a^2 c S}{2 J_y} C_{m_{\delta_e}} \right) \delta_e$$

$$+ \left\{ \Gamma_6(r^2 - p^2) + \Gamma_5 pr + \frac{\rho V_a^2 c S}{J_y} \left[C_{m_0} - C_{m_\alpha} \gamma - C_{m_q} \frac{c}{2V_a} d_{\theta_1} \right] + \dot{d}_{\theta_1} \right\}$$

$$= -a_{\theta_1} \dot{\theta} - a_{\theta_2} \theta + a_{\theta_3} \delta_e + d_{\theta_2},$$

Linear Design Models

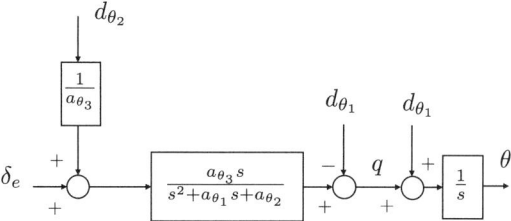

Figure 5.5 Block diagram for the transfer function from the elevator to the pitch angle. The pitch rate q is shown explicitly because it is available from the rate gyros and will be used as a feedback signal in chapter 6.

where

$$a_{\theta_1} \triangleq -\frac{\rho V_a^2 cS}{2J_y} C_{m_q} \frac{c}{2V_a}$$

$$a_{\theta_2} \triangleq -\frac{\rho V_a^2 cS}{2J_y} C_{m_\alpha}$$

$$a_{\theta_3} \triangleq \frac{\rho V_a^2 cS}{2J_y} C_{m_{\delta_e}}$$

$$d_{\theta_2} \triangleq \Gamma_6(r^2 - p^2) + \Gamma_5 pr + \frac{\rho V_a^2 cS}{2J_y} \left[C_{m_0} - C_{m_\alpha} \gamma - C_{m_q} \frac{c}{2V_a} d_{\theta_1} \right] + \dot{d}_{\theta_1}.$$

We have derived a linear model for the evolution of the pitch angle. Taking the Laplace transform, we have

$$\theta(s) = \left(\frac{a_{\theta_3}}{s^2 + a_{\theta_1} s + a_{\theta_2}} \right) \left(\delta_e(s) + \frac{1}{a_{\theta_3}} d_{\theta_2}(s) \right). \quad (5.29)$$

Note that in straight and level flight, $r = p = \phi = \gamma = 0$. In addition, airframes are usually designed so that $C_{m_0} = 0$, which implies that $d_{\theta_2} = 0$. Using the fact that $\dot{\theta} = q + d_{\theta_1}$, we get the block diagram shown in figure 5.5. The model shown in figure 5.5 is useful because the pitch rate q is directly available from the rate gyros for feedback and therefore needs to be accessible in the model.

Altitude

For a constant airspeed, the pitch angle directly influences the climb rate of the aircraft. Therefore, we can develop a transfer function from

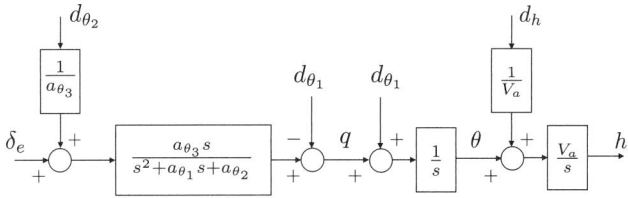

Figure 5.6 Block diagram for longitudinal dynamics.

pitch angle to altitude. From equation (5.3), we have

$$\dot{h} = u \sin \theta - v \sin \phi \cos \theta - w \cos \phi \cos \theta$$
$$= V_a \theta + (u \sin \theta - V_a \theta) - v \sin \phi \cos \theta - w \cos \phi \cos \theta$$
$$= V_a \theta + d_h, \tag{5.30}$$

where

$$d_h \triangleq (u \sin \theta - V_a \theta) - v \sin \phi \cos \theta - w \cos \phi \cos \theta.$$

Note that in straight and level flight, where $v \approx 0$, $w \approx 0$, $u \approx V_a$, $\phi \approx 0$ and θ is small, we have $d_h \approx 0$.

If we assume that V_a is constant and that θ is the input, then in the Laplace domain equation (5.30) becomes

$$h(s) = \frac{V_a}{s} \left(\theta + \frac{1}{V_a} d_h \right), \tag{5.31}$$

and the resulting block diagram for the longitudinal dynamics from the elevator to the altitude is shown in figure 5.6. Alternatively, if the pitch angle is held constant, then increasing the airspeed will result in increased lift over the wings, resulting in a change in altitude. To derive the transfer function from airspeed to altitude, hold θ constant in equation (5.30) and consider V_a as the input to obtain

$$h(s) = \frac{\theta}{s} \left(V_a + \frac{1}{\theta} d_h \right). \tag{5.32}$$

The altitude controllers discussed in chapter 6 will regulate altitude using both pitch angle and airspeed. Similarly, the airspeed will be regulated using both the throttle setting and the pitch angle. For example, when the pitch angle is constant, increasing the throttle will increase the thrust, which increases the airspeed of the vehicle. On the other hand, if the throttle is held constant, then pitching the nose down will

Linear Design Models

decrease the lift causing the aircraft to accelerate downward under the influence of gravity, thereby increasing its airspeed.

Airspeed

To complete the longitudinal models, we derive the transfer functions from throttle and pitch angle to airspeed. Toward that objective, note that if wind speed is zero, then $V_a = \sqrt{u^2 + v^2 + w^2}$, which implies that

$$\dot{V}_a = \frac{u\dot{u} + v\dot{v} + w\dot{w}}{V_a}.$$

Using equation (2.7), we get

$$\dot{V}_a = \dot{u}\cos\alpha\cos\beta + \dot{v}\sin\beta + \dot{w}\sin\alpha\cos\beta$$
$$= \dot{u}\cos\alpha + \dot{w}\sin\alpha + d_{V_1}, \quad (5.33)$$

where

$$d_{V_1} = -\dot{u}(1 - \cos\beta)\cos\alpha - \dot{w}(1 - \cos\beta)\sin\alpha + \dot{v}\sin\beta.$$

Note that when $\beta = 0$, we have $d_{V1} = 0$. Substituting equations (5.4) and (5.6) in equation (5.33), we obtain

$$\dot{V}_a = \cos\alpha \left\{ rv - qw + r - g\sin\theta + \frac{\rho V_a^2 S}{2m}\left[-C_D(\alpha)\cos\alpha \right. \right.$$

$$+ C_L(\alpha)\sin\alpha + (-C_{D_q}\cos\alpha + C_{L_q}\sin\alpha)\frac{cq}{2V_a}$$

$$\left. + (-C_{D_{\delta_e}}\cos\alpha + C_{L_{\delta_e}}\sin\alpha)\delta_e \right] + \frac{\rho S_{\text{prop}} C_{\text{prop}}}{2m}\left[(k\delta_t)^2 - V_a^2\right] \right\}$$

$$+ \sin\alpha \left\{ qu_r - pv_r + g\cos\theta\cos\phi + \frac{\rho V_a^2 S}{2m} \right.$$

$$\times \left[-C_D(\alpha)\sin\alpha - C_L(\alpha)\cos\alpha + (-C_{D_q}\sin\alpha - C_{L_q}\cos\alpha)\frac{cq}{2V_a} \right.$$

$$\left.\left. + (-C_{D_{\delta_e}}\sin\alpha - C_{L_{\delta_e}}\cos\alpha)\delta_e \right] \right\} + d_{V_1}.$$

Using equations (2.7) and the linear approximation $C_D(\alpha) \approx C_{D_0}+C_{D_\alpha}\alpha$, and simplifying, we get

$$\dot{V}_a = r V_a \cos\alpha \sin\beta - p V_a \sin\alpha \sin\beta$$

$$- g \cos\alpha \sin\theta + g \sin\alpha \cos\theta \cos\phi$$

$$+ \frac{\rho V_a^2 S}{2m}\left[-C_D(\alpha) - C_{D_\alpha}\alpha - C_{D_q}\frac{cq}{2V_a} - C_{D_{\delta_e}}\delta_e\right]$$

$$+ \frac{\rho S_{\text{prop}} C_{\text{prop}}}{2m}\left[(k\delta_t)^2 - V_a^2\right] \cos\alpha + d_{V_1}$$

$$= (r V_a \cos\alpha - p V_a \sin\alpha) \sin\beta$$

$$- g \sin(\theta - \alpha) - g \sin\alpha \cos\theta(1 - \cos\phi)$$

$$+ \frac{\rho V_a^2 S}{2m}\left[-C_{D_0} - C_{D_\alpha}\alpha - C_{D_q}\frac{cq}{2V_a} - C_{D_{\delta_e}}\delta_e\right]$$

$$+ \frac{\rho S_{\text{prop}} C_{\text{prop}}}{2m}\left[(k\delta_t)^2 - V_a^2\right] \cos\alpha + d_{V_1}$$

$$= -g \sin\gamma + \frac{\rho V_a^2 S}{2m}\left[-C_{D_0} - C_{D_\alpha}\alpha - C_{D_q}\frac{cq}{2V_a} - C_{D_{\delta_e}}\delta_e\right]$$

$$+ \frac{\rho S_{\text{prop}} C_{\text{prop}}}{2m}\left[(k\delta_t)^2 - V_a^2\right] + d_{V_2}, \tag{5.34}$$

where

$$d_{V_2} = (r V_a \cos\alpha - p V_a \sin\alpha) \sin\beta - g \sin\alpha \cos\theta(1 - \cos\phi)$$

$$+ \frac{\rho S_{\text{prop}} C_{\text{prop}}}{2m}[(k\delta_t)^2 - V_a^2](\cos\alpha - 1) + d_{V_1}.$$

Again note that in level flight $d_{V_2} \approx 0$.

When considering airspeed V_a, there are two inputs of interest: the throttle setting δ_t and the pitch angle θ. Since equation (5.34) is nonlinear in V_a and δ_t, we must first linearize before we can find the desired transfer functions. Following the approach outlined in section 5.5.1, we can linearize equation (5.34) by letting $\bar{V}_a \triangleq V_a - V_a^*$ be the deviation of V_a from trim, $\bar{\theta} \triangleq \theta - \theta^*$ be the deviation of θ from trim, and $\bar{\delta}_t \triangleq \delta_t - \delta_t^*$ be the deviation of the throttle from trim. Equation (5.34) can then

Linear Design Models

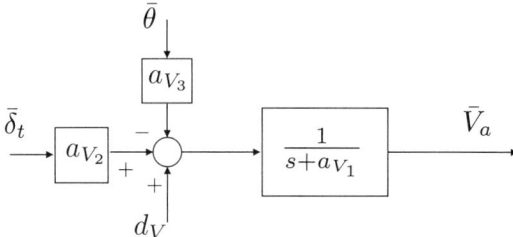

Figure 5.7 Block diagram for airspeed dynamics linearized about trim conditions. The inputs are either the deviation of the pitch angle from trim, or the deviation of the throttle from trim.

be linearized around the wings-level, constant-altitude ($\gamma^* = 0$) trim condition to give

$$\dot{\bar{V}}_a = -g\cos(\theta^* - \alpha^*)\bar{\theta} + \left\{ \frac{\rho V_a^* S}{m} \left[-C_{D_0} - C_{D_\alpha}\alpha^* - C_{D_{\delta_e}}\delta_e^* \right] \right.$$
$$\left. - \frac{\rho S_{\text{prop}}}{m} C_{\text{prop}} V_a^* \right\} \bar{V}_a + \left[\frac{\rho S_{\text{prop}}}{m} C_{\text{prop}} k^2 \delta_t^* \right] \bar{\delta}_t + d_V$$
$$= -a_{V_1} \bar{V}_a + a_{V_2} \bar{\delta}_t - a_{V_3} \bar{\theta} + d_V, \tag{5.35}$$

where

$$a_{V_1} = \frac{\rho V_a^* S}{m} \left[C_{D_0} + C_{D_\alpha}\alpha^* + C_{D_{\delta_e}}\delta_e^* \right] + \frac{\rho S_{\text{prop}}}{m} C_{\text{prop}} V_a^*$$

$$a_{V_2} = \frac{\rho S_{\text{prop}}}{m} C_{\text{prop}} k^2 \delta_t^*$$

$$a_{V_3} = g\cos(\theta^* - \chi^*),$$

and d_V includes d_{V_2} as well as the linearization error. In the Laplace domain we have

$$\bar{V}_a(s) = \frac{1}{s + a_{V_1}} (a_{V_2} \bar{\delta}_t(s) - a_{V_3} \bar{\theta}(s) + d_V(s)). \tag{5.36}$$

The associated block diagram is shown in figure 5.7.

5.5 Linear State-space Models

In this section we will derive linear state-space models for both longitudinal and lateral motion by linearizing equations (5.1)–(5.12) about trim conditions. Section 5.5.1 discusses general linearization techniques. Section 5.5.2 derives the state-space equations for the lateral dynamics, and section 5.5.3 derives the state-space equations for the longitudinal dynamics. Finally, section 5.6 describes the reduced-order modes including the short-period mode, the phugoid mode, the dutch-roll mode, and the spiral-divergence mode.

5.5.1 Linearization

Given the general nonlinear system of equations

$$\dot{x} = f(x, u),$$

where $x \in \mathbb{R}^n$ is the state and $u \in \mathbb{R}^m$ is the control vector, and suppose that, using the techniques discussed in section 5.3, a trim input u^* and state x^* can be found such that

$$\dot{x}^* = f(x^*, u^*) = 0.$$

Letting $\bar{x} \triangleq x - x^*$, we get

$$\dot{\bar{x}} = \dot{x} - \dot{x}^*$$
$$= f(x, u) - f(x^*, u^*)$$
$$= f(x + x^* - x^*, u + u^* - u^*) - f(x^*, u^*)$$
$$= f(x^* + \bar{x}, u^* + \bar{u}) - f(x^*, u^*).$$

Taking the Taylor series expansion of the first term about the trim state, we get

$$\dot{\bar{x}} = f(x^*, u^*) + \frac{\partial f(x^*, u^*)}{\partial x}\bar{x} + \frac{\partial f(x^*, u^*)}{\partial u}\bar{u} + H.O.T. - f(x^*, u^*)$$
$$\approx \frac{\partial f(x^*, u^*)}{\partial x}\bar{x} + \frac{\partial f(x^*, u^*)}{\partial u}\bar{u}. \qquad (5.37)$$

Therefore, the linearized dynamics are determined by finding $\frac{\partial f}{\partial x}$ and $\frac{\partial f}{\partial u}$, evaluated at the trim conditions.

5.5.2 Lateral State-space Equations

For the lateral state-space equations, the state is given by

$$\dot{x}_{\text{lat}} \triangleq (v, p, r, \phi, \psi)^\top,$$

and the input vector is defined as
$$u_{\text{lat}} \triangleq (\delta_a, \delta_r)^\top.$$
Expressing equations (5.5), (5.10), (5.12), (5.7), and (5.9) in terms of x_{lat} and u_{lat}, we get

$$\dot{v} = pw - ru + g\cos\theta\sin\phi + \frac{\rho\sqrt{u^2 + v^2 + w^2}Sb}{2m}\frac{b}{2}\left[C_{Y_p}p + C_{Y_r}r\right]$$

$$+ \frac{\rho(u^2 + v^2 + w^2)S}{2m}\left[C_{Y_0} + C_{Y_\beta}\tan^{-1}\left(\frac{v}{\sqrt{u^2 + w^2}}\right)\right.$$

$$\left. + C_{Y_{\delta_a}}\delta_a + C_{Y_{\delta_r}}\delta_r\right] \tag{5.38}$$

$$\dot{p} = \Gamma_1 pq - \Gamma_2 qr + \frac{\rho\sqrt{u^2 + v^2 + w^2}Sb^2}{2}\frac{b^2}{2}\left[C_{p_p}p + C_{p_r}r\right]$$

$$+ \frac{1}{2}\rho(u^2 + v^2 + w^2)Sb\left[C_{p_0} + C_{p_\beta}\tan^{-1}\left(\frac{v}{\sqrt{u^2 + w^2}}\right)\right.$$

$$\left. + C_{p_{\delta_a}}\delta_a + C_{p_{\delta_r}}\delta_r\right] \tag{5.39}$$

$$\dot{r} = \Gamma_7 pq - \Gamma_1 qr + \frac{\rho\sqrt{u^2 + v^2 + w^2}Sb^2}{2}\frac{b^2}{2}\left[C_{r_p}p + C_{r_r}r\right]$$

$$+ \frac{1}{2}\rho(u^2 + v^2 + w^2)Sb\left[C_{r_0} + C_{r_\beta}\tan^{-1}\left(\frac{v}{\sqrt{u^2 + w^2}}\right)\right.$$

$$\left. + C_{r_{\delta_a}}\delta_a + C_{r_{\delta_r}}\delta_r\right] \tag{5.40}$$

$$\dot{\phi} = p + q\sin\phi\tan\theta + r\cos\phi\tan\theta \tag{5.41}$$

$$\dot{\psi} = q\sin\phi\sec\theta + r\cos\phi\sec\theta, \tag{5.42}$$

where we have utilized the zero-wind expressions
$$\beta = \tan^{-1}\left(\frac{v}{\sqrt{u^2 + w^2}}\right)$$
$$V_a = \sqrt{u^2 + v^2 + w^2}.$$

The Jacobians of equations (5.38)–(5.42) are given by

$$\frac{\partial f_{\text{lat}}}{\partial x_{\text{lat}}} = \begin{pmatrix} \frac{\partial \dot{v}}{\partial v} & \frac{\partial \dot{v}}{\partial p} & \frac{\partial \dot{v}}{\partial r} & \frac{\partial \dot{v}}{\partial \phi} & \frac{\partial \dot{v}}{\partial \psi} \\ \frac{\partial \dot{p}}{\partial v} & \frac{\partial \dot{p}}{\partial p} & \frac{\partial \dot{p}}{\partial r} & \frac{\partial \dot{p}}{\partial \phi} & \frac{\partial \dot{p}}{\partial \psi} \\ \frac{\partial \dot{r}}{\partial v} & \frac{\partial \dot{r}}{\partial p} & \frac{\partial \dot{r}}{\partial r} & \frac{\partial \dot{r}}{\partial \phi} & \frac{\partial \dot{r}}{\partial \psi} \\ \frac{\partial \dot{\phi}}{\partial v} & \frac{\partial \dot{\phi}}{\partial p} & \frac{\partial \dot{\phi}}{\partial r} & \frac{\partial \dot{\phi}}{\partial \phi} & \frac{\partial \dot{\phi}}{\partial \psi} \\ \frac{\partial \dot{\psi}}{\partial v} & \frac{\partial \dot{\psi}}{\partial p} & \frac{\partial \dot{\psi}}{\partial r} & \frac{\partial \dot{\psi}}{\partial \phi} & \frac{\partial \dot{\psi}}{\partial \psi} \end{pmatrix}$$

$$\frac{\partial f_{\text{lat}}}{\partial u_{\text{lat}}} = \begin{pmatrix} \frac{\partial \dot{v}}{\partial \delta_a} & \frac{\partial \dot{v}}{\partial \delta_r} \\ \frac{\partial \dot{p}}{\partial \delta_a} & \frac{\partial \dot{p}}{\partial \delta_r} \\ \frac{\partial \dot{r}}{\partial \delta_a} & \frac{\partial \dot{r}}{\partial \delta_r} \\ \frac{\partial \dot{\phi}}{\partial \delta_a} & \frac{\partial \dot{\phi}}{\partial \delta_r} \\ \frac{\partial \dot{\psi}}{\partial \delta_a} & \frac{\partial \dot{\psi}}{\partial \delta_r} \end{pmatrix}.$$

Toward that end, note that

$$\frac{\partial}{\partial v} \tan^{-1}\left(\frac{v}{\sqrt{u^2+w^2}}\right) = \frac{\sqrt{u^2+w^2}}{u^2+v^2+w^2} = \frac{\sqrt{u^2+w^2}}{V_a^2}.$$

Working out the derivatives, we get that the linearized state-space equations are

$$\begin{pmatrix} \dot{\bar{v}} \\ \dot{\bar{p}} \\ \dot{\bar{r}} \\ \dot{\bar{\phi}} \\ \dot{\bar{\psi}} \end{pmatrix} = \begin{pmatrix} Y_v & Y_p & Y_r & g\cos\theta^*\cos\phi^* & 0 \\ L_v & L_p & L_r & 0 & 0 \\ N_v & N_p & N_r & 0 & 0 \\ 0 & 1 & \cos\phi^*\tan\theta^* & q^*\cos\phi^*\tan\theta^* - r^*\sin\phi^*\tan\theta^* & 0 \\ 0 & 0 & \cos\phi^*\sec\theta^* & p^*\cos\phi^*\sec\theta^* - r^*\sin\phi^*\sec\theta^* & 0 \end{pmatrix}$$

$$\times \begin{pmatrix} \bar{v} \\ \bar{p} \\ \bar{r} \\ \bar{\phi} \\ \bar{\psi} \end{pmatrix} + \begin{pmatrix} Y_{\delta_a} & Y_{\delta_r} \\ L_{\delta_a} & L_{\delta_r} \\ N_{\delta_a} & N_{\delta_r} \\ 0 & 0 \\ 0 & 0 \end{pmatrix} \begin{pmatrix} \bar{\delta}_a \\ \bar{\delta}_r \end{pmatrix}, \qquad (5.43)$$

where the coefficients are given in table 5.1.

TABLE 5.1
Lateral State-space Model Coefficients

Lateral	Formula
Y_v	$\frac{\rho S b v^*}{4 m V_a^*}[C_{Y_p} p^* + C_{Y_r} r^*] + \frac{\rho S v^*}{m}\left[C_{Y_0} + C_{Y_\beta}\beta^* + C_{Y_{\delta_a}}\delta_a^* + C_{Y_{\delta_r}}\delta_r^*\right]$
	$+ \frac{\rho S C_{Y_\beta}}{2m}\sqrt{u^{*2} + w^{*2}}$
Y_p	$w^* + \frac{\rho V_a^* S b}{4m} C_{Y_p}$
Y_r	$-u^* + \frac{\rho V_a^* S b}{4m} C_{Y_r}$
Y_{δ_a}	$\frac{\rho V_a^{*2} S}{2m} C_{Y_{\delta_a}}$
Y_{δ_r}	$\frac{\rho V_a^{*2} S}{2m} C_{Y_{\delta_r}}$
L_v	$\frac{\rho S b^2 v^*}{4 V_a^*}[C_{p_p} p^* + C_{p_r} r^*] + \rho S b v^*[C_{p_0} + C_{p_\beta}\beta^* + C_{p_{\delta_a}}\delta_a^* + C_{p_{\delta_r}}\delta_r^*]$
	$+ \frac{\rho S b C_{p_\beta}}{2}\sqrt{u^{*2} + w^{*2}}$
L_p	$\Gamma_1 q^* + \frac{\rho V_a^* S b^2}{4} C_{p_p}$
L_r	$-\Gamma_2 q^* + \frac{\rho V_a^* S b^2}{4} C_{p_r}$
L_{δ_a}	$\frac{\rho V_a^{*2} S b}{2} C_{p_{\delta_a}}$
L_{δ_r}	$\frac{\rho V_a^{*2} S b}{2} C_{p_{\delta_r}}$
N_v	$\frac{\rho S b^2 v^*}{4 V_a^*}\left[C_{r_p} p^* + C_{r_r} r^*\right] + \rho S b v^*[C_{r_0} + C_{r_\beta}\beta^* + C_{r_{\delta_a}}\delta_a^* + C_{r_{\delta_r}}\delta_r^*]$
	$+ \frac{\rho S b C_{r_\beta}}{2}\sqrt{u^{*2} + w^{*2}}$
N_p	$\Gamma_7 q^* + \frac{\rho V_a^* S b^2}{4} C_{r_p}$
N_r	$-\Gamma_1 q^* + \frac{\rho V_a^* S b^2}{4} C_{r_r}$
N_{δ_a}	$\frac{\rho V_a^{*2} S b}{2} C_{r_{\delta_a}}$
N_{δ_r}	$\frac{\rho V_a^{*2} S b}{2} C_{r_{\delta_r}}$

The lateral equations are often given in terms of $\bar{\beta}$ instead of $\bar{v}$. From equation (2.7), we have

$$v = V_a \sin \beta.$$

Linearizing around $\beta = \beta^*$, we get

$$\bar{v} = V_a^* \cos \beta^* \bar{\beta},$$

which implies that

$$\dot{\bar{\beta}} = \frac{1}{V_a^* \cos \beta^*} \dot{\bar{v}}.$$

Therefore, we can write the state-space equations in terms of $\bar{\beta}$ instead of $\bar{v}$ as

$$\begin{pmatrix} \dot{\bar{\beta}} \\ \dot{\bar{p}} \\ \dot{\bar{r}} \\ \dot{\bar{\phi}} \\ \dot{\bar{\psi}} \end{pmatrix} = \begin{pmatrix} Y_v & \frac{Y_p}{V_a^* \cos \beta^*} & \frac{Y_r}{V_a^* \cos \beta^*} & \frac{g \cos \theta^* \cos \phi^*}{V_a^* \cos \beta^*} & 0 \\ L_v V_a^* \cos \beta^* & L_p & L_r & 0 & 0 \\ N_v V_a^* \cos \beta^* & N_p & N_r & 0 & 0 \\ 0 & 1 & \cos \phi^* \tan \theta^* & q^* \cos \phi^* \tan \theta^* & 0 \\ & & & -r^* \sin \phi^* \tan \theta^* & \\ 0 & 0 & \cos \phi^* \sec \theta^* & p^* \cos \phi^* \sec \theta^* & 0 \\ & & & -r^* \sin \phi^* \sec \theta^* & \end{pmatrix} \begin{pmatrix} \bar{\beta} \\ \bar{p} \\ \bar{r} \\ \bar{\phi} \\ \bar{\psi} \end{pmatrix}$$

$$+ \begin{pmatrix} \frac{Y_{\delta_a}}{V_a^* \cos \beta^*} & \frac{Y_{\delta_r}}{V_a^* \cos \beta^*} \\ L_{\delta_a} & L_{\delta_r} \\ N_{\delta_a} & N_{\delta_r} \\ 0 & 0 \\ 0 & 0 \end{pmatrix} \begin{pmatrix} \bar{\delta}_a \\ \bar{\delta}_r \end{pmatrix}. \tag{5.44}$$

5.5.3 Longitudinal State-space Equations

For the longitudinal state-space equations, the state is given by

$$\dot{x}_{\text{lon}} \triangleq (u, w, q, \theta, h)^\top,$$

and the input vector is defined as

$$u_{\text{lon}} \triangleq (\delta_e, \delta_t)^\top.$$

Linear Design Models

Expressing equations (5.4), (5.6), (5.11), (5.8), and (5.3) in terms of x_{lon} and u_{lon}, we get

$$\dot{u} = rv - qw - g\sin\theta + \frac{\rho V_a^2 S}{2m}\left[C_{X_0} + C_{X_\alpha}\alpha + C_{X_q}\frac{cq}{2V_a} + C_{X_{\delta_e}}\delta_e\right]$$

$$+ \frac{\rho S_{\text{prop}}}{2m}C_{\text{prop}}[(k\delta_t)^2 - V_a^2]$$

$$\dot{w} = qu - pv + g\cos\theta\cos\phi + \frac{\rho V_a^2 S}{2m}\left[C_{Z_0} + C_{Z_\alpha}\alpha + C_{Z_q}\frac{cq}{2V_a} + C_{Z_{\delta_e}}\delta_e\right]$$

$$\dot{q} = \frac{J_{xz}}{J_y}(r^2 - p^2) + \frac{J_z - J_x}{J_y}pr + \frac{1}{2J_y}\rho V_a^2 cS$$

$$\times \left[C_{m_0} + C_{m_\alpha}\alpha + C_{m_q}\frac{cq}{2V_a} + C_{m_{\delta_e}}\delta_e\right]$$

$$\dot{\theta} = q\cos\phi - r\sin\phi$$

$$\dot{h} = u\sin\theta - v\sin\phi\cos\theta - w\cos\phi\cos\theta.$$

Assuming that the lateral states are zero (i.e., $\phi = p = r = \beta = v = 0$) and the windspeed is zero, substituting

$$\alpha = \tan^{-1}\left(\frac{w}{u}\right)$$

$$V_a = \sqrt{u^2 + w^2}$$

from equation (2.7) gives

$$\dot{u} = -qw - g\sin\theta + \frac{\rho(u^2 + w^2)S}{2m}$$

$$\times \left[C_{X_0} + C_{X_\alpha}\tan^{-1}\left(\frac{w}{u}\right) + C_{X_{\delta_e}}\delta_e\right] + \frac{\rho\sqrt{u^2 + w^2}S}{4m}C_{X_q}cq$$

$$+ \frac{\rho S_{\text{prop}}}{2m}C_{\text{prop}}[(k\delta_t)^2 - (u^2 + w^2)] \qquad (5.45)$$

$$\dot{w} = qu + g\cos\theta + \frac{\rho(u^2+w^2)S}{2m}$$

$$\times \left[C_{Z_0} + C_{Z_\alpha}\tan^{-1}\left(\frac{w}{u}\right) + C_{Z_{\delta_e}}\delta_e\right] + \frac{\rho\sqrt{u^2+w^2}S}{4m}C_{Z_q}cq \quad (5.46)$$

$$\dot{q} = \frac{1}{2J_y}\rho(u^2+w^2)cS\left[C_{m_0} + C_{m_\alpha}\tan^{-1}\left(\frac{w}{u}\right) + C_{m_{\delta_e}}\delta_e\right]$$

$$+ \frac{1}{4J_y}\rho\sqrt{u^2+w^2}SC_{m_q}c^2q \quad (5.47)$$

$$\dot{\theta} = q \quad (5.48)$$

$$\dot{h} = u\sin\theta - w\cos\theta. \quad (5.49)$$

The Jacobians of equations (5.45)–(5.49) are given by

$$\frac{\partial f_{\text{lon}}}{\partial x_{\text{lon}}} = \begin{pmatrix} \frac{\partial \dot{u}}{\partial u} & \frac{\partial \dot{u}}{\partial w} & \frac{\partial \dot{u}}{\partial q} & \frac{\partial \dot{u}}{\partial \theta} & \frac{\partial \dot{u}}{\partial h} \\ \frac{\partial \dot{w}}{\partial u} & \frac{\partial \dot{w}}{\partial w} & \frac{\partial \dot{w}}{\partial q} & \frac{\partial \dot{w}}{\partial \theta} & \frac{\partial \dot{w}}{\partial h} \\ \frac{\partial \dot{q}}{\partial u} & \frac{\partial \dot{q}}{\partial w} & \frac{\partial \dot{q}}{\partial q} & \frac{\partial \dot{q}}{\partial \theta} & \frac{\partial \dot{q}}{\partial h} \\ \frac{\partial \dot{\theta}}{\partial u} & \frac{\partial \dot{\theta}}{\partial w} & \frac{\partial \dot{\theta}}{\partial q} & \frac{\partial \dot{\theta}}{\partial \theta} & \frac{\partial \dot{\theta}}{\partial h} \\ \frac{\partial \dot{h}}{\partial u} & \frac{\partial \dot{h}}{\partial w} & \frac{\partial \dot{h}}{\partial q} & \frac{\partial \dot{h}}{\partial \theta} & \frac{\partial \dot{h}}{\partial h} \end{pmatrix}$$

$$\frac{\partial f_{\text{lon}}}{\partial u_{\text{lon}}} = \begin{pmatrix} \frac{\partial \dot{u}}{\partial \delta_e} & \frac{\partial \dot{u}}{\partial \delta_t} \\ \frac{\partial \dot{w}}{\partial \delta_e} & \frac{\partial \dot{w}}{\partial \delta_t} \\ \frac{\partial \dot{q}}{\partial \delta_e} & \frac{\partial \dot{q}}{\partial \delta_t} \\ \frac{\partial \dot{\theta}}{\partial \delta_e} & \frac{\partial \dot{\theta}}{\partial \delta_t} \\ \frac{\partial \dot{h}}{\partial \delta_e} & \frac{\partial \dot{h}}{\partial \delta_t} \end{pmatrix}.$$

Note that

$$\frac{\partial}{\partial u}\tan^{-1}\left(\frac{w}{u}\right) = \frac{1}{1+\frac{w^2}{u^2}}\left(\frac{-w}{u^2}\right) = \frac{-w}{u^2+w^2} = \frac{-w}{V_a^2}$$

Linear Design Models

$$\frac{\partial}{\partial w} \tan^{-1}\left(\frac{w}{u}\right) = \frac{1}{1+\frac{w^2}{u^2}}\left(\frac{1}{u}\right) = \frac{u}{u^2+w^2} = \frac{u}{V_a^2},$$

where we have used equation (2.8) and the fact that $v = 0$. Calculating the derivatives, we get the linearized state-space equations

$$\begin{pmatrix} \dot{\bar{u}} \\ \dot{\bar{w}} \\ \dot{\bar{q}} \\ \dot{\bar{\theta}} \\ \dot{\bar{h}} \end{pmatrix} = \begin{pmatrix} X_u & X_w & X_q & -g\cos\theta^* & 0 \\ Z_u & Z_w & Z_q & -g\sin\theta^* & 0 \\ M_u & M_w & M_q & 0 & 0 \\ 0 & 0 & 1 & 0 & 0 \\ \sin\theta^* & -\cos\theta^* & 0 & u^*\cos\theta^* + w^*\sin\theta^* & 0 \end{pmatrix} \begin{pmatrix} \bar{u} \\ \bar{w} \\ \bar{q} \\ \bar{\theta} \\ \bar{h} \end{pmatrix}$$

$$+ \begin{pmatrix} X_{\delta_e} & X_{\delta_t} \\ Z_{\delta_e} & 0 \\ M_{\delta_e} & 0 \\ 0 & 0 \\ 0 & 0 \end{pmatrix} \begin{pmatrix} \bar{\delta}_e \\ \bar{\delta}_t \end{pmatrix}, \qquad (5.50)$$

where the coefficients are given in table 5.2.

The longitudinal equations are often given in terms of $\bar{\alpha}$ instead of $\bar{w}$. From equation (2.7), we have

$$w = V_a \sin\alpha \cos\beta = V_a \sin\alpha,$$

where we have set $\beta = 0$. Linearizing around $\alpha = \alpha^*$, we get

$$\bar{w} = V_a^* \cos\alpha^* \bar{\alpha},$$

which implies that

$$\dot{\bar{\alpha}} = \frac{1}{V_a^* \cos\alpha^*} \dot{\bar{w}}.$$

TABLE 5.2
Longitudinal State-space Model Coefficients

Longitudinal	Formula
X_u	$\frac{u^*\rho S}{m}\left[C_{X_0}+C_{X_\alpha}\alpha^*+C_{X_{\delta_e}}\delta_e^*\right]-\frac{\rho S w^* C_{X_\alpha}}{2m}$
	$+\frac{\rho S c C_{X_q} u^* q^*}{4mV_a^*}-\frac{\rho S_{\text{prop}} C_{\text{prop}} u^*}{m}$
X_w	$-q^*+\frac{w^*\rho S}{m}\left[C_{X_0}+C_{X_\alpha}\alpha^*+C_{X_{\delta_e}}\delta_e^*\right]+\frac{\rho S c C_{X_q} w^* q^*}{4mV_a^*}$
	$+\frac{\rho S C_{X_\alpha} u^*}{2m}-\frac{\rho S_{\text{prop}} C_{\text{prop}} w^*}{m}$
X_q	$-w^*+\frac{\rho V_a^* S C_{X_q} c}{4m}$
X_{δ_e}	$\frac{\rho V_a^{*2} S C_{X_{\delta_e}}}{2m}$
X_{δ_t}	$\frac{\rho S_{\text{prop}} C_{\text{prop}} k^2 \delta_t^*}{m}$
Z_u	$q^*+\frac{u^*\rho S}{m}\left[C_{Z_0}+C_{Z_\alpha}\alpha^*+C_{Z_{\delta_e}}\delta_e^*\right]-\frac{\rho S C_{Z_\alpha} w^*}{2m}$
	$+\frac{u^*\rho S C_{Z_q} c q^*}{4mV_a^*}$
Z_w	$\frac{w^*\rho S}{m}\left[C_{Z_0}+C_{Z_\alpha}\alpha^*+C_{Z_{\delta_e}}\delta_e^*\right]+\frac{\rho S C_{Z_\alpha} u^*}{2m}$
	$+\frac{\rho w^* S c C_{Z_q} q^*}{4mV_a^*}$
Z_q	$u^*+\frac{\rho V_a^* S C_{Z_q} c}{4m}$
Z_{δ_e}	$\frac{\rho V_a^{*2} S C_{Z_{\delta_e}}}{2m}$
M_u	$\frac{u^*\rho S c}{J_y}\left[C_{m_0}+C_{m_\alpha}\alpha^*+C_{m_{\delta_e}}\delta_e^*\right]-\frac{\rho S c C_{m_\alpha} w^*}{2J_y}$
	$+\frac{\rho S c^2 C_{m_q} q^* u^*}{4J_y V_a^*}$
M_w	$\frac{w^*\rho S c}{J_y}\left[C_{m_0}+C_{m_\alpha}\alpha^*+C_{m_{\delta_e}}\delta_e^*\right]+\frac{\rho S c C_{m_\alpha} u^*}{2J_y}$
	$+\frac{\rho S c^2 C_{m_q} q^* w^*}{4J_y V_a^*}$
M_q	$\frac{\rho V_a^* S c^2 C_{m_q}}{4J_y}$
M_{δ_e}	$\frac{\rho V_a^{*2} S c C_{m_{\delta_e}}}{2J_y}$

Linear Design Models

We can, therefore, write the the state-space equations in terms of $\bar{\alpha}$ instead of $\bar{w}$ as

$$\begin{pmatrix} \dot{\bar{u}} \\ \dot{\bar{\alpha}} \\ \dot{\bar{q}} \\ \dot{\bar{\theta}} \\ \dot{\bar{h}} \end{pmatrix} = \begin{pmatrix} X_u & X_w V_a^* \cos\alpha^* & X_q & -g\cos\theta^* & 0 \\ \frac{Z_u}{V_a^* \cos\alpha^*} & Z_w & \frac{Z_q}{V_a^* \cos\alpha^*} & \frac{-g\sin\theta^*}{V_a^* \cos\alpha^*} & 0 \\ M_u & M_w V_a^* \cos\alpha^* & M_q & 0 & 0 \\ 0 & 0 & 1 & 0 & 0 \\ \sin\theta^* & -V_a^* \cos\theta^* \cos\alpha^* & 0 & u^*\cos\theta^* + w^*\sin\theta^* & 0 \end{pmatrix}$$

$$\times \begin{pmatrix} \bar{u} \\ \bar{\alpha} \\ \bar{q} \\ \bar{\theta} \\ \bar{h} \end{pmatrix} + \begin{pmatrix} X_{\delta_e} & X_{\delta_t} \\ \frac{Z_{\delta_e}}{V_a^* \cos\alpha^*} & 0 \\ M_{\delta_e} & 0 \\ 0 & 0 \\ 0 & 0 \end{pmatrix} \begin{pmatrix} \bar{\delta}_e \\ \bar{\delta}_t \end{pmatrix}. \tag{5.51}$$

5.6 Reduced-order Modes

The traditional literature on aircraft dynamics and control define several open-loop aircraft dynamic modes. These include the short-period mode, the phugoid mode, the rolling mode, the spiral-divergence mode, and the dutch-roll mode. In this section we will briefly describe each of these modes and show how to approximate the eigenvalues associated with these modes.

Short-period Mode
If we assume a constant altitude and a constant thrust input, then we can simplify the longitudinal state-space model in equation (5.51) to

$$\begin{pmatrix} \dot{\bar{u}} \\ \dot{\bar{\alpha}} \\ \dot{\bar{q}} \\ \dot{\bar{\theta}} \end{pmatrix} = \begin{pmatrix} X_u & X_w V_a^* \cos\alpha^* & X_q & -g\cos\theta^* \\ \frac{Z_u}{V_a^* \cos\alpha^*} & Z_w & \frac{Z_q}{V_a^* \cos\alpha^*} & \frac{-g\sin\theta^*}{V_a^* \cos\alpha^*} \\ M_u & M_w V_a^* \cos\alpha & M_q & 0 \\ 0 & 0 & 1 & 0 \end{pmatrix} \begin{pmatrix} \bar{u} \\ \bar{\alpha} \\ \bar{q} \\ \bar{\theta} \end{pmatrix} + \begin{pmatrix} X_{\delta_e} \\ \frac{Z_{\delta_e}}{V_a^* \cos\alpha^*} \\ M_{\delta_e} \\ 0 \end{pmatrix} \bar{\delta}_e. \tag{5.52}$$

If we compute the eigenvalues of the state matrix, we will find that there is one fast, damped mode and one slow, lightly damped mode. The fast mode is called the *short period* mode. The slow, lightly damped mode is called the *phugoid* mode.

For the short-period mode, we will assume that u is constant (i.e., $\bar{u} = \dot{\bar{u}} = 0$). It follows that the state-space equations in equation (5.52) can be written as

$$\dot{\bar{\alpha}} = Z_w \bar{\alpha} + \frac{Z_q}{V_a^* \cos \alpha^*} \dot{\bar{\theta}} - \frac{g \sin \theta^*}{V_a^* \cos \alpha^*} \bar{\theta} + \frac{Z_{\delta_e}}{V_a^* \cos \alpha^*} \bar{\delta}_e$$

$$\ddot{\bar{\theta}} = M_w V_a^* \cos \alpha^* \bar{\alpha} + M_q \dot{\bar{\theta}},$$

where we have substituted $\bar{q} = \dot{\bar{\theta}}$. Taking the Laplace transform of these equations gives

$$\begin{pmatrix} s - Z_w & -\frac{Z_q s}{V_a^* \cos \alpha^*} + \frac{g \sin \theta^*}{V_a^* \cos \alpha^*} \\ -M_w V_a^* \cos \alpha^* & s^2 - M_q s \end{pmatrix} \begin{pmatrix} \bar{\alpha}(s) \\ \bar{\theta}(s) \end{pmatrix} = \begin{pmatrix} \frac{Z_{\delta_e}}{V_a^* \cos \alpha^*} \\ 0 \end{pmatrix} \bar{\delta}_e(s),$$

which implies that

$$\begin{pmatrix} \bar{\alpha}(s) \\ \bar{\theta}(s) \end{pmatrix} = \frac{\begin{pmatrix} s^2 - M_q s & \frac{Z_q s}{V_a^* \cos \alpha^*} - \frac{g \sin \theta^*}{V_a^* \cos \alpha^*} \\ M_w V_a^* \cos \alpha^* & s - Z_w \end{pmatrix}}{(s^2 - M_q s)(s - Z_w) + M_w V_a^* \cos \alpha^* \left(-\frac{Z_q s}{V_a^* \cos \alpha^*} + \frac{g \sin \theta^*}{V_a^* \cos \alpha^*}\right)}$$

$$\times \begin{pmatrix} \frac{Z_{\delta_e}}{V_a^* \cos \alpha^*} \\ 0 \end{pmatrix} \bar{\delta}_e(s).$$

Assuming that we have linearized around level flight (i.e., $\theta^* = 0$), the characteristic equation becomes

$$s \left(s^2 + (-Z_w - M_q) s + M_q Z_w - M_w Z_q\right) = 0.$$

Therefore, the short-period poles are approximately equal to

$$\lambda_{\text{short}} = \frac{Z_w + M_q}{2} \pm \sqrt{\left(\frac{Z_w + M_q}{2}\right)^2 - M_q Z_w + M_w Z_q}.$$

Phugoid Mode

Assuming that α is constant (i.e., $\bar{\alpha} = \dot{\bar{\alpha}} = 0$), then $\alpha = \alpha^*$ and equation (5.52) becomes

$$\begin{pmatrix} \dot{\bar{u}} \\ 0 \\ \dot{\bar{q}} \\ \dot{\bar{\theta}} \end{pmatrix} = \begin{pmatrix} X_u & X_w V_a^* \sin\alpha^* & X_q & -g\cos\theta^* \\ \frac{Z_u}{V_a^* \cos\alpha^*} & Z_w & \frac{Z_q}{V_a^* \cos\alpha^*} & \frac{-g\sin\theta^*}{V_a^* \cos\alpha^*} \\ M_u & M_w V_a^* \cos\alpha^* & M_q & 0 \\ 0 & 0 & 1 & 0 \\ -\sin\theta^* & -V_a^* \cos\theta^* \cos\alpha^* & 0 & u^*\cos\theta^* + w^*\sin\theta^* \end{pmatrix}$$

$$\times \begin{pmatrix} \bar{u} \\ 0 \\ \bar{q} \\ \bar{\theta} \end{pmatrix} + \begin{pmatrix} X_{\delta_e} \\ \frac{Z_{\delta_e}}{V_a^* \cos\alpha^*} \\ M_{\delta_e} \\ 0 \\ 0 \end{pmatrix} \bar{\delta}_e.$$

Taking the Laplace transform of the first two equations gives

$$\begin{pmatrix} s - X_u & -X_q s + g\cos\theta^* \\ -Z_u & -Z_q s + g\sin\theta^* \end{pmatrix} \begin{pmatrix} \bar{u}(s) \\ \bar{\theta}(s) \end{pmatrix} = \begin{pmatrix} X_{\delta_e} \\ Z_{\delta_e} \end{pmatrix} \bar{\delta}_e.$$

Again assuming that $\theta^* = 0$, we get that the characteristic equation is given by

$$s^2 + \left(\frac{Z_u X_q - X_u Z_q}{Z_q}\right) s - \frac{g Z_u}{Z_q} = 0.$$

The poles of the phugoid mode are approximately given by

$$\lambda_{\text{phugoid}} = -\frac{Z_u X_q - X_u Z_q}{2 Z_q} \pm \sqrt{\left(\frac{Z_u X_q - X_u Z_q}{2 Z_q}\right)^2 + \frac{g Z_u}{Z_q}}.$$

Roll Mode

If we ignore the heading dynamics and assume a constant pitch angle (i.e., $\bar{\theta} = 0$), then equation (5.44) becomes

$$\begin{pmatrix} \dot{\bar{\beta}} \\ \dot{\bar{p}} \\ \dot{\bar{r}} \\ \dot{\bar{\phi}} \end{pmatrix} = \begin{pmatrix} Y_v & \frac{Y_p}{V_a^* \cos \beta^*} & \frac{Y_r}{V_a^* \cos \beta^*} & \frac{g \cos \theta^* \cos \phi^*}{V_a^* \cos \beta^*} \\ L_v V_a^* \cos \beta^* & L_p & L_r & 0 \\ N_v V_a^* \cos \beta^* & N_p & N_r & 0 \\ 0 & 1 & 0 & 0 \end{pmatrix} \begin{pmatrix} \bar{\beta} \\ \bar{p} \\ \bar{r} \\ \bar{\phi} \end{pmatrix}$$

$$+ \begin{pmatrix} \frac{Y_{\delta_a}}{V_a^* \cos \beta^*} & \frac{Y_{\delta_r}}{V_a^* \cos \beta^*} \\ L_{\delta_a} & L_{\delta_r} \\ N_{\delta_a} & N_{\delta_r} \\ 0 & 0 \end{pmatrix} \begin{pmatrix} \bar{\delta}_a \\ \bar{\delta}_r \end{pmatrix}. \quad (5.53)$$

The dynamics for $\bar{p}$ are obtained from equation (5.53) as

$$\dot{\bar{p}} = L_v V_a^* \cos \beta^* \bar{\beta} + L_p \bar{p} + L_r \bar{r} + L_{\delta_a} \bar{\delta}_a + L_{\delta_r} \bar{\delta}_r.$$

The rolling mode is obtained by assuming that $\bar{\beta} = \bar{r} = \bar{\delta}_r = 0$:

$$\dot{\bar{p}} = +L_p \bar{p} + L_{\delta_a} \bar{\delta}_a.$$

The transfer function is therefore

$$\bar{p}(s) = \frac{L_{\delta_a}}{s - L_p} \bar{\delta}_a(s).$$

An approximation of the eigenvalue for the rolling mode is therefore given by

$$\lambda_{\text{rolling}} = L_p.$$

Spiral-divergence Mode

For the spiral-divergence mode we assume that $\dot{\bar{p}} = \bar{p} = 0$, and that the rudder command is negligible. Therefore, from the second and third equations in equation (5.53), we get

$$0 = L_v V_a^* \cos \beta^* \bar{\beta} + L_r \bar{r} + L_{\delta_a} \bar{\delta}_a \quad (5.54)$$

$$\dot{\bar{r}} = N_v V_a^* \cos \beta^* \bar{\beta} + N_r \bar{r} + N_{\delta_a} \bar{\delta}_a. \quad (5.55)$$

Solving equation (5.54) for $\bar{\beta}$ and substituting into equation (5.55), we obtain

$$\dot{\bar{r}} = \left(\frac{N_r L_v - N_v L_r}{L_v}\right) \bar{r} + \left(\frac{N_{\delta_a} L_v - N_v L_{\delta_a}}{L_v}\right) \bar{\delta}_a.$$

In the frequency domain, we have

$$\bar{r}(s) = \frac{\left(\frac{N_{\delta_a} L_v - N_v L_{\delta_a}}{L_v}\right)}{s - \left(\frac{N_r L_v - N_v L_r}{L_v}\right)} \bar{\delta}_a(s).$$

From this, the pole of the spiral mode is approximately

$$\lambda_{\text{spiral}} = \frac{N_r L_v - N_v L_r}{L_v},$$

which is typically in the right half of the complex plane and is, therefore, an unstable mode.

Dutch-roll Mode
For the dutch-roll mode, we neglect the rolling motions and focus on the equations for sideslip and yaw. From equation (5.53), we have

$$\begin{pmatrix} \dot{\bar{\beta}} \\ \dot{\bar{r}} \end{pmatrix} = \begin{pmatrix} Y_v & \frac{Y_r}{V_a^* \cos \beta^*} \\ N_v V_a^* \cos \beta^* & N_r \end{pmatrix} \begin{pmatrix} \bar{\beta} \\ \bar{r} \end{pmatrix} + \begin{pmatrix} \frac{Y_{\delta_r}}{V_a^* \cos \beta^*} \\ N_{\delta_r} \end{pmatrix} \bar{\delta}_r.$$

The characteristic equation is given by

$$\det\left(sI - \begin{pmatrix} Y_v & \frac{Y_r}{V_a^* \cos \beta^*} \\ N_v V_a^* \cos \beta^* & N_r \end{pmatrix}\right) = s^2 + (-Y_v - N_r)s + (Y_v N_r - N_v Y_r) = 0.$$

Therefore, the poles of the dutch-roll mode are approximated by

$$\lambda_{\text{dutch roll}} = \frac{Y_v + N_r}{2} \pm \sqrt{\left(\frac{Y_v + N_r}{2}\right)^2 - (Y_v N_r - N_v Y_r)}.$$

5.7 Chapter Summary

The objective of this chapter is to develop design models that can be used to develop low-level autopilot models for a fixed-wing miniature air vehicle. In particular we focus on linear models about trim conditions. In section 5.1 we summarized the nonlinear equations of motion developed in chapters 3 and 4. In section 5.2 we introduced the notion of a coordinated turn, which was used later in the chapter to model the relationship between roll angle and course rate. In section 5.3 we introduced the notion of trim states and inputs. In section 5.4 we linearized the nonlinear model and developed transfer functions that model the dominant relationships. The motion of the aircraft was decomposed into lateral and longitudinal dynamics. The transfer functions for the lateral dynamics are given by equations (5.26) and (5.27), which express the relationship between the aileron deflection and the roll angle, and the relationship between the roll angle and the course angle, respectively. For aircraft that have a rudder and the ability to measure the side slip angle, equation (5.28) expresses the relationship between the rudder deflection and the side slip angle. The transfer functions for the longitudinal dynamics are given by equations (5.29), (5.31), (5.32), and (5.36), which model the relationship between the elevator deflection and the pitch angle, the pitch angle and the altitude, the airspeed and the altitude, and the throttle and pitch angle to the airspeed, respectively. In section 5.5 we developed state-space models linearized about the trim condition. The state-space model for the lateral dynamics is given in equation (5.43). The state-space model for the longitudinal dynamics is given in equation (5.50). In section 5.6 we discussed the modes associated with the linear models developed in this chapter, as they are defined in the traditional aeronautics literature. For the lateral dynamics, the modes are the roll mode, the dutch-roll mode, and the spiral-divergence mode. For the longitudinal dynamics, the modes are the short-period mode and the phugoid mode.

Notes and References

The models that we have developed in this chapter are standard. An excellent discussion of trim including algorithms for computing trim is contained in [7]. The transfer function models are discussed in detail in [4]. The state-space models are derived in [1, 2, 5, 6, 7, 12], which also discuss the reduced-order modes derived in section 5.6. The derivation of the coordinated turn in equation (5.13) follows [130].

5.8 Design Project

5.1. Read appendix F and familiarize yourself with the Simulink `trim` and `linmod` commands.

5.2. Copy and rename your current Simulink diagram to `mavsim_trim.mdl` and modify the file so that it has the proper input-output structure, as shown in figure F.1.

5.3. Create a Matlab script that computes the trim values for the Simulink simulation developed in chapters 2 through 4. The input to the Matlab script should be the desired airspeed V_a, the desired path angle $\pm \gamma$, and the desired turn radius $\pm R$, where $+R$ indicates a right-hand turn and $-R$ indicates a left-hand turn.

5.4. Use the Matlab script to compute the trimmed state and controls for wings-level flight with $V_a = 10$ m/s and $\gamma = 0$ rad. Set the initial states in your original Simulink simulation to the trim state, and the inputs to the trim controls. If the trim algorithm is correct, the MAV states will remain constant during the simulation. Run the trim algorithm for various values of γ. The only variable that should change is the altitude h. Convince yourself that the climb rate is correct.

5.5. Use the Matlab script to compute the trimmed state and controls for constant turns with $V_a = 10$ m/s and $R = 50$ m. Set the initial states in your original Simulink simulation to the trim state, and the inputs to the trim controls. If the trim algorithm is correct, the UAV states will remain constant during the simulation except for the heading ψ.

5.6. Create a Matlab script that uses the trim values computed in the previous problem to create the transfer functions listed in section 5.4.

5.7. Create a Matlab script that uses the trim values and the `linmod` command to linearize the Simulink model about the trim condition to produce the state-space models given in equations (5.50) and (5.43).

5.8. Compute eigenvalues of `A_lon` and notice that one of the eigenvalues will be zero and that there are two complex conjugate pairs. Using the formula

$$(s + \lambda)(s + \lambda^*) = s^2 + 2\Re\lambda s + |\lambda|^2 = s^2 + 2\zeta\omega_n s + \omega_n^2,$$

extract ω_n and ζ from the two complex conjugate pairs of poles. The pair with the larger ω_n correspond to the short-period mode, and the pair with the smaller ω_n correspond to the phugoid mode. The phugoid and short-period modes can be excited by starting the simulation in a wings-level, constant-altitude trim condition, and placing an impulse on the elevator. The file

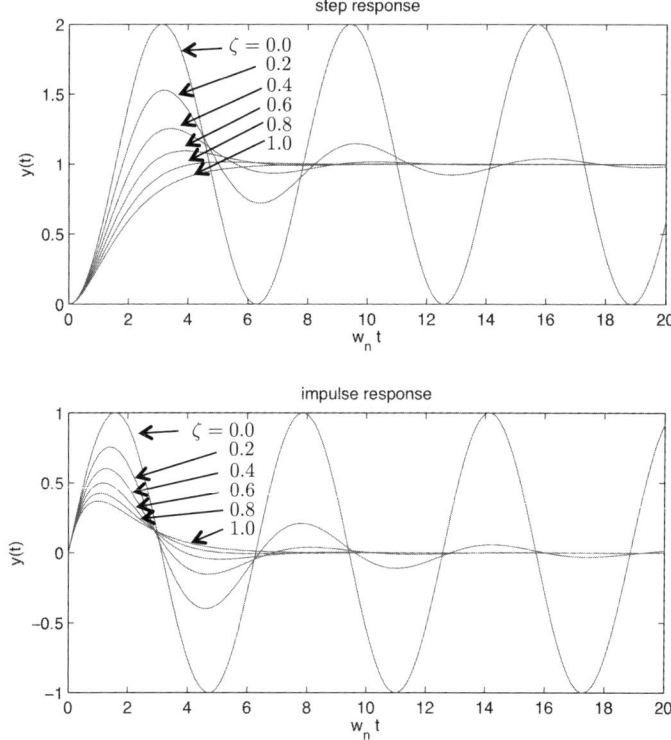

Figure 5.8 Step and Impulse response for a second order system with transfer function equal to $T(s) = \omega_n^2/(s^2 + 2\zeta\omega_n s + \omega_n^2)$.

mavsim_chap5.mdl on the website shows how to implement an impulse and doublet. Using figure 5.8 convince yourself that the eigenvalues of A_lon adequately predict the short period and phugoid modes.

5.9. Compute eigenvalues of A_lat and notice that there is an eigenvalue at zero, a real eigenvalue in the right half plane, a real eigenvalue in the left half plane, and a complex conjugate pair. The real eigenvalue in the right half plane is the spiral-divergence mode, the real eigenvalue in the left half plane is the roll mode, and the complex eigenvalues are the dutch-roll mode. The lateral modes can be excited by starting the simulation in a wings-level, constant-altitude trim condition, and placing a unit doublet on the aileron or on the rudder. Using figure 5.8 convince yourself that the eigenvalues of A_lat adequately predict the roll, spiral-divergence, and dutch-roll modes.

6

Autopilot Design Using Successive Loop Closure

In general terms, an autopilot is a system used to guide an aircraft without the assistance of a pilot. For manned aircraft, the autopilot can be as simple as a single-axis wing-leveling autopilot, or as complicated as a full flight control system that controls position (altitude, latitude, longitude) and attitude (roll, pitch, yaw) during the various phases of flight (e.g., take-off, ascent, level flight, descent, approach, landing). For MAVs, the autopilot is in complete control of the aircraft during all phases of flight. While some control functions may reside in the ground control station, the autopilot portion of the MAV control system resides on board the MAV.

This chapter presents an autopilot design suitable for the sensors and computational resources available on board MAVs. We will utilize a method called successive loop closure to design lateral and longitudinal autopilots. The successive loop closure approach is discussed generally in section 6.1. Because the lifting surfaces of aircraft have limited range, we discuss actuator saturation and the limit it imposes on performance in section 6.2. Lateral and longitudinal autopilot designs are presented in sections 6.3 and 6.4. The chapter concludes with a discussion of discrete-time implementation of proportion-integral-derivative (PID) feedback control laws in section 6.5.

6.1 Successive Loop Closure

The primary goal in autopilot design is to control the inertial position (p_n, p_e, h) and attitude (ϕ, θ, χ) of the MAV. For most flight maneuvers of interest, autopilots designed on the assumption of decoupled dynamics yield good performance. In the discussion that follows, we will assume that the longitudinal dynamics (forward speed, pitching, climbing/descending motions) are decoupled from the lateral dynamics (rolling, yawing motions). This simplifies the development of the autopilot significantly and allows us to utilize a technique commonly used for autopilot design called successive loop closure.

The basic idea behind successive loop closure is to close several simple feedback loops in succession around the open-loop plant dynamics

Figure 6.1 Open-loop transfer function modeled as a cascade of three transfer functions.

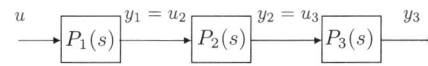

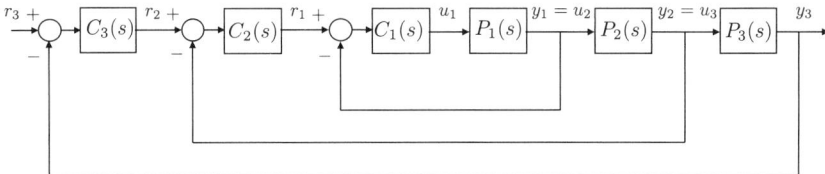

Figure 6.2 Three-stage successive loop closure design.

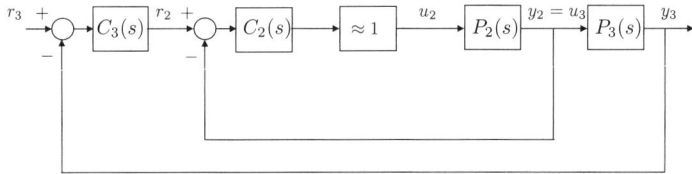

Figure 6.3 Successive loop closure design with inner loop modeled as a unity gain.

rather than designing a single (presumably more complicated) control system. To illustrate how this approach can be applied, consider the open-loop system shown in figure 6.1. The open-loop dynamics are given by the product of three transfer functions in series: $P(s) = P_1(s)P_2(s)P_3(s)$. Each of the transfer functions has an output (y_1, y_2, y_3) that can be measured and used for feedback. Typically, each of the transfer functions, $P_1(s)$, $P_2(s)$, $P_3(s)$, is of relatively low order—usually first or second order. In this case, we are interested in controlling the output y_3. Instead of closing a single feedback loop with y_3, we will instead close feedback loops around y_1, y_2, and y_3 in succession, as shown in figure 6.2. We will design the compensators $C_1(s)$, $C_2(s)$, and $C_3(s)$ in succession. A necessary condition in the design process is that the inner loop has the highest bandwidth, with each successive loop bandwidth a factor of 5 to 10 times smaller in frequency.

Examining the inner loop shown in figure 6.2, the goal is to design a closed-loop system from r_1 to y_1 having a bandwidth ω_{BW1}. The key assumption we make is that for frequencies well below ω_{BW1}, the closed-loop transfer function $y_1(s)/r_1(s)$ can be modeled as a gain of 1. This is depicted schematically in figure 6.3. With the inner-loop transfer function modeled as a gain of 1, design of the second loop is simplified because it includes only the plant transfer function $P_2(s)$ and the compensator $C_2(s)$. The critical step in closing the loops successively

Autopilot Design

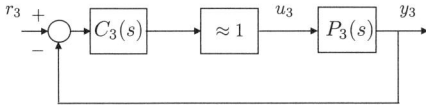

Figure 6.4 Successive-loop-closure design with two inner loops modeled as a unity gain.

is to design the bandwidth of the next loop so that it is a factor of S smaller than the preceding loop, where S is typically in the range of 5 to 10. In this case, we require $\omega_{BW2} < \frac{1}{S}\omega_{BW1}$ thus ensuring that the unity gain assumption on the inner loop is not violated over the range of frequencies that the middle loop operates.

With the two inner loops operating as designed, $y_2(s)/r_2(s) \approx 1$ and the transfer function from $r_2(s)$ to $y_2(s)$ can be replaced with a gain of 1 for the design of the outermost loop, as shown in figure 6.4. Again, there is a bandwidth constraint on the design of the outer loop: $\omega_{BW3} < \frac{1}{S_2}\omega_{BW2}$. Because each of the plant models $P_1(s)$, $P_2(s)$, and $P_3(s)$ is first or second order, conventional PID or lead-lag compensators can be employed effectively. Transfer-function-based design methods such as root-locus or loop-shaping approaches are commonly used.

The following sections discuss the design of a lateral autopilot and a longitudinal autopilot. Transfer functions modeling the lateral and longitudinal dynamics were developed in section 5.4 and will be used to design the autopilots in this chapter.

6.2 Saturation Constraints and Performance

The successive-loop-closure design process implies that performance of the system is limited by the performance of the inner-most loop. The performance of the inner-most loop is often limited by saturation constraints. For example, in the design of the lateral autopilot, the fact that the ailerons have physical limits on their angular deflection implies that the roll rate of the aircraft is limited. The goal is to design the bandwidth of the inner loop to be as large as possible, without violating the saturation constraints, and then design the outer loops to ensure bandwidth separation of the successive loops. In this section, we briefly describe how knowledge of the plant and controller transfer functions and the actuator saturation constraints can be used to develop performance specifications for the inner-most loops. We will use a second-order system to illustrate the process.

Given the second-order system shown in figure 6.5 with proportional feedback on the output error and derivative feedback on the output, the closed-loop transfer function is

$$\frac{y}{y^c} = \frac{b_0 k_p}{s^2 + (a_1 + b_0 k_d)s + (a_0 + b_0 k_p)}. \tag{6.1}$$

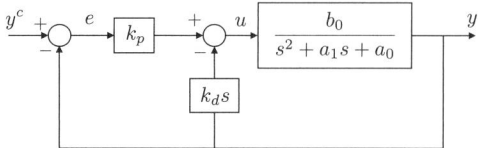

Figure 6.5 Control system example.

We can see that the closed-loop poles of the system are defined by the selection of the control gains k_p and k_d. Note also that the actuator effort u can be expressed as $u = k_p e - k_d \dot{y}$. When $\dot{y}$ is zero or small, the size of the actuator effort u is primarily governed by the size of the control error e and the control gain k_p. If the system is stable, the largest control effort in response to a step input will occur immediately after the step, where $u^{\max} = k_p e^{\max}$. Rearranging this expression, we find that the proportional control gain can be determined from the maximum anticipated output error and the saturation limits of the actuator as

$$k_p = \frac{u^{\max}}{e^{\max}}, \tag{6.2}$$

where $u^{\max}$ is the maximum control effort the system can provide, and $e^{\max}$ is the step error that results from a step input of nominal size.

The canonical second-order transfer function with no zeros is given by the standard form

$$\frac{y}{y^c} = \frac{\omega_n^2}{s^2 + 2\zeta\omega_n s + \omega_n^2}, \tag{6.3}$$

where y^c is the commanded value, ζ is the damping ratio, and ω_n is the natural frequency. If $0 \leq \zeta < 1$, then the system is said to be *under damped*, and the poles are complex and given by

$$\text{poles} = -\zeta\omega_n \pm j\omega_n\sqrt{1 - \zeta^2}. \tag{6.4}$$

By comparing the coefficients of the denominator polynomials of the transfer function of the closed-loop system in equation (6.1) and the canonical second-order system transfer function in equation (6.3), and taking into account the saturation limits of the actuator, we can derive an expression for the achievable bandwidth of the closed-loop system.

Autopilot Design

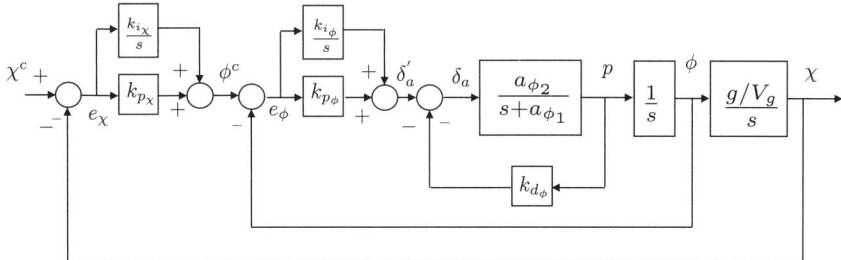

Figure 6.6 Autopilot for lateral control using successive loop closure.

Equating the coefficients of the s^0 terms gives

$$\omega_n = \sqrt{a_0 + b_0 k_p}$$
$$= \sqrt{a_0 + b_0 \frac{u^{\max}}{e^{\max}}},$$

which is an upper limit on the bandwidth of the closed-loop system, ensuring that saturation of the actuator is avoided. We will utilize this approach in sections 6.3.1 and 6.4.1 to determine the bandwidth of the roll and pitch loops.

6.3 Lateral-directional Autopilot

Figure 6.6 shows the block diagram for a lateral autopilot using successive loop closure. There are five gains associated with the lateral autopilot. The derivative gain k_{d_ϕ} provides roll rate damping for the innermost loop. The roll attitude is regulated with the proportional and integral gains k_{p_ϕ} and k_{i_ϕ}. The course angle is regulated with the proportional and integral gains k_{p_χ} and k_{i_χ}. The idea with successive loop closure is that the gains are successively chosen beginning with the inner loop and working outward. In particular, k_{d_ϕ} and k_{p_ϕ} are usually selected first, k_{i_ϕ} second, and finally k_{p_χ} and k_{i_χ}.

6.3.1 Roll Attitude Loop Design

The inner loop of the lateral autopilot is used to control roll angle and roll rate, as shown in figure 6.7. If the transfer function coefficients a_{ϕ_1} and a_{ϕ_2} are known, then there is a systematic method for selecting the control gains k_{d_ϕ} and k_{p_ϕ} based on the desired response of closed-loop dynamics. From figure 6.7, the transfer function from ϕ^c to ϕ is given by

$$H_{\phi/\phi^c}(s) = \frac{k_{p_\phi} a_{\phi_2}}{s^2 + (a_{\phi_1} + a_{\phi_2} k_{d_\phi})s + k_{p_\phi} a_{\phi_2}}.$$

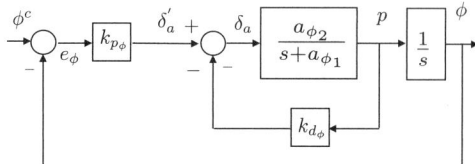

Figure 6.7 Roll attitude hold control loops.

Note that the DC gain is equal to one. If the desired response is given by the canonical second-order transfer function

$$\frac{\phi(s)}{\phi^c(s)} = \frac{\omega_{n_\phi}^2}{s^2 + 2\zeta_\phi \omega_{n_\phi} s + \omega_{n_\phi}^2},$$

then equating denominator polynomial coefficients, we get

$$\omega_{n_\phi}^2 = k_{p_\phi} a_{\phi_2} \tag{6.5}$$

$$2\zeta_\phi \omega_{n_\phi} = a_{\phi_1} + a_{\phi_2} k_{d_\phi}. \tag{6.6}$$

According to equation (6.2), the proportional gain is selected so that the ailerons saturate when the roll error is $e_\phi^{\max}$, where $e_\phi^{\max}$ is a design parameter. Therefore from equation (6.2) we get

$$k_{p_\phi} = \frac{\delta_a^{\max}}{e_\phi^{\max}} \operatorname{sign}(a_{\phi_2}). \tag{6.7}$$

The natural frequency of the roll loop is therefore given by

$$\omega_{n_\phi} = \sqrt{|a_{\phi_2}| \frac{\delta_a^{\max}}{e_\phi^{\max}}}. \tag{6.8}$$

Solving equation (6.6) for k_{d_ϕ} gives

$$k_{d_\phi} = \frac{2\zeta_\phi \omega_{n_\phi} - a_{\phi_1}}{a_{\phi_2}}, \tag{6.9}$$

where the damping ratio ζ_ϕ is a design parameter.

Integrator on Roll
Note that the open-loop transfer function in figure 6.7 is a type one system, which implies that zero steady-state tracking error in roll should be

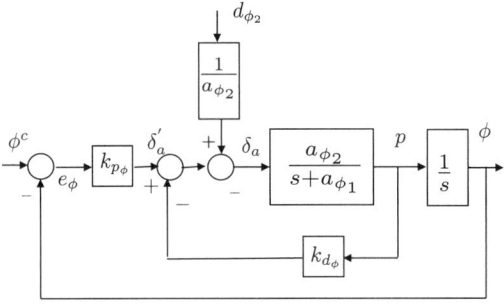

Figure 6.8 Roll attitude hold loop with input disturbance.

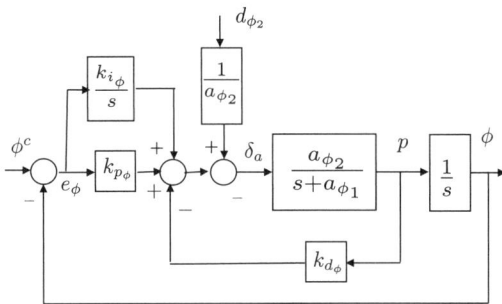

Figure 6.9 Integrator for roll attitude hold.

achievable without an integrator. From figure 5.2, however, we see that there is a disturbance that enters at the summing junction before δ_a. This disturbance represents the terms in the dynamics that were neglected in the process of creating the linear, reduced-order model of the roll dynamics. It can also represent physical perturbations to the system, such as those from gusts or turbulence. Figure 6.8 shows the roll loop with the disturbance. Solving for $\phi(s)$ in figure 6.8, we get

$$\phi = \left(\frac{1}{s^2 + (a_{\phi_1} + a_{\phi_2} k_{d_\phi})s + a_{\phi_2} k_{p_\phi}} \right) d_{\phi_2}$$
$$+ \left(\frac{a_{\phi_2} k_{d_\phi}}{s^2 + (a_{\phi_1} + a_{\phi_2} k_{d_\phi})s + a_{\phi_2} k_{p_\phi}} \right) \phi^c.$$

Note that if d_{ϕ_2} is a constant disturbance (i.e., $d_{\phi_2} = A/s$) then from the final value theorem, the steady-state error due to d_{ϕ_2} is $\frac{A}{a_{\phi_2} k_{p_\phi}}$. In a constant orbit, p, q, and r will be constants, so d_{ϕ_2} will also be constant, as can be seen from equation (5.25). Therefore, it is desirable to remove the steady-state error using an integrator. Figure 6.9 shows the roll attitude hold loop with an integrator added to reject the disturbance d_{ϕ_2}.

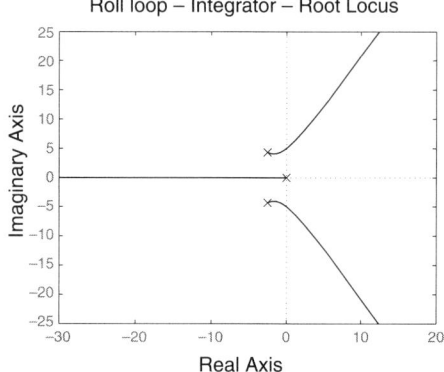

Figure 6.10 Roll loop root locus as a function of the integral gain k_{i_ϕ}.

Solving for $\phi(s)$ in figure 6.9, we get

$$\phi = \left(\frac{s}{s^3 + (a_{\phi_1} + a_{\phi_2}k_{d_\phi})s^2 + a_{\phi_2}k_{p_\phi}s + a_{\phi_2}k_{i_\phi}}\right) d_{\phi_2}$$

$$+ \left(\frac{a_{\phi_2}k_{p_\phi}\left(s + \frac{k_{i_\phi}}{k_{p_\phi}}\right)}{s^3 + (a_{\phi_1} + a_{\phi_2}k_{d_\phi})s^2 + a_{\phi_2}k_{p_\phi}s + a_{\phi_2}k_{i_\phi}}\right) \phi^c.$$

Note that in this case, the final-value theorem predicts zero steady-state error for a constant d_{ϕ_2}. If d_{ϕ_2} is a ramp (i.e., $d_{\phi_2} = A/s^2$), then the steady-state error is given by $\frac{A}{a_{\phi_2}k_{i_\phi}}$. If a_{ϕ_1} and a_{ϕ_2} are known, then k_{i_ϕ} can be effectively selected using root locus techniques. The closed-loop poles of the system are given by

$$s^3 + (a_{\phi_1} + a_{\phi_2}k_{d_\phi})s^2 + a_{\phi_2}k_{p_\phi}s + a_{\phi_2}k_{i_\phi} = 0,$$

which can be placed in Evans form as

$$1 + k_{i_\phi} \left(\frac{a_{\phi_2}}{s\left(s^2 + (a_{\phi_1} + a_{\phi_2}k_{d_\phi})s + a_{\phi_2}k_{p_\phi}\right)}\right) = 0.$$

Figure 6.10 shows the root locus of the characteristic equation plotted as a function of k_{i_ϕ}. For small values of gain, the system remains stable.

The output of the roll attitude hold loop is

$$\delta_a = k_{p_\phi}(\phi^c - \phi) + \frac{k_{i_\phi}}{s}(\phi^c - \phi) - k_{d_\phi}p.$$

6.3.2 Course Hold

The next step in the successive-loop-closure design of the lateral autopilot is to design the course-hold outer loop. If the inner loop from ϕ^c to ϕ

Autopilot Design

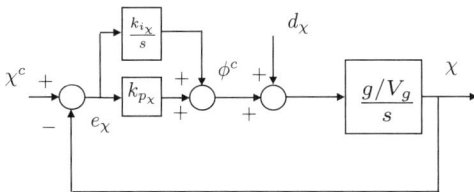

Figure 6.11 Course hold outer feedback loop.

has been adequately tuned, then $H_{\phi/\phi^c} \approx 1$ over the range of frequencies from 0 to ω_{n_ϕ}. Under this condition, the block diagram of figure 6.6 can be simplified to the block diagram in figure 6.11 for the purposes of designing the outer loop.

The objective of the course hold design is to select k_{p_χ} and k_{i_χ} in figure 6.6 so that the course χ asymptotically tracks steps in the commanded course χ^c. From the simplified block diagram, the transfer functions from the inputs χ^c and d_χ to the output χ are given by

$$\chi = \frac{g/V_g s}{s^2 + k_{p_\chi} g/V_g s + k_{i_\chi} g/V_g} d_\chi + \frac{k_{p_\chi} g/V_g s + k_{i_\chi} g/V_g}{s^2 + k_{p_\chi} g/V_g s + k_{i_\chi} g/V_g} \chi^c. \quad (6.10)$$

Note that if d_χ and χ^c are constants, then the final value theorem implies that $\chi \to \chi^c$. The transfer function from χ^c to χ has the form

$$H_\chi = \frac{2\zeta_\chi \omega_{n_\chi} s + \omega_{n_\chi}^2}{s^2 + 2\zeta_\chi \omega_{n_\chi} s + \omega_{n_\chi}^2}. \quad (6.11)$$

As with the inner feedback loops, we can choose the natural frequency and damping of the outer loop and from those values calculate the feedback gains k_{p_χ} and k_{i_χ}. Figure 6.12 shows the frequency response and the step response for H_χ. Note that because of the numerator zero, the standard intuition for the selection of ζ does not hold for this transfer function. Larger ζ results in larger bandwidth and smaller overshoot.

Comparing coefficients in equations (6.10) and (6.11), we find

$$\omega_{n_\chi}^2 = g/V_g k_{i_\chi}$$

$$2\zeta_\chi \omega_{n_\chi} = g/V_g k_{p_\chi}.$$

Solving these expressions for k_{p_χ} and k_{i_χ}, we get

$$k_{p_\chi} = 2\zeta_\chi \omega_{n_\chi} V_g/g \quad (6.12)$$

$$k_{i_\chi} = \omega_{n_\chi}^2 V_g/g. \quad (6.13)$$

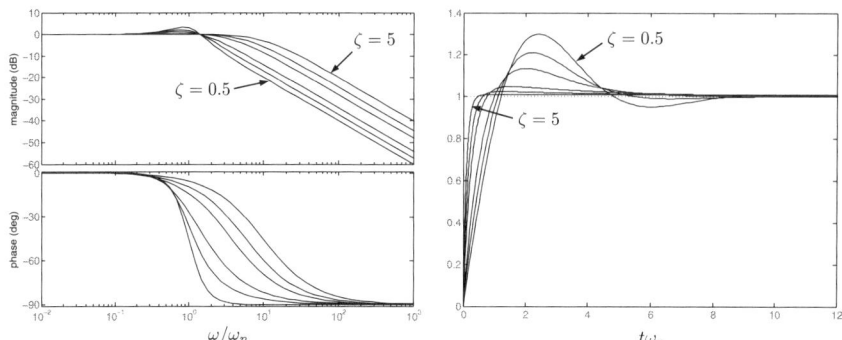

Figure 6.12 Frequency and step response for a second-order system with a transfer function zero for $\zeta = 0.5, 0.7, 1, 2, 3, 5$.

To ensure proper function of this successive-loop-closure design, it is essential that there be sufficient bandwidth separation between the inner and outer feedback loops. Adequate separation can be achieved by letting

$$\omega_{n_\chi} = \frac{1}{W_\chi}\omega_{n_\phi},$$

where the separation W_χ is a design parameter that is usually chosen to be greater than five. Generally, more bandwidth separation is better. More bandwidth separation requires either slower response in the χ loop (lower ω_{n_χ}), or faster response in the ϕ loop (higher ω_{n_ϕ}). Faster response usually comes at the cost of requiring more actuator control authority, which may not be possible given the physical constraints of the actuators.

The output of the course hold loop is

$$\phi^c = k_{p_\chi}(\chi^c - \chi) + \frac{k_{i_\chi}}{s}(\chi^c - \chi).$$

6.3.3 Sideslip Hold

If the aircraft is equipped with a rudder, the rudder can be used to maintain zero sideslip angle, $\beta(t) = 0$. The sideslip hold loop is shown in figure 6.13, and the transfer function from β^c to β is given by

$$H_{\beta/\beta^c}(s) = \frac{a_{\beta_2}k_{p_\beta}s + a_{\beta_2}k_{i_\beta}}{s^2 + (a_{\beta_1} + a_{\beta_2}k_{p_\beta})s + a_{\beta_2}k_{i_\beta}}.$$

Note that the DC gain is equal to one. If the desired closed poles are the roots of

$$s^2 + 2\zeta_\beta\omega_{n_\beta}s + \omega_{n_\beta}^2 = 0,$$

Autopilot Design

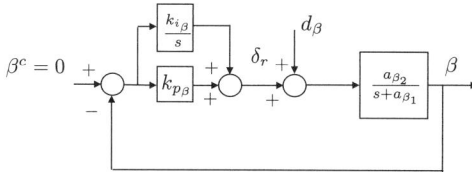

Figure 6.13 Sideslip hold control loop.

then equating coefficients gives

$$\omega_{n_\beta}^2 = a_{\beta_2} k_{i_\beta} \tag{6.14}$$

$$2\zeta_\beta \omega_{n_\beta} = a_{\beta_1} + a_{\beta_2} k_{p_\beta}. \tag{6.15}$$

Suppose that the maximum error in sideslip is given by $e_\beta^{\max}$ and that the maximum allowable rudder deflection is given by $\delta_r^{\max}$. Then by the approach of section 6.2, we obtain

$$k_{p_\beta} = \frac{\delta_r^{\max}}{e_\beta^{\max}} \text{sign}(a_{\beta_2}). \tag{6.16}$$

By choosing a value for ζ_β to give the desired damping, we can solve equations (6.14) and (6.15) to give

$$k_{i_\beta} = \frac{1}{a_{\beta_2}} \left(\frac{a_{\beta_1} + a_{\beta_2} k_{p_\beta}}{2\zeta_\beta} \right)^2. \tag{6.17}$$

The output of the sideslip hold loop is

$$\delta_r = -k_{p_\beta} \beta - \frac{k_{i_\beta}}{s} \beta.$$

6.4 Longitudinal Autopilot

The longitudinal autopilot is more complicated than the lateral autopilot because airspeed plays a significant role in the longitudinal dynamics. Our objective in designing the longitudinal autopilot will be to regulate airspeed and altitude using the throttle and the elevator as actuators. The method used to regulate altitude and airspeed depends on the altitude error. The flight regimes are shown in figure 6.14.

In the *take-off zone*, full throttle is commanded and the pitch attitude is regulated to a fixed pitch angle θ^c using the elevator. The objective in the *climb zone* is to maximize the climb rate given the current atmospheric conditions. To maximize the climb rate, full

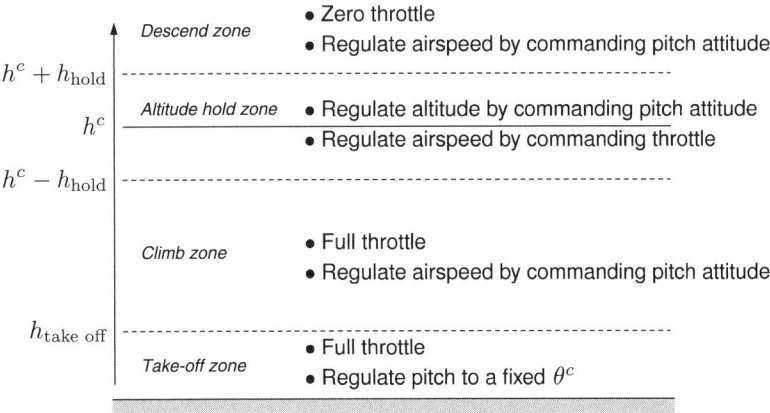

Figure 6.14 Flight regimes for the longitudinal autopilot.

throttle is commanded and the airspeed is regulated using the pitch angle. If the airspeed increases above its nominal value, then the aircraft is caused to pitch up, which results in an increase in climb rate and a decrease in airspeed. Similarly, if the airspeed drops below the nominal value, the aircraft is pitched down, thereby increasing the airspeed but also decreasing the climb rate. Regulating the airspeed using pitch attitude effectively avoids stall conditions. Note, however, that we do not regulate airspeed with pitch attitude immediately after take-off. After take-off the aircraft is attempting to increase its airspeed and doing so by pitching down would drive the aircraft into the ground.

The *descend zone* is similar to the climb zone except that the throttle is commanded to zero. Again, stall conditions are avoided by regulating airspeed using the pitch angle, thus maximizing the descent rate at a given airspeed. In the *altitude hold zone*, the airspeed is regulated by adjusting the throttle, and the altitude is regulated by commanding the pitch attitude.

To implement the longitudinal autopilot shown in figure 6.14, we need the following feedback loops: (1) pitch attitude hold using elevator, (2) airspeed hold using throttle, (3) airspeed hold using pitch attitude, and (4) altitude hold using pitch attitude. The design of ach of these loops will be discussed in the next four subsections. Finally, the complete longitudinal autopilot will be presented in section 6.4.5.

6.4.1 Pitch Attitude Hold

The pitch attitude hold loop is similar to the roll attitude hold loop, and we will follow a similar line of reasoning in its development. From

Autopilot Design

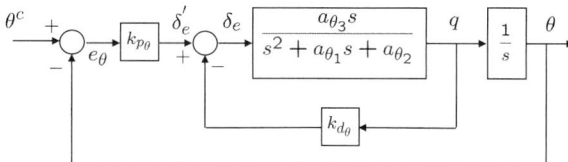

Figure 6.15 Pitch attitude hold feedback loops.

figure 6.15, the transfer function from θ^c to θ is given by

$$H_{\theta/\theta^c}(s) = \frac{k_{p_\theta} a_{\theta_3}}{s^2 + (a_{\theta_1} + k_{d_\theta} a_{\theta_3})s + (a_{\theta_2} + k_{p_\theta} a_{\theta_3})}. \quad (6.18)$$

Note that in this case, the DC gain is not equal to one.

If the desired response is given by the canonical second-order transfer function

$$\frac{K_{\theta_{DC}} \omega_{n_\theta}^2}{s^2 + 2\zeta_\theta \omega_{n_\theta} s + \omega_{n_\theta}^2},$$

then, equating denominator coefficients, we get

$$\omega_{n_\theta}^2 = a_{\theta_2} + k_{p_\theta} a_{\theta_3} \quad (6.19)$$

$$2\zeta_\theta \omega_{n_\theta} = a_{\theta_1} + k_{d_\theta} a_{\theta_3}. \quad (6.20)$$

If we set the proportional gain to avoid saturation when the maximum input error is experienced, we get

$$k_{p_\theta} = \frac{\delta_e^{\max}}{e_\theta^{\max}} \text{sign}(a_{\theta_3}),$$

where the sign of a_{θ_3} is taken since a_{θ_3} is based on $C_{m_{\delta_e}}$, which is typically negative. To ensure stability, k_{p_θ} and a_{θ_3} need to be of the same sign. From equation (6.19), the bandwidth limit of the pitch loop can be calculated as

$$\omega_{n_\theta} = \sqrt{a_{\theta_2} + \frac{\delta_e^{\max}}{e_\theta^{\max}} |a_{\theta_3}|}, \quad (6.21)$$

and solving equation (6.20) for k_{d_θ}, we get

$$k_{d_\theta} = \frac{2\zeta_\theta \omega_{n_\theta} - a_{\theta_1}}{a_{\theta_3}}. \quad (6.22)$$

In summary, knowing the actuator saturation limit $\delta_e^{\max}$ and the maximum anticipated pitch error, $e_\theta^{\max}$ can be selected to determine the proportional gain k_{p_θ} and the bandwidth of the pitch loop. Selecting the desired damping ratio ζ_θ fixes the derivative gain value k_{d_θ}.

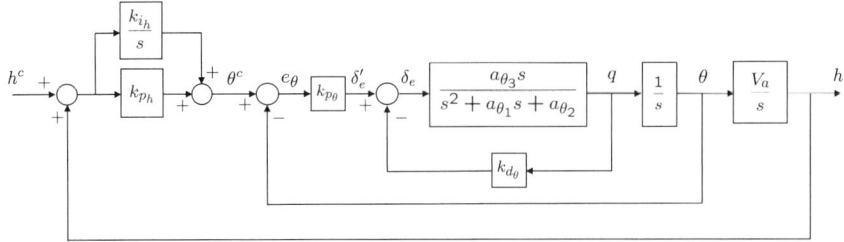

Figure 6.16 Successive loop feedback structure for altitude-hold autopilot.

The DC gain of this inner-loop transfer function approaches one as the $k_{p_\theta} \to \infty$. The DC gain is given by

$$K_{\theta_{DC}} = \frac{k_{p_\theta} a_{\theta_3}}{(a_{\theta_2} + k_{p_\theta} a_{\theta_3})}, \quad (6.23)$$

which for typical gain values is significantly less than one. The design of the outer loops will use this DC gain to represent the gain of the inner loop over its full bandwidth. An integral feedback term could be employed to ensure unity DC gain on the inner loop. The addition of an integral term, however, can severely limit the bandwidth of the inner loop. For this reason, we have chosen not to use integral control on the pitch loop. Note however, that in the design project, the actual pitch angle will not converge to the commanded pitch angle. This fact will be taken into account in the development of the outer loops. The output of the pitch attitude-hold loop is

$$\delta_e = k_{p_\theta}(\theta^c - \theta) - k_{d_\theta} q.$$

6.4.2 Altitude Hold Using Commanded Pitch

The altitude-hold autopilot utilizes a successive-loop-closure strategy with the pitch-attitude-hold autopilot as an inner loop, as shown in figure 6.16. Assuming that the pitch loop functions as designed and that $\theta \approx K_{\theta_{DC}} \theta^c$, the altitude-hold loop using the commanded pitch can be approximated by the block diagram shown in figure 6.17.

In the Laplace domain, we have

$$h(s) = \left(\frac{K_{\theta_{DC}} V_a k_{p_h} \left(s + \frac{k_{i_h}}{k_{p_h}}\right)}{s^2 + K_{\theta_{DC}} V_a k_{p_h} s + K_{\theta_{DC}} V_a k_{i_h}} \right) h^c(s)$$

$$+ \left(\frac{s}{s^2 + K_{\theta_{DC}} V_a k_{p_h} s + K_{\theta_{DC}} V_a k_{i_h}} \right) d_h(s),$$

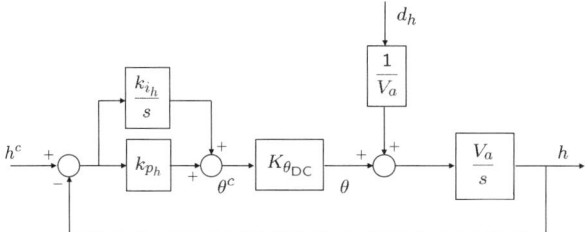

Figure 6.17 The altitude-hold loop using the commanded pitch angle.

where again we see that the DC gain is equal to one, and constant disturbances are rejected. The closed-loop transfer function is again independent of aircraft parameters and is dependent only on the known airspeed. The gains k_{p_h} and k_{i_h} should be chosen such that the bandwidth of the altitude-from-pitch loop is less than the bandwidth of the pitch-attitude-hold loop. Similar to the course loop, let

$$\omega_{n_h} = \frac{1}{W_h}\omega_{n_\theta},$$

where the bandwidth separation W_h is a design parameter that is usually between five and fifteen. If the desired response of the altitude-hold loop is given by the canonical second-order transfer function

$$\frac{\omega_{n_h}^2}{s^2 + 2\zeta_h \omega_{n_h} s + \omega_{n_h}^2},$$

then, equating denominator coefficients, we get

$$\omega_{n_h}^2 = K_{\theta_{DC}} V_a k_{i_h}$$
$$2\zeta_h \omega_{n_h} = K_{\theta_{DC}} V_a k_{p_h}.$$

Solving these expressions for k_{i_h} and k_{p_h}, we get

$$k_{i_h} = \frac{\omega_{n_h}^2}{K_{\theta_{DC}} V_a} \tag{6.24}$$

$$k_{p_h} = \frac{2\zeta_h \omega_{n_h}}{K_{\theta_{DC}} V_a}. \tag{6.25}$$

Therefore, selecting the desired damping ratio ζ_h and the bandwidth separation W_h fixes the value for k_{p_h} and k_{i_h}.

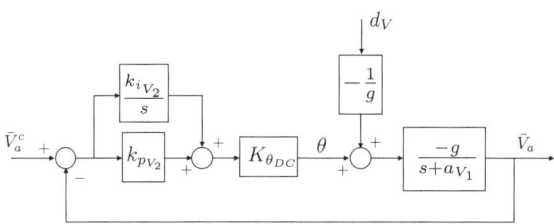

Figure 6.18 PI controller to regulate airspeed using the pitch angle.

The output of the altitude-hold-with-pitch loop is

$$\theta^c = k_{p_h}(h^c - h) + \frac{k_{i_h}}{s}(h^c - h).$$

6.4.3 Airspeed Hold Using Commanded Pitch

The dynamic model for airspeed using pitch angle is shown in figure 5.7. Disturbance rejection again requires a PI controller. The resulting block diagram is shown in figure 6.18.

In the Laplace domain, we have

$$\bar{V}_a(s) = \left(\frac{-K_{\theta_{DC}} g k_{p_{V_2}} \left(s + \frac{k_{i_{V_2}}}{k_{p_{V_2}}}\right)}{s^2 + (a_{V_1} - K_{\theta_{DC}} g k_{p_{V_2}})s - K_{\theta_{DC}} g k_{i_{V_2}}} \right) \bar{V}_a^c(s)$$

$$+ \left(\frac{s}{s^2 + (a_{V_1} - K_{\theta_{DC}} g k_{p_{V_2}})s - K_{\theta_{DC}} g k_{i_{V_2}}} \right) d_V(s). \quad (6.26)$$

Note that the DC gain is equal to one and that step disturbances are rejected. To hold a constant airspeed, the pitch angle must approach a non-zero angle of attack. The integrator will wind up to command the appropriate angle of attack.

The gains $k_{p_{V_2}}$ and $k_{i_{V_2}}$ should be chosen so that the bandwidth of the airspeed-from-pitch loop is less than the bandwidth of the pitch-attitude-hold loop. Let

$$\omega_{n_{V_2}} = \frac{1}{W_{V_2}} \omega_{n_\theta},$$

where the bandwidth separation W_{V_2} is a design parameter. Following a similar procedure to what we have done previously, we can determine values for the feedback gains by matching the denominator coefficients in equation (6.26) with those of a canonical second-order transfer function. Denoting the desired natural frequency and damping ratio we seek to achieve with feedback as $\omega_{n_{V_2}}^2$ and ζ_{V_2}, respectively, matching coefficients gives

$$\omega_{n_{V_2}}^2 = -K_{\theta_{DC}} g k_{i_{V_2}}$$

$$2\zeta_{V_2}\omega_{n_{V_2}} = a_{V_1} - K_{\theta_{DC}} g k_{p_{V_2}}.$$

Solving for the control gains gives

$$k_{i_{V_2}} = -\frac{\omega_{n_{V_2}}^2}{K_{\theta_{DC}} g} \tag{6.27}$$

$$k_{p_{V_2}} = \frac{a_{V_1} - 2\zeta_{V_2}\omega_{n_{V_2}}}{K_{\theta_{DC}} g}. \tag{6.28}$$

Thus, selecting the damping ratio ζ_{V_2} and the bandwidth separation W_{V_2} fixes the control gains $k_{i_{V_2}}$ and $k_{p_{V_2}}$. The output of the airspeed hold with pitch loop is

$$\theta^c = k_{p_{V_2}}(V_a^c - V_a) + \frac{k_{i_{V_2}}}{s}(V_a^c - V_a).$$

6.4.4 Airspeed Hold Using Throttle

The dynamic model for airspeed using the throttle as an input is shown in figure 5.7. The associated closed-loop system is shown in figure 6.19. If we use proportional control, then

$$\bar{V}_a(s) = \left(\frac{a_{V_2} k_{pv}}{s + (a_{V_1} + a_{V_2} k_{pv})}\right) \bar{V}_a^c(s) + \left(\frac{1}{s + (a_{V_1} + a_{V_2} k_{pv})}\right) d_V(s).$$

Note that the DC gain is not equal to one and that step disturbances are not rejected. If, on the other hand, we use proportional-integral control, then

$$\bar{V}_a = \left(\frac{a_{V_2}(k_{pv} s + k_{iv})}{s^2 + (a_{V_1} + a_{V_2} k_{pv})s + a_{V_2} k_{iv}}\right) \bar{V}_a^c$$

$$+ \left(\frac{1}{s^2 + (a_{V_1} + a_{V_2} k_{pv})s + a_{V_2} k_{iv}}\right) d_V.$$

It is clear that using a PI controller results in a DC gain of one, with step disturbance rejection. If a_{V_1} and a_{V_2} are known, then the gains k_{pv} and k_{iv} can be determined using the same technique we have used previously. Equating the closed-loop transfer function denominator coefficients

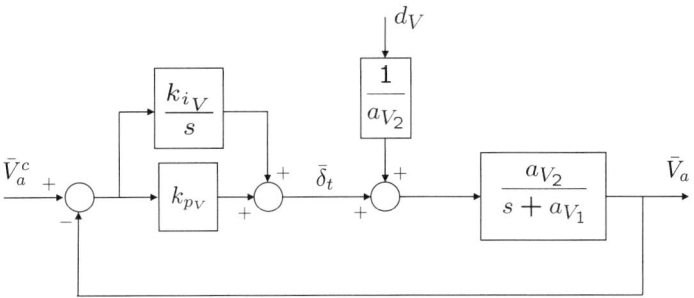

Figure 6.19 Airspeed hold using throttle.

with those of a canonical second-order transfer function, we get

$$\omega_{n_V}^2 = a_{V_2} k_{i_V}$$
$$2\zeta_V \omega_{n_V} = a_{V_1} + a_{V_2} k_{p_V}.$$

Inverting these expressions gives the control gains

$$k_{i_V} = \frac{\omega_{n_V}^2}{a_{V_2}} \tag{6.29}$$

$$k_{p_V} = \frac{2\zeta_V \omega_{n_V} - a_{V_1}}{a_{V_2}}. \tag{6.30}$$

The design parameters for this loop are the damping coefficient ζ_V and the natural frequency ω_{n_V}.

Note that since $\bar{V}_a^c = V_a^c - V_a^*$ and $\bar{V}_a = V_a - V_a^*$, the error signal in figure 6.19 is

$$e = \bar{V}_a^c - \bar{V}_a = V_a^c - V_a.$$

Therefore, the control loop shown in figure 6.19 can be implemented without knowledge of the trim velocity V_a^*. If the throttle trim value δ_t^* is known, then the throttle command is

$$\delta_t = \delta_t^* + \bar{\delta}_t.$$

However, if δ_t^* is not precisely known, then the error in δ_t^* can be thought of as a step disturbance, and the integrator will wind up to reject the disturbances.

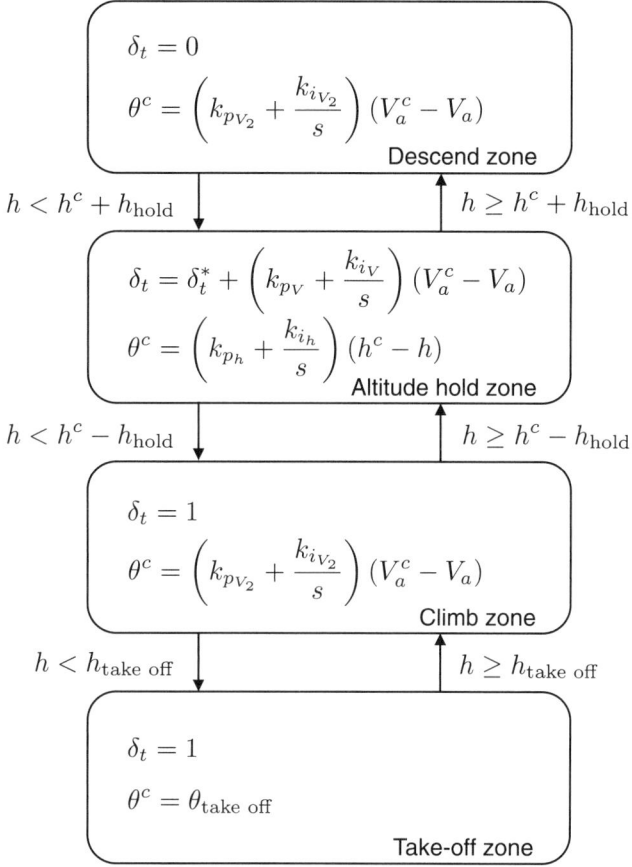

Figure 6.20 Altitude-control state machine.

The output of the airspeed hold with throttle loop is

$$\delta_t = \delta_t^* + k_{p_V}(V_a^c - V_a) + \frac{k_{i_V}}{s}(V_a^c - V_a).$$

6.4.5 Altitude-control State Machine

The longitudinal autopilot deals with the control of the longitudinal motions in the body $\mathbf{i}^b$-$\mathbf{k}^b$ plane: pitch angle, altitude, and airspeed. Up to this point, we have described four different longitudinal autopilot modes: (1) pitch attitude hold, (2) altitude hold using commanded pitch, (3) airspeed hold using commanded pitch, and (4) airspeed hold using throttle. These longitudinal control modes can be combined to create the altitude control state machine shown in figure 6.20. In the climb zone, the throttle is set to its maximum value ($\delta_t = 1$) and the airspeed hold from commanded pitch mode is used to control the airspeed, thus ensuring that the aircraft avoids stall conditions. In simple terms, this causes the MAV to climb at its maximum possible climb rate until it is close to the altitude set point. Similarly, in the descend zone, the throttle is set to its minimum value ($\delta_t = 0$) and the airspeed hold from commanded pitch mode is again used to control airspeed. In this way, the MAV descends at a steady rate until it reaches the altitude hold zone. In the altitude hold zone, the airspeed-from-throttle mode is used to regulate the airspeed around V_a^c, and the altitude-from-pitch mode is used to regulate the altitude around h^c. The pitch attitude control loop is active in all four zones.

6.5 Digital Implementation of PID Loops

The longitudinal and lateral control strategies presented in this chapter consist of several proportional-integral-derivative (PID) control loops. In this section we briefly describe how PID loops can be implemented in discrete time. A general PID control signal is given by

$$u(t) = k_p e(t) + k_i \int_{-\infty}^{t} e(\tau) d\tau + k_d \frac{de}{dt}(t),$$

where $e(t) = y^c(t) - y(t)$ is the error between the commanded output $y^c(t)$ and the current output $y(t)$. In the Laplace domain, we have

$$U(s) = k_p E(s) + k_i \frac{E(s)}{s} + k_d s E(s).$$

Since a pure differentiator is not causal, the standard approach is to use a band-limited differentiator so that

$$U(s) = k_p E(s) + k_i \frac{E(s)}{s} + k_d \frac{s}{\tau s + 1} E(s).$$

To convert to discrete time, we use the Tustin or trapezoidal rule, where the Laplace variable s is replaced with the z-transform approximation

$$s \mapsto \frac{2}{T_s} \left(\frac{1 - z^{-1}}{1 + z^{-1}} \right),$$

where T_s is the sample period [28]. Letting $I(s) \triangleq E(s)/s$, an integrator in the z domain becomes

$$I(z) = \frac{T_s}{2} \left(\frac{1 + z^{-1}}{1 - z^{-1}} \right) E(z).$$

Transforming to the time domain, we have

$$I[n] = I[n-1] + \frac{T_s}{2}\left(E[n] + E[n-1]\right). \tag{6.31}$$

A formula for discrete implementation of a differentiator can be derived in a similar manner. Letting $D(s) \triangleq (s/(\tau s + 1))E(s)$, the differentiator in the z domain is

$$D(z) = \frac{\frac{2}{T_s}\left(\frac{1-z^{-1}}{1+z^{-1}}\right)}{\frac{2\tau}{T_s}\left(\frac{1-z^{-1}}{1+z^{-1}}\right) + 1} E(z)$$

$$= \frac{\left(\frac{2}{2\tau+T_s}\right)(1-z^{-1})}{1 - \left(\frac{2\tau-T_s}{2\tau+T_s}\right)z^{-1}} E(z).$$

Transforming to the time domain, we have

$$D[n] = \left(\frac{2\tau - T_s}{2\tau + T_s}\right) D[n-1] + \left(\frac{2}{2\tau + T_s}\right)(E[n] - E[n-1]). \tag{6.32}$$

Matlab code that implements a general PID loop is shown below.

```
function u = pidloop(y_c, y, flag, kp, ki, kd, limit, Ts, tau)
persistent integrator;
persistent differentiator;
persistent error_d1;
if flag==1, % reset (initialize) persistent variables
            % when flag==1
  integrator = 0;
  differentiator = 0;
  error_d1  = 0; % _d1 means delayed by one time step
end
error = y_c - y; % compute the current error
integrator = integrator + (Ts/2)*(error + error_d1);
% update integrator
differentiator = (2*tau-Ts)/(2*tau+Ts)*differentiator...
  + 2/(2*tau+Ts)*(error - error_d1);
% update differentiator
error_d1 = error; % update the error for next time through
                  % the loop
u = sat(...             % implement PID control
  kp * error +...       % proportional term
  ki * integrator +...  % integral term
  kd * differentiator,... % derivative term
  limit...              % ensure abs(u)<=limit
  );
% implement integrator anti-windup
if ki~=0
  u_unsat = kp*error + ki*integrator + kd*differentiator;
```

```
28      integrator = integrator + Ts/ki * (u - u_unsat);
29   end
30
31   function out = sat(in, limit)
32      if in > limit,      out = limit;
33      elseif in < -limit; out = -limit;
34      else                out = in;
35      end
```

The inputs on line 1 are the commanded output y_c; the current output y; a flag used to reset the integrator; the PID gains k_p, k_i, and k_d; the limit of the saturation command; the sample time T_s; and the time constant τ of the differentiator. Line 11 implements equation (6.31), and lines 12–13 implement equation (6.32).

A potential problem with a straight-forward implementation of PID controllers is integrator wind up. When the error $y_c - y$ is large and a large error persists for an extended period of time, the value of the integrator, as computed in line 11, can become large, or "wind up." A large integrator will cause u, as computed in line 15–20, to saturate, which will cause the system to push with maximum effort in the direction needed to correct the error. Since the value of the integrator will continue to wind up until the error signal changes sign, the control signal may not come out of saturation until well after the error has changed sign, which can cause a large overshoot and may potentially destabilize the system.

Since integrator wind up can destabilize the autopilot loops, it is important that each loop have an anti-wind-up scheme. A number of different anti-wind-up schemes are possible. A particularly simple scheme, which is shown in Lines 22–25, is to subtract from the integrator exactly the amount needed to keep u at the saturation bound. In particular, let

$$u^-_{\text{unsat}} = k_p e + k_d D + k_i I^-$$

denote the unsaturated control value before updating the integrator, where I^- is the value of the integrator before applying the anti-wind-up scheme and let

$$u^+_{\text{unsat}} = k_p e + k_d D + k_i I^+$$

denote the unsaturated control value after updating the integrator, where

$$I^+ = I^- + \Delta I,$$

and ΔI is the update. The objective is to find ΔI so that $u^+_{\text{unsat}} = u$, where u is value of the control after the saturation command is applied. Noting that

$$u^+_{\text{unsat}} = u^-_{\text{unsat}} + k_i \Delta I,$$

we can solve for ΔI to obtain

$$\Delta I = \frac{1}{k_i}(u - u^-_{\text{unsat}}).$$

The multiplication by T_s in line 24 is to account for the sampled-data implementation.

6.6 Chapter Summary

In this chapter, we utilized the technique of successive loop closure to develop lateral and longitudinal autopilots for a MAV. The lateral autopilot includes roll-attitude hold as an inner loop and course-angle hold as an outer loop. The longitudinal autopilot is more complicated and depends on the altitude zone. A pitch-attitude-hold loop is used as an inner loop in every zone. In the take-off zone, the MAV is given full throttle, and the MAV is regulated to maintain a fixed take-off pitch angle. In the climb zone, the MAV is given full throttle, and the airspeed is regulated with an airspeed-hold-with-pitch autopilot loop. The descend zone is similar to the climb zone except the MAV is given minimum throttle. In the altitude hold zone, the altitude is regulated with an altitude-using-pitch autopilot loop, and the airspeed is regulated with an airspeed-using-throttle autopilot loop.

6.6.1 Summary of Design Process for Lateral Autopilot

Input: The transfer function coefficients a_{ϕ_1} and a_{ϕ_2}, the nominal airspeed V_a, and the aileron limit $\delta_a^{\max}$.

Tuning Parameters: The roll angle limit $\phi^{\max}$, the damping coefficients ζ_ϕ and ζ_χ, the roll integrator gain k_{i_ϕ}, and the bandwidth separation $W_\chi > 1$, where $\omega_{n_\phi} = W_\chi \omega_{n_\chi}$.

Compute Natural Frequencies: Compute the natural frequency of the inner loop ω_{n_ϕ} using equation (6.8), and the natural frequency of the outer loop using $\omega_{n_\chi} = \omega_{n_\phi}/W_\chi$.

Compute Gains: Compute the gains k_{p_ϕ}, k_{d_ϕ}, k_{p_χ}, and k_{i_χ} using equations (6.7), (6.9), (6.12), and (6.13).

6.6.2 Summary of Design Process for Longitudinal Autopilot

Input: The transfer function coefficients a_{θ_1}, a_{θ_2}, a_{θ_3}, a_{V_1}, and a_{V_2}; the nominal airspeed V_a; and the elevator limit $\delta_e^{\max}$.

Tuning Parameters: The pitch angle limit $e_\theta^{\max}$; the damping coefficients ζ_θ, ζ_h, ζ_V, and ζ_{V_2}; the natural frequency ω_{n_V}; and the bandwidth separation for the altitude loop W_h and the airspeed-using-pitch loop W_{V_2}.

Compute Natural Frequencies: Compute the natural frequency of the pitch loop ω_{n_θ} using equation (6.21). Compute the natural frequency of the altitude loop using $\omega_{n_h} = \omega_{n_\theta} / W_h$, and of the airspeed-using-pitch loop using $\omega_{n_{V_2}} = \omega_{n_\theta} / W_{V_2}$.

Compute Gains: Compute the gains k_{p_θ} and k_{d_θ} using equation (6.22). Compute the DC gain of the pitch loop using equation (6.23). Compute k_{p_h} and k_{i_h} using equations (6.25) and (6.24). Compute the gains $k_{p_{V_2}}$ and $K_{i_{V_2}}$ using equations (6.28) and (6.27). Compute the gains k_{p_V} and k_{i_V} using equations (6.30) and (6.29).

Notes and References

Portions of the design process that we outline in this chapter appear in [29]. Similar techniques using root locus are given in [1, 2, 5, 6]. A standard reference for digital implementation of PID controllers is [28]. Simple anti-wind-up schemes are discussed in [28, 30].

6.7 Design Project

In this assignment you will use simplified design models to tune the gains of the PID loops for the lateral and longitudinal autopilot. To do this, you will need to create some auxiliary Simulink models that implement the design models. The final step will be to implement the control loops on the full simulation model. Simulink models that will help you in this process are included on the book website.

> 6.1. Create a Matlab script that computes the gains for the roll attitude hold loop. Assume that the maximum aileron deflection is $\delta_a^{\max} = 45$ degrees, the saturation limit is achieved for a step size of $\phi^{\max} = 15$ degrees, and the nominal airspeed is $V_a = 10$ m/s. Use the Simulink file `roll_loop.mdl` on the website to tune the values of ζ_ϕ and k_{i_ϕ} to get acceptable performance.

6.2. Augment your Matlab script to compute the gains for the course hold loop. The Simulink file `course_loop.mdl` implements the course hold loop with the roll hold as an inner loop. Tune the bandwidth separation and the damping ratio ζ_χ to get acceptable performance for step inputs in course angle of 25 degrees.

6.3. Augment your Matlab script to compute the gains for sideslip hold. Use the Simulink model `sideslip_loop.mdl` on the website to tune the value of ζ_β.

6.4. Augment your Matlab script to compute the gains for the pitch attitude hold loop. Assume that the maximum elevator deflection is $\delta_e^{\max} = 45$ degrees, and that the saturation limit is achieved for a step size of $e_\theta^{\max} = 10$ degrees. Use the Simulink file `pitch_loop.mdl` on the website to tune the value of ζ_θ.

6.5. Augment your Matlab script to compute the gains for altitude hold using pitch as an input. Use the Simulink model `altitude_from_pitch_loop.mdl` on the website to tune the value of ζ_h and the bandwidth separation.

6.6. Augment your Matlab script to compute the gains for airspeed hold using pitch as an input. Use the Simulink model `airspeed_from_pitch_loop.mdl` on the website to tune the value of ζ and the bandwidth separation.

6.7. Augment your Matlab script to compute the gains for airspeed hold using throttle as an input. Use the Simulink model `airspeed_from_throttle_loop.mdl` on the website to tune the value of ζ and ω_n.

6.8. The final step of the design is to implement the lateral-directional and longitudinal autopilot on the simulation model. Modify your simulation model so that the aircraft is in its own subsystem. To ensure that the autopilot code can be easily transferred to embedded code written in, for example, C/C++, write the autopilot function using a Matlab script. An example of how to organize your simulation is given in the Simulink model `mavsim_chap6.mdl` on the website. A gutted version of the autopilot code is also given on the web site. To implement the longitudinal autopilot, you will need to implement a state machine using, for example, the **switch** statement.

7

Sensors for MAVs

Critical to the creation and realization of small unmanned air vehicles has been the development of small, lightweight solid-state sensors. Based on microelectromechanical systems (MEMS) technology, small but accurate sensors such as accelerometers, angular rate sensors, and pressure sensors have enabled the development of increasingly smaller and more capable autonomous aircraft. Coupled with the development of small global positioning systems (GPS), computationally capable microcontrollers, and more powerful batteries, the capabilities of MAVs have gone from being purely radio controlled (RC) by pilots on the ground to highly autonomous systems in less than 20 years. The objective of this chapter is to describe the onboard sensors typically used on MAVs and to quantify what they measure. We will focus on sensors used for guidance, navigation, and control of the aircraft. Payload sensors, such as cameras, and their use will be described in chapter 13.

The following sensors are often found on MAVs:

- Accelerometers
- Rate gyros
- Pressure sensors
- Magnetometers
- GPS

The following sections will discuss each of these sensors, describe their sensing characteristics, and propose models that describe their behavior for analysis and simulation purposes.

7.1 Accelerometers

Acceleration transducers (accelerometers) typically employ a proof mass held in place by a compliant suspension as shown in figure 7.1. When the case of the accelerometer experiences an acceleration, the proof mass moves relative to the case through a distance proportional to the acceleration. The acceleration experienced by the proof mass is converted to a displacement by the springs in the suspension. A simple

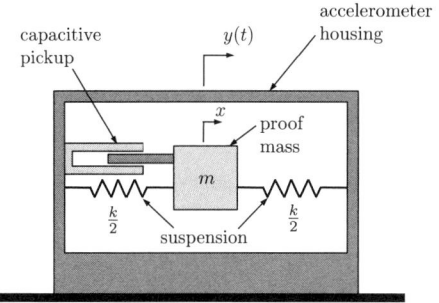

Figure 7.1 Conceptual depiction of MEMS accelerometer.

force balance analysis of the proof mass yields the relationship

$$m\ddot{x} + kx = ky(t),$$

where x is the inertial position of the proof mass and $y(t)$ is the inertial position of the housing—the acceleration of which we want to sense. Given that the deflection of the suspension is $\delta = y(t) - x$, this relation can be expressed as

$$\ddot{x} = \frac{k}{m}\delta.$$

Thus, the acceleration of the proof mass is proportional to the deflection of the suspension. At frequencies below the resonant frequency, the acceleration of the proof mass is the same as the acceleration of the housing. This can be seen by examining the transfer function from the housing position input to the proof mass position output

$$\frac{X(s)}{Y(s)} = \frac{1}{\frac{m}{k}s^2 + 1},$$

or equivalently, the transfer function from the housing acceleration input to the proof mass acceleration output

$$\frac{A_X(s)}{A_Y(s)} = \frac{1}{\frac{m}{k}s^2 + 1}.$$

At frequencies corresponding to $\omega < \sqrt{k/m}$, the transfer function $A_X(s)/A_Y(s) \approx 1$ and the displacement of the proof mass is an accurate indicator of the acceleration of the body to which the accelerometer is attached.

The accelerometer in figure 7.1 is shown with a capacitive transducer to convert the proof mass displacement into a voltage output as is common in many MEMS devices. Other approaches to convert the displacement to a usable signal include piezoelectric, reluctive, and strain-based designs. As with other analog devices, accelerometer measurements are subject to signal bias and random uncertainty. The output of an accelerometer can be modeled as

$$\Upsilon_{\text{accel}} = k_{\text{accel}} A + \beta_{\text{accel}} + \eta'_{\text{accel}},$$

where Υ_{accel} is in volts, k_{accel} is a gain, A is the acceleration in meters per second squared, β_{accel} is a bias term, and η'_{accel} is zero-mean Gaussian noise. The gain k_{accel} may be found on the data sheet of the sensor. Due to variations in manufacturing, however, it is imprecisely known. A one-time lab calibration is usually done to accurately determine the calibration constant, or gain, of the sensor. The bias term β_{accel} is dependent on temperature and should be calibrated prior to each flight.

In aircraft applications, three accelerometers are commonly used. The accelerometers are mounted near the center of mass, with the sensitive axis of one accelerometer aligned with each of the body axes. Accelerometers measure the specific force in the body frame of the vehicle. Another interpretation is that they measure the difference between the acceleration of the aircraft and the gravitational acceleration. To understand this phenomena, imagine that the device shown in figure 7.1 were to be turned ninety degrees and set on a table. The forces acting on the casing will be gravity pulling down, and an equal and opposite normal force pushing up to keep the casing on the table. Therefore, the total acceleration on the casing will be zero. However, since the normal force of the table does not act on the proof mass, it will deflect under the force of gravity and the sensor will measure an acceleration equal to one g. Therefore the measured acceleration is the total acceleration of the casing minus gravity. Mathematically we have

$$\begin{pmatrix} a_x \\ a_y \\ a_z \end{pmatrix} = \frac{d\mathbf{v}}{dt_b} + \boldsymbol{\omega}_{b/i} \times \mathbf{v} - \mathcal{R}_v^b \begin{pmatrix} 0 \\ 0 \\ g \end{pmatrix},$$

which can be expressed in component form as

$$a_x = \dot{u} + qw - rv + g \sin \theta$$
$$a_y = \dot{v} + ru - pw - g \cos \theta \sin \phi$$
$$a_z = \dot{w} + pv - qu - g \cos \theta \cos \phi.$$

It can be seen that each accelerometer measures elements of linear acceleration, Coriolis acceleration, and gravitational acceleration.

The voltage output of an accelerometer is converted into a number corresponding to the voltage inside the autopilot microcontroller by an analog-to-digital converter at a sample rate T_s. Through calibration, this voltage can be converted to a numerical representation of the acceleration in meters per second squared. Assuming that the biases can be removed through the calibration process, the accelerometer signals inside the autopilot can be modeled as

$$y_{accel,x} = \dot{u} + qw - rv + g \sin\theta + \eta_{accel,x}$$
$$y_{accel,y} = \dot{v} + ru - pw - g \cos\theta \sin\phi + \eta_{accel,y} \quad (7.1)$$
$$y_{accel,z} = \dot{w} + pv - qu - g \cos\theta \cos\phi + \eta_{accel,z},$$

where $\eta_{accel,x}$, $\eta_{accel,y}$, and $\eta_{accel,z}$ are zero-mean Gaussian processes with variance $\sigma^2_{accel,x}$, $\sigma^2_{accel,y}$, and $\sigma^2_{accel,z}$ respectively. Because of the calibration, the units of $y_{accel,x}$, $y_{accel,y}$, and $y_{accel,z}$ are in m/s².

Depending on the organization of the simulation software, the terms $\dot{u}$, $\dot{v}$, and $\dot{w}$ (state derivatives), may be inconvenient to calculate for inclusion in equation (7.1). As an alternative, we can substitute from equations (5.4), (5.5), and (5.6) to obtain

$$y_{accel,x} = \frac{\rho V_a^2 S}{2m}\left[C_X(\alpha) + C_{X_q}(\alpha)\frac{\bar{c}q}{2V_a} + C_{X_{\delta_e}}(\alpha)\delta_e\right]$$
$$\quad + \frac{\rho S_{prop} C_{prop}}{2m}[(k_{motor}\delta_t)^2 - V_a^2] + \eta_{accel,x}$$
$$y_{accel,y} = \frac{\rho V_a^2 S}{2m}\left[C_{Y_0} + C_{Y_\beta}\beta + C_{Y_p}\frac{bp}{2V_a} + C_{Y_r}\frac{br}{2V_a}\right.$$
$$\quad \left. + C_{Y_{\delta_a}}\delta_a + C_{Y_{\delta_r}}\delta_r\right] + \eta_{accel,y} \quad (7.2)$$
$$y_{accel,z} = \frac{\rho V_a^2 S}{2m}\left[C_Z(\alpha) + C_{Z_q}(\alpha)\frac{\bar{c}q}{2V_a} + C_{Z_{\delta_e}}(\alpha)\delta_e\right] + \eta_{accel,z}.$$

However, since the forces are already calculated as part of the dynamics, the best way to organize the simulation files is to use the forces to compute the output of the accelerometers. The resulting equations are

$$y_{accel,x} = \frac{f_x}{m} + g\sin\theta + \eta_{accel,x}$$
$$y_{accel,y} = \frac{f_y}{m} - g\cos\theta\sin\phi + \eta_{accel,y} \quad (7.3)$$
$$y_{accel,z} = \frac{f_z}{m} - g\cos\theta\cos\phi + \eta_{accel,z},$$

where f_x, f_y, and f_z are given in equation (4.18). With the exception of the noise terms, the terms on the right hand sides of equation (7.3) represent the specific force experienced by the aircraft. The acceleration of the aircraft is commonly expressed in units of g, the gravitational constant. To express the acceleration measurements in g's, equation (7.3) can be divided by g. The choice of units is up to the preference of the engineer, however, maintaining consistent units reduces the potential for mistakes in implementation.

7.2 Rate Gyros

MEMS rate gyros typically operate based on the principle of the Coriolis acceleration. In the early 19th century, French scientist G.G. de Coriolis discovered that a point translating on a rotating rigid body experiences an acceleration, now called Coriolis acceleration, that is proportional to the velocity of the point and the rate of rotation of the body

$$\mathbf{a}_C = 2\mathbf{\Omega} \times \mathbf{v}, \qquad (7.4)$$

where $\mathbf{\Omega}$ is the angular velocity of the body in an inertial reference frame, and $\mathbf{v}$ is the velocity of the point in the reference frame of the body. In this case, $\mathbf{\Omega}$ and $\mathbf{v}$ are both vector quantities and $\times$ represents the vector cross product.

MEMS rate gyros commonly consist of a vibrating proof mass as depicted in figure 7.2. In this figure, the cantilever and proof mass are actuated at their resonant frequency to cause oscillation in the vertical plane. The cantilever is actuated so that the velocity of the proof mass due to these oscillations is a constant amplitude sinusoid

$$\mathbf{v} = A\omega_n \sin(\omega_n t),$$

where A is the amplitude of the oscillation and ω_n is the natural frequency of the oscillation. If the sensitive axis of the rate gyro is configured to be the longitudinal axis of the undeflected cantilever, then rotation about this axis will result in a Coriolis acceleration in the horizontal plane described by equation (7.4) and shown in figure 7.2. Similar to the accelerometer, the Coriolis acceleration of the proof mass results in a lateral deflection of the cantilever. This lateral deflection of the cantilever can be detected in several ways: by capacitive coupling, through a piezoelectrically generated charge, or through a change in piezoresistance of the cantilever. Whatever the transduction method, a voltage proportional to the lateral Coriolis acceleration is produced.

Sensors for MAVs

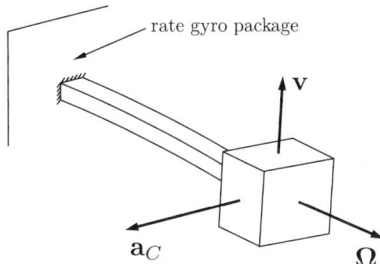

Figure 7.2 Conceptual depiction of proof mass rate gyro. ω is the angular velocity of the sensor package to be measured. $\mathbf{v}$ is the actuated vibration velocity of the cantilever. $\mathbf{a}_C$ is the Coriolis acceleration that results as the sensor package undergoes an angular velocity.

With the sensing axis orthogonal to the direction of vibration, the ideal output voltage of the rate gyro is proportional to the amplitude of Coriolis acceleration, and is given by

$$V_{\text{gyro}} = k_C |\mathbf{a}_C|$$
$$= 2k_C |\mathbf{\Omega} \times \mathbf{v}|.$$

Since $\mathbf{\Omega}$, the angular rate of rotation about the sensitive axis of the gyro, and $\mathbf{v}$ are orthogonal

$$|\mathbf{\Omega} \times \mathbf{v}| = \Omega |\mathbf{v}|,$$

and

$$V_{\text{gyro}} = 2k_C \Omega |A\omega_n \sin(\omega_n t)|$$
$$= 2k_C A\omega_n \Omega$$
$$= K_C \Omega,$$

where K_C is a calibration constant and Ω represents the magnitude and direction (sign) of the angular velocity about the sensitive axis.

The output of a rate gyro can be modeled as

$$\Upsilon_{\text{gyro}} = k_{\text{gyro}} \Omega + \beta_{\text{gyro}} + \eta'_{\text{gyro}},$$

where Υ_{gyro} corresponds to the measured rate of rotation in volts, k_{gyro} is a gain converting the rate in radians per second to volts, Ω is the angular rate in radians per second, β_{gyro} is a bias term, and η'_{gyro} is zero-mean Gaussian noise. An approximate value for the gain k_{gyro} should be given on the spec sheet of the sensor. To ensure accurate measurements, the value of this gain should be determined through experimental calibration. The bias term β_{gyro} is strongly dependent on temperature and should be calibrated prior to each flight. For low-cost MEMS gyros, drift in this bias term can be significant and care must be taken to zero the gyro bias periodically during flight. This is done by flying a straight and level path ($\Omega = 0$) and resetting the gyro bias so that Υ_{gyro} averages zero over a period of 100 or so samples.

For simulation purposes, we are interested in modeling the calibrated gyro signals inside the autopilot. The rate gyro signals are converted from analog voltages coming out of the sensor to numerical representations of angular rates (in units of rad/s) inside the autopilot. We assume that the gyros have been calibrated so that in the nominal case 1 rad/s of angular rate experienced by the sensor results in a numerical measurement inside the autopilot of 1 rad/s (i.e., the gain from the physical rate to its numerical representation inside the autopilot is one) and that the biases have been estimated and subtracted from the measurements. It is common to measure the angular rates about each of the body axes using three gyros by aligning the sensitive axis of a gyro along each of the $\mathbf{i}^b$, $\mathbf{j}^b$, and $\mathbf{k}^b$ axes of the MAV. These rate gyro measurements of angular body rates p, q, and r can be modeled as

$$y_{\text{gyro},x} = p + \eta_{\text{gyro},x}$$
$$y_{\text{gyro},y} = q + \eta_{\text{gyro},y} \qquad (7.5)$$
$$y_{\text{gyro},z} = r + \eta_{\text{gyro},z},$$

where $y_{\text{gyro},x}$, $y_{\text{gyro},y}$, and $y_{\text{gyro},z}$ are angular rate measurements with units of rad/s. The variables $\eta_{\text{gyro},x}$, $\eta_{\text{gyro},y}$, and $\eta_{\text{gyro},z}$ represent zero-mean Gaussian processes with variances $\sigma^2_{\text{gyro},x}$, $\sigma^2_{\text{gyro},y}$, and $\sigma^2_{\text{gyro},z}$, respectively. MEMS gyros are analog devices that are sampled by the autopilot microcontroller. We will assume that the sample rate is given by T_s.

7.3 Pressure Sensors

Pressure, a quantity commonly associated with fluids, is defined as the force per unit area acting on a surface. Pressure acts in a direction normal to the surface of the body to which it is applied. We will use measurements of pressure to provide indications of the altitude of the aircraft and the airspeed of the aircraft. To measure altitude, we will use an absolute pressure sensor. To measure airspeed, we will use a differential pressure sensor.

7.3.1 Altitude Measurement

Measurements of altitude can be inferred from measurements of atmospheric pressure. The basic equation of hydrostatics, given by

$$P_2 - P_1 = \rho g (z_2 - z_1), \qquad (7.6)$$

states that for a static fluid, the pressure at a point of interest changes with the depth of the point below the surface of the fluid. This relationship assumes that the density of the fluid is constant between the points of interest. Although the air in the atmosphere is compressible and its density changes significantly over altitudes from sea level to altitudes commonly flown by modern aircraft, the hydrostatic relationship of equation (7.6) can be useful over small altitude changes where the air density remains essentially constant.

We are typically interested in the altitude or height of the aircraft above a ground station and the corresponding change in pressure between the ground and the altitude of interest. From equation (7.6), the change in pressure due to a change in altitude is given by

$$P - P_{\text{ground}} = -\rho g(h - h_{\text{ground}}) \tag{7.7}$$
$$= -\rho g h_{\text{AGL}},$$

where h is the absolute altitude of the aircraft, h_{ground} is the absolute altitude of the ground, $h_{\text{AGL}} = h - h_{\text{ground}}$, and h and h_{ground} are measured with respect to sea level and P is the corresponding absolute pressure measurement. The change in sign between equations (7.6) and (7.7) comes from the fact that depth, z, is measured positive down while altitude h is measured positive up. A decrease in altitude above the ground results in an increase in measured pressure. In implementation, P_{ground} is the atmospheric pressure measured at ground level prior to take off and ρ is the air density at the flight location.

Equation (7.7) assumes that the air density is constant over the altitude range of interest. In truth, it varies with both weather conditions and altitude. Assuming that weather conditions are invariant over the flight duration, we must consider the effects of changing air density due to changes in pressure and temperature that occur with altitude.

Below altitudes of 11,000 m above sea level, the pressure of the atmosphere can be calculated using the barometric formula [31]. This formula takes into account the change in density and pressure due to decreasing temperature with altitude and is given by

$$P = P_0 \left[\frac{T_0}{T_0 + L_0 h_{\text{ASL}}} \right]^{\frac{gM}{RL_0}}, \tag{7.8}$$

where $P_0 = 101,325 \text{ N/m}^2$ is the standard pressure at sea level, $T_0 = 288.15 \text{ K}$ is the standard temperature at sea level, $L_0 = -0.0065 \text{ K/m}$ is the lapse rate or the rate of temperature decrease in the lower atmosphere, $g = 9.80665 \text{ m/s}^2$ is the gravitational constant, $R = 8.31432 \text{ N-m/(mol-K)}$ is the universal gas constant for air, and

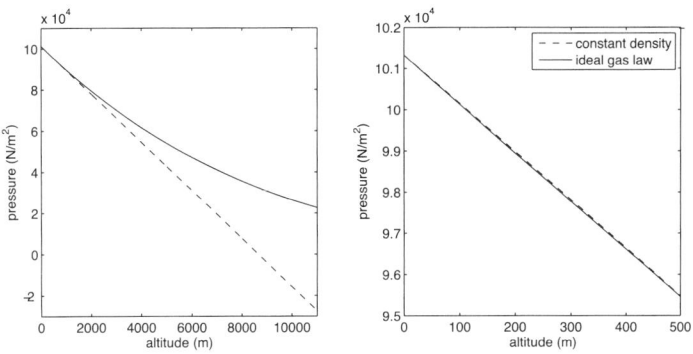

Figure 7.3 Comparison of atmospheric pressure calculations using constant-density and variable-density models.

$M = 0.0289644$ kg/mol is the standard molar mass of atmospheric air. The altitude h_{ASL} is referenced to sea level.

The relative significance of the constant-density assumption can be seen by comparing pressures calculated using equations (7.7) and (7.8) as shown in figure 7.3. It can be seen that over the full range of altitudes for which the barometric formula is valid (0 to 11,000 m above sea level), the pressure versus altitude relationship is not linear and that the linear approximation of equation (7.7) is not valid. The plot on the right of figure 7.3, however, shows that over narrower altitude ranges, such as those common to small unmanned aircraft, a linear approximation can be used with reasonable accuracy. For this particular plot, equation (7.7) was employed with $h_{\text{ground}} = 0$ and air density calculated at sea level.

One key to accurately calculating altitude from pressure using equation (7.7) is to have an accurate measure of air density at the flight location. This can be determined from the ideal gas formula, with measurements of the local temperature and barometric pressure at flight time according to

$$\rho = \frac{MP}{RT},$$

using values for the universal gas constant and molar mass of air specified above. Notice that in this formula, temperature is expressed in units of Kelvin. The conversion from Fahrenheit to Kelvin is given by

$$T\,[K] = \frac{5}{9}(T\,[F] - 32) + 273.15.$$

The atmospheric pressure is expressed in N/m². Typical weather data reports pressure in inches of mercury (Hg). The conversion factor is 1 N/m² = 3385 inches of Hg.

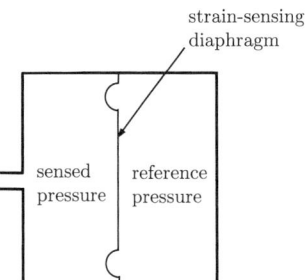

Figure 7.4 Schematic of an absolute pressure sensor.

In practice, we will utilize the measurement of absolute pressure to give an indication of altitude above ground level of the aircraft. Figure 7.4 shows an example of an absolute pressure sensor in schematic form. The pressure sensor consists of two volumes separated by a diaphragm. The volume on the right is closed and at a constant reference pressure. The volume at the left is open to the ambient air. Changes in the pressure of the ambient air cause the diaphragm to deflect. These deflections are measured and produce a signal proportional to the sensed pressure.

Following equation (7.7), the output of the absolute pressure sensor of interest is given by

$$y_{\text{abs pres}} = (P_{\text{ground}} - P) + \beta_{\text{abs pres}} + \eta_{\text{abs pres}}$$
$$= \rho g h_{\text{AGL}} + \beta_{\text{abs pres}} + \eta_{\text{abs pres}} \quad (7.9)$$

where h_{AGL} is the altitude above ground level, β_{abspres} is a temperature-related bias drift, and η_{abspres} is zero-mean Gaussian noise with variance $\sigma^2_{\text{abspres}}$. P_{ground} is the pressure measured at ground level prior to take-off and held in the memory of the autopilot microcontroller. P is the absolute pressure measured by the sensor during flight. The difference between these two measurements is proportional to the altitude of the aircraft above ground level.

7.3.2 Airspeed Sensor

Airspeed can be measured using a pitot-static probe in conjunction with a differential pressure transducer as depicted schematically in figure 7.5. The pitot-static tube has two ports: one that is exposed to the total pressure and another that is exposed to the static pressure. The total pressure is also known as the stagnation pressure or pitot pressure. It is the pressure at the tip of the probe, which is open to the oncoming flow. The flow is stagnant or stopped at the tip. As a result, pressure is built up so that the pressure at the tip is higher than that of the surrounding fluid. The static pressure is simply the ambient pressure

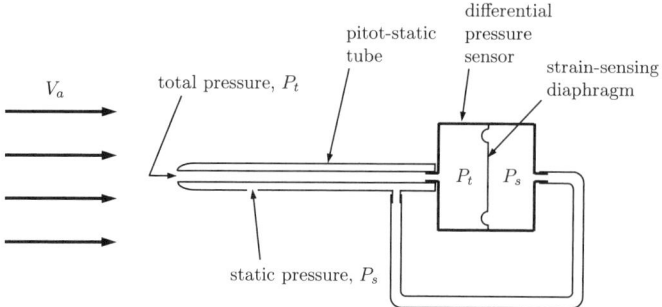

Figure 7.5 Schematic of pitot-static tube and differential pressure sensor. Not to scale.

of the surrounding fluid (or atmosphere). The difference in pressures on each side of the diaphragm in the differential pressure sensor cause the diaphragm to deflect causing a strain in the diaphragm proportional to the pressure difference. This strain is measured, producing a voltage output representing the differential pressure.

Bernoulli's equation states that total pressure is the sum of the static pressure and dynamic pressure. In equation form, we can write this as

$$P_t = P_s + \frac{\rho V_a^2}{2},$$

where ρ is the air density and V_a is the airspeed of the MAV. Rearranging, we have

$$\frac{\rho V_a^2}{2} = P_t - P_s,$$

which is the quantity that the differential pressure sensor measures. With proper calibration to convert the sensor output from volts to a number inside the microcontroller representing pressure in units of N/m^2, we can model the output of the differential pressure sensor as

$$y_{\text{diff pres}} = \frac{\rho V_a^2}{2} + \beta_{\text{diff pres}} + \eta_{\text{diff pres}}, \qquad (7.10)$$

where $\beta_{\text{diff pres}}$ is a temperature-related bias drift, $\eta_{\text{diff pres}}$ is zero-mean Gaussian noise with variance $\sigma^2_{\text{diff pres}}$. The absolute and differential pressure sensors are analog devices that are sampled by the onboard processor at the same update rate as the main autopilot control loop.

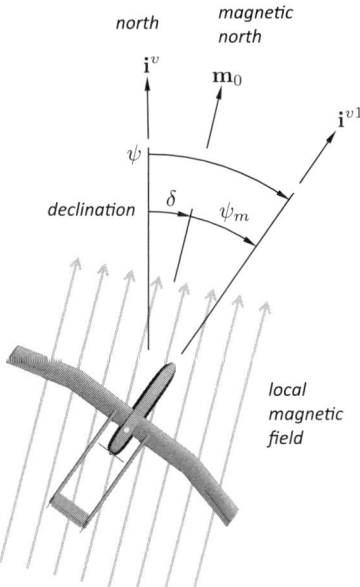

Figure 7.6 Magnetic field and compass measurement.

7.4 Digital Compasses

The earth's magnetic field has been used as a navigational aid for centuries. The first magnetic compasses are believed to have originated with the Chinese around the first century AD Compasses appeared in Europe around the 11th century AD and were used by Christopher Columbus and other world explorers in the late 15th century. The earth's magnetic field continues to provide a means for navigation for a variety of vehicles, including unmanned aircraft.

The magnetic field around the earth behaves similarly to that of a common magnetic dipole with the magnetic field lines running normal to the earth's surface at the poles and parallel to the earth's surface near the equator. Except near the poles, the earth's magnetic field points to magnetic north. A compass measures the direction of the magnetic field locally and provides an indication of heading relative to magnetic north, ψ_m. This is depicted schematically in figure 7.6. The declination angle δ is the angle between true north and magnetic north.

The earth's magnetic field is three dimensional, with north, east, and down components that vary with location along the earth's surface. For example, in Provo, Utah, the north component (X) of the magnetic field is 21,053 nT, the east component (Y) is 4520 nT, and the down component (Z) is 47,689 nT. The declination angle is 12.12 degrees. Figure 7.7 shows the declination angle over the surface of the earth and illustrates the significant dependence of the magnetic north direction on location. The inclination angle is the angle that the

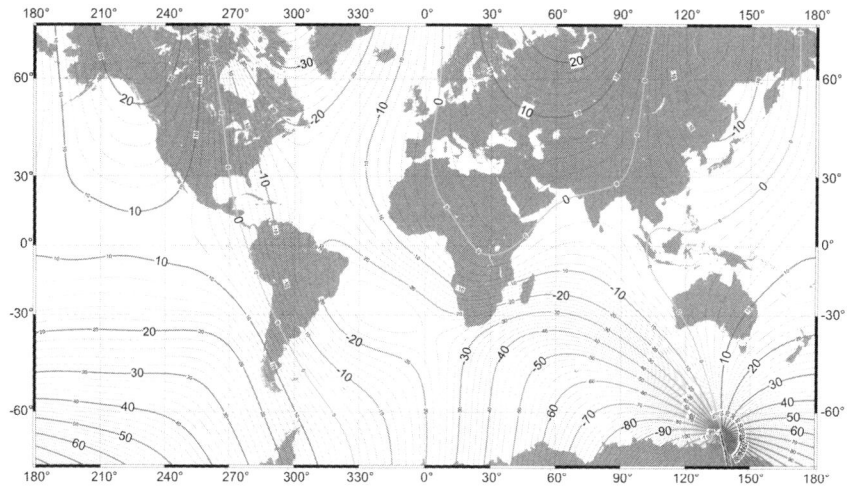

Figure 7.7 Magnetic field declination according to the US/UK World Magnetic Model. Adapted from [32].

magnetic field makes with the horizontal plane. In Provo, the inclination is 65.7 degrees.

Modern digital compasses use three-axis magnetometers to measure the strength of the magnetic field along three orthogonal axes. In UAV applications, these axes of measurement are usually aligned with the body axes of the aircraft. Although only two sensing axes are required to measure the magnetic heading if the aircraft is in level flight, a third sensing axis is necessary if the aircraft is pitching or rolling out of the horizontal plane.

From figure 7.6, we can see that the heading angle is the sum of the declination angle and the magnetic heading measurement

$$\psi = \delta + \psi_m. \tag{7.11}$$

The declination angle for a given latitude and longitude can be calculated using models such as the World Magnetic Model (WMM), available from the National Geophysical Data Center (NGDC) [32]. The magnetic heading can be determined from measurements of the magnetic field strength along the body-frame axes. To do so, we project the body-frame measurements of the magnetic field onto the horizontal plane. The angle between the horizontal component of the magnetic field and the $\mathbf{i}^{v1}$ axis (the heading) is the magnetic heading. Mathematically, we can calculate the magnetic heading from the following expressions as

$$\mathbf{m}_0^{v1} = \begin{pmatrix} m_{0x}^{v1} \\ m_{0y}^{v1} \\ m_{0z}^{v1} \end{pmatrix} = \mathcal{R}_b^{v1}(\phi, \theta) \mathbf{m}_0^b$$

$$= \mathcal{R}_{v2}^{v1}(\theta) \mathcal{R}_b^{v2}(\phi) \mathbf{m}_0^b$$

$$= \begin{pmatrix} \cos\theta & 0 & \sin\theta \\ 0 & 1 & 0 \\ -\sin\theta & 0 & \cos\theta \end{pmatrix} \begin{pmatrix} 1 & 0 & 0 \\ 0 & \cos\phi & -\sin\phi \\ 0 & \sin\phi & \cos\phi \end{pmatrix} \mathbf{m}_0^b$$

$$\begin{pmatrix} m_{0x}^{v1} \\ m_{0y}^{v1} \\ m_{0z}^{v1} \end{pmatrix} = \begin{pmatrix} c_\theta & s_\theta s_\phi & s_\theta c_\phi \\ 0 & c_\phi & -s_\phi \\ -s_\theta & c_\theta s_\phi & c_\theta c_\phi \end{pmatrix} \mathbf{m}_0^b, \quad (7.12)$$

and

$$\psi_m = -\text{atan2}(m_{0y}^{v1}, m_{0x}^{v1}). \quad (7.13)$$

In these equations, $\mathbf{m}_0^b$ is a vector containing the body-frame measurements of the magnetic field taken onboard the vehicle. The four-quadrant inverse tangent function atan2(y,x) returns the arctangent of y/x in the range $[-\pi, \pi]$ using the signs of both arguments to determine the quadrant of the return value. The components m_{0x}^{v1} and m_{0y}^{v1} are the horizontal components of the magnetic field measurement that result when $\mathbf{m}_0^b$ is projected onto the horizontal plane. Notice that in level flight ($\phi = \theta = 0$), $\mathbf{m}_0^{v1} = \mathbf{m}_0^b$.

In practice, the use of magnetometers and digital compasses can be challenging. This is primarily due to the sensitivity of the sensor to electromagnetic interference. Careful placement of the sensor on the aircraft is essential to avoid interference from electric motors, servos, and power wiring. Magnetometers can also be sensitive to interference from power lines and weather systems. Some of the challenges associated with magnetometers have been addressed by manufacturers that package magnetometers, signal conditioning, and a microcontroller into a single-chip solution called a digital compass. These digital compasses vary in their sophistication with full-featured versions incorporating tilt compensation and automatic declination/inclination calculation from latitude and longitude data.

To create a sensor model for simulation purposes, a reasonable approach to modeling a digital compass is to assume that the compass gives a measure of the true heading with a bias error, due to uncertainty in the declination angle, and sensor noise from the magnetometers.

Mathematically, this can be represented as

$$y_{\text{mag}} = \psi + \beta_{\text{mag}} + \eta_{\text{mag}}, \qquad (7.14)$$

where η_{mag} is a zero-mean Gaussian process with variance σ_{mag}^2 and β_{mag} is a bias error. A digital compass communicates over a serial link with the autopilot at sample rate T_s.

7.5 Global Positioning System

The Global Positioning System (GPS) is a satellite-based navigation system that provides 3-D position information for objects on or near the earth's surface. The NAVSTAR GPS system was developed by the U.S. Department of Defense and has been fully operational since 1993. For unmanned aircraft, it would be difficult to overstate the significance of the development and availability of the GPS system. It was, and continues to be, a critical enabling technology for small UAVs. The GPS system and global navigation satellite systems have been described in detail in numerous texts (e.g., [33, 34, 35, 36]). In this section, we will provide a brief overview of GPS position sensing and present a model for GPS position sensing suitable for simulation.

The key component of the GPS system is the constellation of 24 satellites that continuously orbit the earth at an altitude of 20,180 km [33]. The configuration of the satellite orbits is designed so that any point on the earth's surface is observable by at least four satellites at all times. By measuring the times of flight of signals from a minimum of four satellites to a receiver on or near the surface of the earth, the location of the receiver in three dimensions can be determined. The time of flight of the radio wave signal is used to determine the range from each satellite to the receiver. Because synchronization errors exist between the satellite clocks and the receiver clock, the range estimate determined from the time of flight measurement is called *pseudorange* to distinguish it from the true range.

Because of clock synchronization errors between the satellites (whose atomic clocks are almost perfectly synchronized) and the receiver, four independent pseudorange measurements are required to triangulate the position of the receiver as depicted in figure 7.8. Why are four pseudorange measurements required? We know that with one range measurement from a known location, we can locate a point on a line (1-D). Two range measurements can locate a point on a plane (2-D) and three range measurements can locate a point on a 3-D surface. To resolve position in three dimensions with a receiver clock offset error, at least four measurements are needed. The geometry associated with

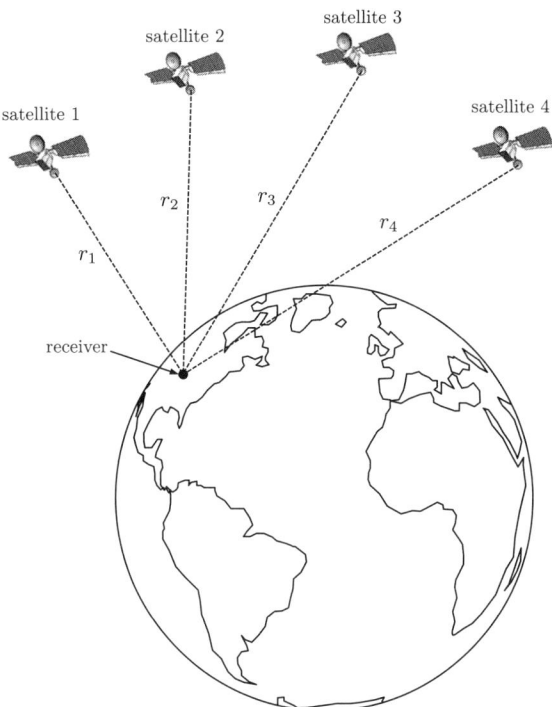

Figure 7.8 Pseudo-range measurements from four satellites used to triangulate the position of a receiver.

pseudorange measurements from four different satellites form a system of four nonlinear algebraic equations in four unknowns: latitude, longitude, and altitude of the GPS receiver, and receiver clock time offset [33].

7.5.1 GPS Measurement Error

The accuracy of a GPS position measurement is affected by the accuracy of the satellite pseudorange measurements and by the geometry of the satellites from which pseudorange measurements are taken. The effect of satellite geometry is taken into consideration through a factor called dilution of precision (DOP). Pseudorange accuracy is affected by errors in the time of flight measurement for each satellite. Given that the electromagnetic radio signals from the satellites travel at the speed of light, small timing errors can cause significant positioning errors. For example, a timing error of only 10 ns can result in a positioning error of about 3 m. Error sources in the time of flight are discussed briefly in the following paragraphs drawing on information presented in [33, 34].

Ephemeris Data

The satellite ephemeris is a mathematical description of its orbit. Calculation of the receiver location requires the satellite locations to be known. Ephemeris errors in the pseudorange calculation are due to uncertainty in the transmitted location of the satellite. Errors in the range of 1 to 5 m are typical.

Satellite Clock

GPS satellites use cesium and rubidium atomic clocks, which over the course of one day, drift about 10 ns, introducing an error of about 3.5 m. Given that the clocks are updated every 12 hours, satellite clock errors introduce a positioning error of 1 to 2 m on average.

Ionosphere

The ionosphere is the uppermost layer of the earth's atmosphere and is characterized by the presence of free electrons that delay the transmission of GPS signals. Although receivers make corrections for this delay based on information in the GPS message, errors caused by variations in the speed of light through the ionosphere are the largest source of ranging errors in GPS measurements. Errors are typically between 2 and 5 m.

Troposphere

The troposphere is the lowest layer of the earth's atmosphere, extending from the earth's surface to an altitude of between 7 and 20 km. Most of the mass of the atmosphere is in the troposphere and almost all of the weather activity around the earth occurs in the troposphere. Variations in the temperature, pressure, and humidity in the troposphere affect the speed of light and thus affect time of flight and pseudorange estimates. The uncertainty in the speed of light through the troposphere introduces range errors of about 1 m.

Multipath Reception

Multipath errors are caused when a GPS receiver receives reflected signals that mask the true signal of interest. Multipath is most significant for static receivers located near large reflecting surfaces, such as might be encountered near large buildings or structures. Multipath errors are below 1 m in most circumstances.

Receiver Measurement

Receiver measurement errors stem from the inherent limits with which the timing of the satellite signal can be resolved. Improvements in

TABLE 7.1
Standard pseudorange error model (1-σ, in meters) [34]

Error source	Bias	Random	Total
Ephemeris data	2.1	0.0	2.1
Satellite clock	2.0	0.7	2.1
Ionosphere	4.0	0.5	4.0
Troposphere monitoring	0.5	0.5	0.7
Multipath	1.0	1.0	1.4
Receiver measurement	0.5	0.2	0.5
UERE, rms	5.1	1.4	5.3
Filtered UERE, rms	5.1	0.4	5.1

signal tracking and processing have resulted in modern receivers that can compute the signal timing with sufficient accuracy to keep ranging errors due to the receiver less than 0.5 m.

Pseudorange errors from the sources described above are treated as statistically uncorrelated and can be added using the root sum of squares. The cumulative effect of each of these error sources on the pseudorange measurment is called the user-equivalent range error (UERE). Parkinson, et al. [34] characterized these errors as a combination of slowly varying biases and random noise. The magnitudes of these errors are tabulated in table 7.1. Recent publications indicate that measurement accuracies have improved in recent years due to improvements in error modeling and receiver technology with total UERE being estimated as approximately 4.0 m (1-σ) [33].

The pseudorange error sources described above contribute to the UERE in the range estimates for individual satellites. An additional source of position error in the GPS system comes from the geometric configuration of the satellites used to compute the position of the receiver. This satellite geometry error is expressed in terms of a single factor called the dilution of precision (DOP) [33]. The DOP value describes the increase in positioning error attributed to the positioning of the satellites in the constellation. In general, a GPS position estimate from a group of visible satellites that are positioned close to one another will result in a higher DOP value, while a position estimate from a group of visible satellites that are spread apart will result in a lower DOP value.

There are a variety of DOP terms defined in the literature. The two DOP terms of greatest interest to us are the horizontal DOP (HDOP) and the vertical DOP (VDOP). HDOP describes the influence of the satellite geometry on the GPS position measurement accuracy in the horizontal plane, while VDOP describes the influence of satellite geometry on position measurement accuracy in altitude. Since DOP depends on the number and configuration of visible satellites, it varies continuously with time. In open areas where satellites are readily visible, a nominal HDOP value would be 1.3, while a nominal VDOP value would be 1.8 [33].

The total error in a GPS position measurement takes into account the UERE and DOP. The standard deviation of the rms error in the northeast plane is given by

$$E_{\text{n-e,rms}} = \text{HDOP} \times \text{UERE}_{\text{rms}}$$
$$= (1.3)(5.1 \text{ m}) \quad (7.15)$$
$$= 6.6 \text{ m}.$$

Similarly, the standard deviation of the rms altitude error is given by

$$E_{\text{h,rms}} = \text{VDOP} \times \text{UERE}_{\text{rms}}$$
$$= (1.8)(5.1 \text{ m}) \quad (7.16)$$
$$= 9.2 \text{ m}.$$

These expressions give an indication of the size of the error that can be anticipated for a single receiver position measurement. As indicated in table 7.1, these errors consist of statistically independent slowly-varying biases and random noise components. Techniques, such as differential GPS, can be used to reduce the bias error components of GPS position measurements to much smaller values.

7.5.2 Transient Characteristics of GPS Positioning Error

The previous discussion has given us a good sense of the root-mean-square magnitude of the positioning errors involved in GPS measurements. For the purposes of simulation, however, we are not only interested in the size of the error, we are also interested in knowing the dynamic characteristics of the error. Referring to equation (7.15) and assuming that the horizontal position error is composed of a north position error and east position error that are independent but of similar size, we can calculate the north and east error magnitudes to be about 4.7 m in size. The north, east, and altitude errors are comprised of a slowly changing bias along with random noise. For example, based on a VDOP of 1.8 and the UERE values of table 7.1, we can approximately model the altitude position error from GPS as having a slowly varying, zero-mean bias of 9.2 m and a random noise component of 0.7 m.

Sensors for MAVs

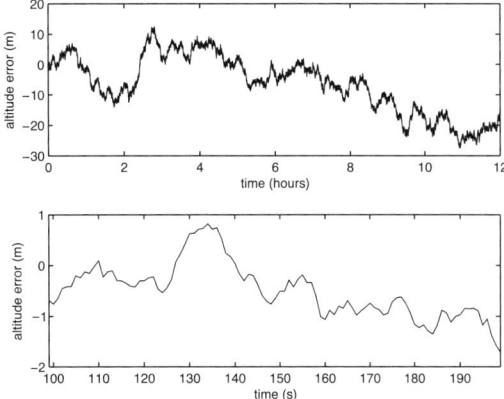

Figure 7.9 Example GPS altitude position error from Gauss-Markov model.

TABLE 7.2
Gauss-Markov error model parameters

Direction	Nominal 1-σ error (m)		Model Parameters		
	Bias	Random	Std. Dev. η_{GPS} (m)	$1/k_{GPS}$ (s)	T_s (s)
North	4.7	0.4	0.21	1100	1.0
East	4.7	0.4	0.21	1100	1.0
Altitude	9.2	0.7	0.40	1100	1.0

To model the transient behavior of the error, we follow the approach of [37] and model the error as a Gauss-Markov process. Gauss-Markov processes are modeled by

$$\nu[n+1] = e^{-k_{GPS} T_s} \nu[n] + \eta_{GPS}[n] \quad (7.17)$$

where $\nu[n]$ is the error being modeled, $\eta_{GPS}[n]$ is zero-mean Gaussian white noise, $1/k_{GPS}$ is the time constant of the process, and T_s is the sample time. Figure 7.9 shows results from the Gauss-Markov GPS altitude error model given by equation (7.17). The error over the 12-hour period shown has a standard deviation of 9.4 m, while the noise component of the error has a standard deviation of 0.69 m. The upper plot shows the error over a 12-hour period, while the lower plot shows the error over a 100-second segment of time. Suitable Gauss-Markov process parameters to model the GPS error are given in table 7.2.

Drawing on the error model in equation (7.17) and the parameters in table 7.2, we can create position error models for the north, east, and altitude measurements from GPS: ν_n, ν_e, and ν_h. Accordingly, a model for GPS measurements that is suitable for simulation purposes is given by

$$y_{GPS,n}[n] = p_n[n] + v_n[n] \tag{7.18}$$
$$y_{GPS,e}[n] = p_e[n] + v_e[n] \tag{7.19}$$
$$y_{GPS,h}[n] = -p_d[n] + v_h[n], \tag{7.20}$$

where p_n, p_e, and h are the actual earth coordinates and altitude above sea level, and n is the sample index. GPS measurements are commonly available from small UAV receivers at 1 Hz. New systems suitable for small UAV implementations provide GPS measurements at 5 Hz updates.

7.5.3 GPS Velocity Measurements

Using carrier phase Doppler measurements from GPS satellite signals, the velocity of the receiver can be calculated to accuracies with standard deviations in the range of 0.01 to 0.05 m/s. Many modern GPS receiver chips provide velocity information as part of their output data packet. In addition, they provide information on horizontal ground speed and course over the ground. Horizontal ground speed and course are calculated from the north and east velocity components from GPS as

$$V_g = \sqrt{V_n^2 + V_e^2} \tag{7.21}$$
$$\chi = \tan^{-1}\left(\frac{V_n}{V_e}\right), \tag{7.22}$$

where $V_n = V_a \cos \psi + w_n$ and $V_e = V_a \sin \psi + w_e$.

Using basic principles of uncertainty analysis [38], the uncertainty in ground speed and course measurements can be estimated to be

$$\sigma_{V_g} = \sqrt{\frac{V_n^2 \sigma_{V_n}^2 + V_e^2 \sigma_{V_e}^2}{V_n^2 + V_e^2}}$$

$$\sigma_\chi = \sqrt{\frac{V_n^2 \sigma_{V_e}^2 + V_e^2 \sigma_{V_n}^2}{(V_n^2 + V_e^2)^2}}.$$

If the uncertainty in the north and east directions have the same magnitude (i.e., $\sigma_{V_n} = \sigma_{V_e} = \sigma_V$), these expressions simplify to be

$$\sigma_{V_g} = \sigma_V \tag{7.23}$$
$$\sigma_\chi = \frac{\sigma_V}{V_g}. \tag{7.24}$$

Notice that the uncertainty in the course measurement scales with the inverse of the ground speed—for high speeds the error is small and for low speeds the error is large. This is not unexpected since course is undefined for a stationary object. Based on equations (7.21)–(7.22) and (7.23)–(7.24), we can model the ground speed and course measurements available from GPS as

$$y_{\text{GPS},V_g} = \sqrt{(V_a \cos \psi + w_n)^2 + (V_a \sin \psi + w_e)^2} + \eta_V \qquad (7.25)$$

$$y_{\text{GPS},\chi} = \text{atan2}(V_a \sin \psi + w_e, V_a \cos \psi + w_n) + \eta_\chi, \qquad (7.26)$$

where η_V and η_χ are zero-mean Gaussian processes with variances $\sigma_{V_g}^2$ and σ_χ^2.

7.6 Chapter Summary

In this chapter, we have described the sensors commonly found on small unmanned aircraft and proposed models that describe their function for the purposes of simulation, analysis, and observer design. The simulation models characterize the errors of the sensors and their effective update rates. We have focused on sensors used for guidance, navigation, and control of the aircraft including accelerometers, rate gyros, absolute pressure sensors, differential pressure sensors, magnetometers, and GPS. Camera sensors will be discussed in chapter 13.

Notes and References

The accelerometers, rate gyros, and pressure sensors used on small unmanned aircraft are usually based on MEMS technology due to their small size and weight. Several references provide excellent overviews of these devices including [39, 40, 41]. The development of the global positioning system has been described in detail in several texts. Details describing its function and modeling of position errors can be found in [33, 34, 35, 36]. Specific information for sensor models can be found in manufacturer data sheets for the devices of interest.

7.7 Design Project

The objective of this project assignment is to add the sensors to the simulation model of the MAV.

   7.1. Download the files associated with this chapter from the book website. Note that we have added a block for the sensors that contains two files: sensors.m and gps.m. The file sensors.m will model all of the sensors that update at rate T_s (gyros, accelerometers, pressure sensors), and `gps.m` will model the GPS sensor, which is updated at rate $T_{s,\text{GPS}}$.

7.2. Using the sensor parameters listed in appendix H, modify `sensors.m` to simulate the output of the rate gyros (eq. (7.5)), the accelerometers (eq. (7.3)), and the pressure sensors (eq (7.9) and (7.10)).

7.3. Using the sensor parameters listed in appendix H, modify `gps.m` to simulate the position measurement output of the GPS sensor (eq. (7.18)–(7.20)) and the ground speed and course output of the GPS sensor (eq. (7.25)–(7.26)).

7.4. Using a Simulink scope, observe the output of each sensor and verify that its sign and magnitude are approximately correct, and that the shape of the waveform is approximately correct.

8

State Estimation

The autopilot designed in chapter 6 assumes that states of the system like roll and pitch angles are available for feedback. However, one of the challenges of MAV flight control is that sensors that directly measure roll and pitch are not available. Therefore, the objective of this chapter is to describe techniques for estimating the state of a small or micro air vehicle from the sensor measurements described in chapter 7. Since rate gyros directly measure roll rates in the body frame, the states p, q, and r can be recovered by low-pass filtering the rate gyros. Therefore, we begin the chapter by discussing digital implementation of low-pass filters in section 8.2. In section 8.3 we describe a simple state-estimation scheme that is based on mathematically inverting the sensor models. However, this scheme does not account for the dynamics of the system and therefore does not perform well over the full range of flight conditions. Accordingly, in section 8.4 we introduce dynamic-observer theory as a precursor to our discussion on the Kalman filter. A mathematical derivation of the Kalman filter is given in section 8.5. For those with limited exposure to stochastic processes, appendix G provides an overview of basic concepts from probability theory with a focus on Gaussian stochastic processes. The last two sections of the chapter describe applications of the Kalman filter. In section 8.6, an extended Kalman filter is designed to estimate the roll and pitch attitude of the MAV, and in section 8.7, an extended Kalman filter is used to estimate the position, ground speed, course, and heading of the MAV, as well as the wind speed and direction.

8.1 Benchmark Maneuver

To illustrate the different estimation schemes presented in this chapter, we will use a maneuver that adequately excites all of the states. Initially the MAV will be in wings-level, trimmed flight at an altitude of 100 m, and an airspeed of 10 m/s. The maneuver is defined by commanding a constant airspeed of 10 m/s and by commanding altitude and heading as shown in figure 8.1. By using the same benchmark maneuver to estimate the performance of the different estimators developed in this chapter, we can evaluate their relative performance. The benchmark maneuver commands longitudinal and lateral motions simultaneously,

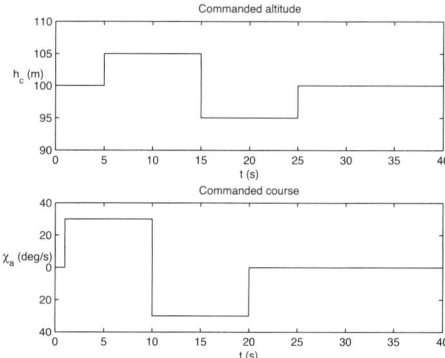

Figure 8.1 Altitude and heading commands that define the benchmark maneuver used to evaluate and tune the state estimation scheme.

thus exposing any significant sensitivities of estimators to assumptions of decoupled dynamics.

8.2 Low-pass Filters

Since some of the estimation schemes described in this chapter require low-pass filtering of the sensor signal, this section describes digital implementation of a single pole low-pass filter. The Laplace transform representation of a simple unity DC-gain low-pass filter with cut-off frequency a is given by

$$Y(s) = \frac{a}{s+a} U(s),$$

where $U(s) = \mathcal{L}\{u(t)\}$ and $u(t)$ is the input of the filter, and where $Y(s) = \mathcal{L}\{y(t)\}$ and $y(t)$ is the output. Taking the inverse Laplace transform gives

$$\dot{y} = -ay + au. \tag{8.1}$$

From linear-systems theory, it is well known that the sampled-data solution to equation (8.1) is given by

$$y(t+T_s) = e^{-aT_s} y(t) + a \int_0^{T_s} e^{-a(T_s - \tau)} u(\tau) \, d\tau.$$

Assuming that $u(t)$ is constant between sample periods results in the expression

$$y[n+1] = e^{-aT_s} y[n] + a \int_0^{T_s} e^{-a(T_s - \tau)} \, d\tau \, u[n]$$
$$= e^{-aT_s} y[n] + (1 - e^{-aT_s}) u[n]. \tag{8.2}$$

If we let $\alpha_{\text{LPF}} = e^{-aT_s}$ then we get the simple form

$$y[n+1] = \alpha_{\text{LPF}} y[n] + (1 - \alpha_{\text{LPF}}) u[n].$$

Note that this equation has a nice physical interpretation: the new value of y (filtered value) is a weighted average of the old value of y and u (unfiltered value). If u is noisy, then $\alpha_{\text{LPF}} \in [0, 1]$ should be close to unity. However, if u is relatively noise free, then α should be close to zero.

We will use the notation $LPF(\cdot)$ to represent the low-pass filter operator. Therefore $\hat{x} = LPF(x)$ is the low-pass filtered version of x.

8.3 State Estimation by Inverting the Sensor Model

In this section we will derive the simplest possible state estimation scheme based on inverting the sensor models derived in chapter 7. While this method is effective for angular rates, altitude, and airspeed, it is not effective for estimating the Euler angles or the position and course of the MAV.

8.3.1 Angular Rates

The angular rates p, q, and r can be estimated by low-pass filtering the rate gyro signals given by equation (7.5) to obtain

$$\hat{p} = LPF(y_{\text{gyro},x}) \tag{8.3}$$

$$\hat{q} = LPF(y_{\text{gyro},y}) \tag{8.4}$$

$$\hat{r} = LPF(y_{\text{gyro},z}). \tag{8.5}$$

For the benchmark maneuver discussed in section 8.1, the estimation error for p, q, and r are shown in figure 8.2. From the figure we see that low-pass filtering the gyro measurements produces acceptable estimates of p, q, and r.

8.3.2 Altitude

A estimate of the altitude can be obtained from the absolute pressure sensor. Applying a low-pass filter to equation (7.9) and dividing by ρg we get

$$\hat{h} = \frac{LPF(y_{\text{static pres}})}{\rho g}. \tag{8.6}$$

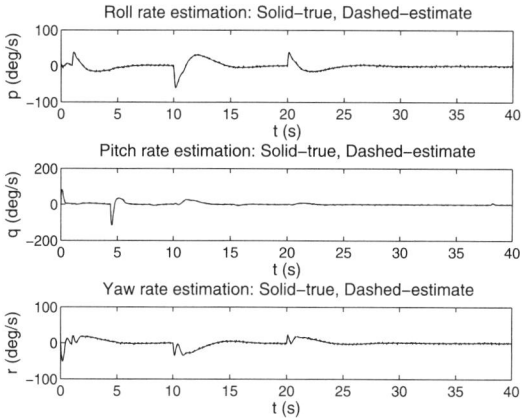

Figure 8.2 Estimation error on the angular rates obtained by low-pass filtering the rate gyros.

8.3.3 Airspeed

The airspeed can be estimated by applying a low-pass filter to the differential pressure sensor represented by equation (7.10), and inverting to obtain

$$\hat{V}_a = \sqrt{\frac{2}{\rho} LPF(y_{\text{diff pres}})}. \tag{8.7}$$

For the benchmark maneuver discussed in section 8.1, the estimates of the altitude and airspeed are shown in figure 8.3, together with truth data. As can be seen from the figure, inverting the sensor model produces a fairly accurate model of the altitude and airspeed.

8.3.4 Roll and Pitch Angles

Roll and pitch angles are the most difficult variables to estimate well on small unmanned aircraft. A simple scheme that works in unaccelerated flight can be derived as follows. Recall from equation (7.1) that

$$y_{\text{accel},x} = \dot{u} + qw - rv + g \sin \theta + \eta_{\text{accel},x}$$
$$y_{\text{accel},y} = \dot{v} + ru - pw - g \cos \theta \sin \phi + \eta_{\text{accel},y}$$
$$y_{\text{accel},z} = \dot{w} + pv - qu - g \cos \theta \cos \phi + \eta_{\text{accel},z}.$$

State Estimation

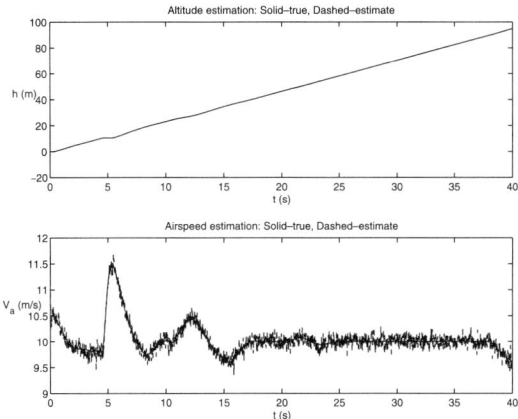

Figure 8.3 Estimation error of the altitude and airspeed obtained by low-pass filtering the pressure sensors and inverting the sensor model. For altitude and airspeed, the accuracy of the simple scheme is adequate.

In unaccelerated flight we have $\dot{u} = \dot{v} = \dot{w} = p = q = r = 0$, which implies that

$$LPF(y_{\text{accel},x}) = g \sin \theta$$
$$LPF(y_{\text{accel},y}) = -g \cos \theta \sin \phi$$
$$LPF(y_{\text{accel},z}) = -g \cos \theta \cos \phi.$$

Solving for ϕ and θ we get

$$\hat{\phi}_{\text{accel}} = \tan^{-1}\left(\frac{LPF(y_{\text{accel},y})}{LPF(y_{\text{accel},z})}\right) \tag{8.8}$$

$$\hat{\theta}_{\text{accel}} = \sin^{-1}\left(\frac{LPF(y_{\text{accel},x})}{g}\right). \tag{8.9}$$

The estimation errors of the roll and pitch angles for the benchmark maneuver discussed in section 8.1 are shown in figure 8.4, where it is clear that the estimation error during accelerated flight is unacceptable. In section 8.6 we will use an extended Kalman filter to provide more accurate estimates of the roll and pitch angles.

8.3.5 Position, Course, and Ground Speed

The position of the MAV can be estimated by low-pass filtering equations (7.18) and (7.19). The biases due to multipath, clock, and satellite geometry will not be removed. The estimates of the position

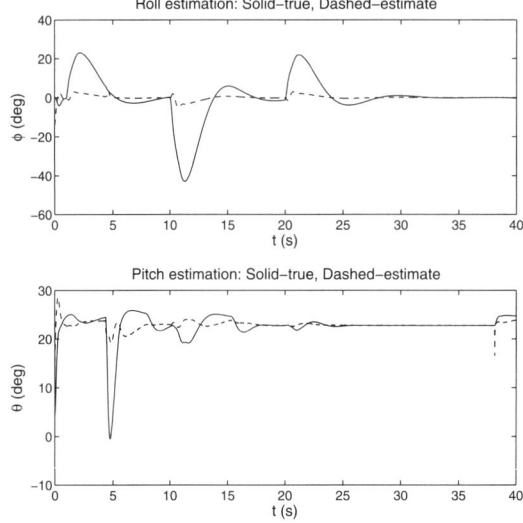

Figure 8.4 Estimation error on the roll and pitch angles obtained by low-pass filtering the accelerometers and inverting the model. Since this scheme assumes unaccelerated flight during maneuvers where acceleration exists, the estimation error can be unacceptably large.

variables are therefore given by

$$\hat{p}_n = LPF(y_{\text{GPS},n}) \tag{8.10}$$

$$\hat{p}_e = LPF(y_{\text{GPS},e}). \tag{8.11}$$

Similarly, estimates of the course angle and ground speed of the MAV can be obtained by low-pass filtering equations (7.26) and (7.25) to obtain

$$\hat{\chi} = LPF(y_{\text{GPS},\chi}) \tag{8.12}$$

$$\hat{V}_g = LPF(y_{\text{GPS},V_g}). \tag{8.13}$$

The primary downside of low-pass filtering GPS signals is that since the sample rate is slow (usually on the order of 1 Hz), there is significant delay in the estimate. The estimation scheme described in section 8.7 will resolve this problem.

For the benchmark maneuver discussed in section 8.1, the estimation error for the north and east position, and for the course and ground speed are shown in figure 8.5. There is significant error due, in part, to the fact that the GPS sensor is only updated at 1 Hz. Clearly, simply low-pass filtering the GPS data does not produce satisfactory results.

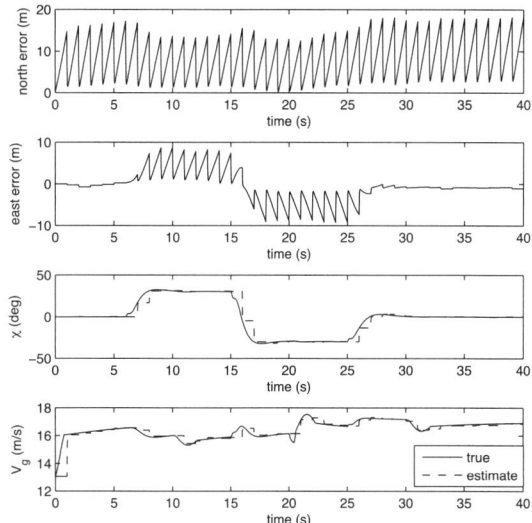

Figure 8.5 Estimation error for the north and east position, course, and ground speed obtained by low-pass filtering the GPS sensors.

We have shown in this section that adequate estimates of the body rates p, q, and r, as well as the altitude and the airspeed can be obtained by low-pass filtering the sensors. However, estimating the roll and pitch angles and the position, course, and ground speed will require more sophisticated techniques. In particular, a simple low pass filter does not account for the underlying dynamics of the system. In the following section we will introduce dynamic observer theory. The most commonly used dynamic observer is the Kalman filter which will be derived in section 8.5. The application of the Kalman filter to attitude estimation is in section 8.6, and the application of the Kalman filter to position, course, and groundspeed estimation is in section 8.7.

8.4 Dynamic-observer Theory

The objective of this section is to briefly review observer theory, which serves as a precursor to our discussion on the Kalman filter. Suppose that we have a linear time-invariant system modeled by the equations

$$\dot{x} = Ax + Bu$$
$$y = Cx.$$

A continuous-time observer for this system is given by the equation

$$\dot{\hat{x}} = \underbrace{A\hat{x} + Bu}_{\text{copy of the model}} + \underbrace{L(y - C\hat{x})}_{\text{correction due to sensor reading}}, \tag{8.14}$$

where $\hat{x}$ is the estimated value of x. Defining the observation error as $\tilde{x} = x - \hat{x}$ we find that

$$\dot{\tilde{x}} = (A - LC)\tilde{x},$$

which implies that the observation error decays exponentially to zero if L is chosen so that the eigenvalues of $A - LC$ are in the open left half of the complex plane.

In practice, the sensors are usually sampled and processed in digital hardware at sample rate T_s. How do we modify the observer equation shown in equation (8.14) to account for sampled sensor readings? One approach is to propagate the system model between samples using the equation

$$\dot{\hat{x}} = A\hat{x} + Bu, \tag{8.15}$$

and then to update the estimate when a measurement is received using

$$\hat{x}^+ = \hat{x}^- + L(y(t_n) - C\hat{x}^-), \tag{8.16}$$

where t_n is the instant in time that the measurement is received and $\hat{x}^-$ is the state estimate produced by equation (8.15) at time t_n. Equation (8.15) is then re-instantiated with initial conditions given by $\hat{x}^+$. If the system is nonlinear, then the propagation and update equations become

$$\dot{\hat{x}} = f(\hat{x}, u) \tag{8.17}$$

$$\hat{x}^+ = \hat{x}^- + L(y(t_n) - h(\hat{x}^-)). \tag{8.18}$$

The observation process is shown graphically in figure 8.6. Note that it is not necessary to have a fixed sample rate.

A pseudo-code implementation of the continuous-discrete observer is shown in algorithm 1. In line 1 the state estimate is initialized to zero. If additional information is known, then the state can be initialized accordingly. The ordinary differential equation in equation (8.17) is propagated between samples with the for-loop in lines 4–6 using and

State Estimation

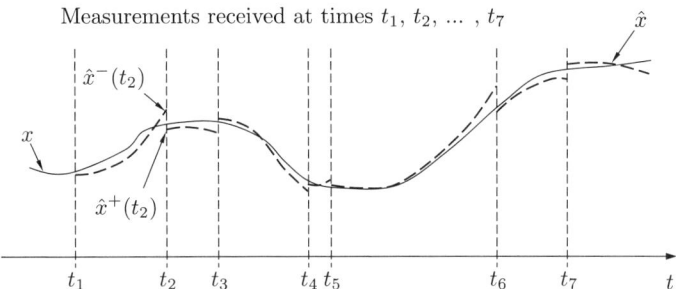

Figure 8.6 Time line for continuous-discrete dynamic observer. The vertical dashed lines indicate sample times at which measurements are received. In between measurements, the state is propagated using equation (8.17). When a measurement is received, the state is updated using equation (8.18).

Euler integration method. When a measurement is received, the state is updated using equation (8.18) in line 8.

Algorithm 1 Continuous-discrete Observer

1: Initialize: $\hat{x} = \chi_0$.
2: Pick an output sample rate T_{out} that is less than the sample rates of the sensors.
3: At each sample time T_{out}:
4: **for** $i = 1$ to N **do** {Propagate the state equation.}
5: $\quad \hat{x} = \hat{x} + \left(\frac{T_{out}}{N}\right) f(\hat{x}, u)$
6: **end for**
7: **if** A measurement has been received from sensor i **then** {Measurement Update}
8: $\quad \hat{x} = \hat{x} + L_i \left(y_i - h_i(\hat{x})\right)$
9: **end if**

8.5 Derivation of the Continuous-discrete Kalman Filter

The key parameter for the dynamic observer discussed in the previous section is the observer gain L. The Kalman filter and extended Kalman filters discussed in the remainder of this chapter are standard techniques for choosing L. If the process and measurement are linear, and the process and measurement noise are zero-mean white Gaussian processes with known covariance matrices, then the Kalman filter gives the optimal gain, where the optimality criteria will be defined later in this section. There are several different forms for the Kalman filter, but the form that is particularly useful for MAV applications is the continuous-propagation, discrete-measurement Kalman filter.

We will assume that the (linear) system dynamics are given by

$$\dot{x} = Ax + Bu + \xi \tag{8.19}$$

$$y[n] = Cx[n] + \eta[n],$$

where $y[n] = y(t_n)$ is the nth sample of y, $x[n] = x(t_n)$ is the nth sample of x, $\eta[n]$ is the measurement noise at time t_n, ξ is a zero-mean Gaussian random process with covariance Q, and $\eta[n]$ is a zero-mean Gaussian random variable with covariance R. The random process ξ is called the process noise and represents modeling error and disturbances on the system. The random variable η is called the measurement noise and represents noise on the sensors. The covariance R can usually be estimated from sensor calibration, but the covariance Q is generally unknown and therefore becomes a system parameter that can be tuned to improve the performance of the observer. Note that the sample rate does not need to be fixed.

Similar to equations (8.15) and (8.16) the continuous-discrete Kalman filter has the form

$$\dot{\hat{x}} = A\hat{x} + Bu$$

$$\hat{x}^+ = \hat{x}^- + L(y(t_n) - C\hat{x}^-).$$

Define the estimation error as $\tilde{x} = x - \hat{x}$. The covariance of the estimation error at time t is given by

$$P(t) \triangleq E\{\tilde{x}(t)\tilde{x}(t)^\top\}. \tag{8.20}$$

Note that $P(t)$ is symmetric and positive semi-definite, therefore, its eigenvalues are real and non-negative. Also, small eigenvalues of $P(t)$ imply small variance, which implies low average estimation error. Therefore, we would like to choose $L(t)$ to minimize the eigenvalues of $P(t)$. Recall that

$$\text{tr}(P) = \sum_{i=1}^{n} \lambda_i,$$

where $\text{tr}(P)$ is the trace of P, and λ_i are the eigenvalues of P. Therefore, minimizing $\text{tr}(P)$ minimizes the estimation error covariance. The Kalman filter is derived by finding L to minimize $\text{tr}(P)$.

Between Measurements

Differentiating $\tilde{x}$ we get

$$\dot{\tilde{x}} = \dot{x} - \dot{\hat{x}}$$
$$= Ax + Bu + \xi - A\hat{x} - Bu$$
$$= A\tilde{x} + \xi.$$

Solving the differential equation with initial conditions $\tilde{x}_0$ we obtain

$$\tilde{x}(t) = e^{At}\tilde{x}_0 + \int_0^t e^{A(t-\tau)}\xi(\tau)\,d\tau.$$

We can compute the evolution of the error covariance P as

$$\dot{P} = \frac{d}{dt}E\{\tilde{x}\tilde{x}^\top\}$$
$$= E\{\dot{\tilde{x}}\tilde{x}^\top + \tilde{x}\dot{\tilde{x}}^\top\}$$
$$= E\left\{A\tilde{x}\tilde{x}^\top + \xi\tilde{x}^\top + \tilde{x}\tilde{x}^\top A^\top + \tilde{x}\xi^\top\right\}$$
$$= AP + PA^\top + E\{\xi\tilde{x}^\top\} + E\{\tilde{x}\xi^\top\}.$$

We can compute $E\{\tilde{x}\xi^\top\}$ as

$$E\{\tilde{x}\xi^\top\} = E\{e^{At}\tilde{x}_0\xi^\top(t) + \int_0^t e^{A(t-\tau)}\xi(\tau)\xi^\top(\tau)\,d\tau$$
$$= \int_0^t e^{A(t-\tau)}Q\delta(t-\tau)\,d\tau$$
$$= \frac{1}{2}Q,$$

where the $\frac{1}{2}$ is because we only use half of the area inside the delta function. Therefore, since Q is symmetric we have that P evolves between measurements as

$$\dot{P} = AP + PA^\top + Q.$$

At Measurements

At a measurement, we have that

$$\tilde{x}^+ = x - \hat{x}^+$$
$$= x - \hat{x}^- - L\left(Cx + \eta - C\hat{x}^-\right)$$
$$= \tilde{x}^- - LC\tilde{x}^- - L\eta.$$

We also have that

$$P^+ = E\{\tilde{x}^+ \tilde{x}^{+T}\}$$
$$= E\left\{\left(\tilde{x}^- - LC\tilde{x}^- - L\eta\right)\left(\tilde{x}^- - LC\tilde{x}^- - L\eta\right)^\top\right\}$$
$$= E\left\{\tilde{x}^-\tilde{x}^{-\top} - \tilde{x}^-\tilde{x}^{-\top}C^\top L^\top - \tilde{x}^-\eta^\top L^\top\right.$$
$$- LC\tilde{x}^-\tilde{x}^{-\top} + LC\tilde{x}^-\tilde{x}^{-\top}C^\top L^\top + LC\tilde{x}^-\eta^\top L^\top$$
$$\left. - L\eta\tilde{x}^{-\top} + L\eta\tilde{x}^{-\top}C^\top L^\top + L\eta\eta^\top L^\top\right\}$$
$$= P^- - P^-C^\top L^\top - LCP^- + LCP^-C^\top L^\top + LRL^\top, \quad (8.21)$$

where we have used the fact that since η and $\tilde{x}^-$ are independent, $E\{\tilde{x}^-\eta^\top L^\top\} = E\{L\eta\tilde{x}^{-\top}\} = 0$.

In the derivation that follows, we will need the following matrix relationships:

$$\frac{\partial}{\partial A}\mathrm{tr}(BAD) = B^\top D^\top \quad (8.22)$$

$$\frac{\partial}{\partial A}\mathrm{tr}(ABA^\top) = 2AB, \text{ if } B = B^\top. \quad (8.23)$$

Our objective is to pick L to minimize $\mathrm{tr}(P^+)$. A necessary condition is that

$$\frac{\partial}{\partial L}\mathrm{tr}(P^+) = -P^-C^\top - P^-C^\top + 2LCP^-C^\top + 2LR = 0$$

$$\implies 2L(R + CP^-C^\top) = 2P^-C^\top$$

$$\implies L = P^-C^\top(R + CP^-C^\top)^{-1}.$$

Substituting into equation (8.21) gives

$$P^+ = P^- + P^-C^\mathsf{T}(R+CP^-C^\mathsf{T})^{-1}CP^- - P^-C^\mathsf{T}(R+CP^-C^\mathsf{T})^{-1}CP^-$$
$$+ P^-C^\mathsf{T}(R+CP^-C^\mathsf{T})^{-1}(CP^-C^\mathsf{T}+R)(R+CP^-C^\mathsf{T})^{-1}CP^-$$
$$= P^- - P^-C^\mathsf{T}(R+CP^-C^\mathsf{T})^{-1}CP^-$$
$$= (I - P^-C^\mathsf{T}(R+CP^-C^\mathsf{T})^{-1}C)P^-$$
$$= (I - LC)P^-.$$

We can therefore summarize the Kalman filter as follows. In between measurements, propagate the equations

$$\dot{\hat{x}} = A\hat{x} + Bu$$
$$\dot{P} = AP + PA^\mathsf{T} + Q,$$

where $\hat{x}$ is the estimate of the state, and P is the symmetric covariance matrix of the estimation error. When a measurement from the ith sensor is received, update the state estimate and error covariance according to the equations

$$L_i = P^-C_i^\mathsf{T}(R_i + C_i P^- C_i^\mathsf{T})^{-1}$$
$$P^+ = (I - L_i C_i)P^-$$
$$\hat{x}^+ = \hat{x}^- + L_i(y_i(t_n) - C_i \hat{x}^-),$$

where L_i is called the Kalman gain for sensor i.

We have assumed that the system propagation model and measurement model are linear. However, for many applications, including the applications discussed later in this chapter, the system propagation model and the measurement model are nonlinear. In other words, the model in equation (8.19) becomes

$$\dot{x} = f(x, u) + \xi$$
$$y[n] = h(x[n], u[n]) + \eta[n].$$

For this case, the state propagation and update laws use the nonlinear model, but the propagation and update of the error covariance use the Jacobian of f for A, and the Jacobian of h for C. The resulting algorithm is called the Extended Kalman Filter (EKF). Pseudo-code for the EKF is shown in algorithm 2. The state is initialized in line 1. The propagation of the ordinary differential equations (ODEs) for $\hat{x}$ and P using an Euler integration scheme are given by the for-loop in lines 4-8. The update equations for the ith sensor are given in lines 9-14. The application of algorithm 2 to roll and pitch angle estimation is described in section 8.6. The application of algorithm 2 to position, heading, ground speed, course, and wind estimation is described in section 8.7.

Algorithm 2 Continuous-discrete Extended Kalman Filter

1: Initialize: $\hat{x} = \chi_0$.
2: Pick an output sample rate T_{out} that is less than the sample rates of the sensors.
3: At each sample time T_{out}:
4: **for** $i = 1$ to N **do** {Prediction Step}
5: $\quad \hat{x} = \hat{x} + \left(\frac{T_{out}}{N}\right) f(\hat{x}, u)$
6: $\quad A = \frac{\partial f}{\partial x}(\hat{x}, u)$
7: $\quad P = P + \left(\frac{T_{out}}{N}\right) \left(AP + PA^\top + Q\right)$
8: **end for**
9: **if** Measurement has been received from sensor i **then** {Measurement Update}
10: $\quad C_i = \frac{\partial h_i}{\partial x}(\hat{x}, u[n])$
11: $\quad L_i = PC_i^\top (R_i + C_i P C_i^\top)^{-1}$
12: $\quad P = (I - L_i C_i) P$
13: $\quad \hat{x} = \hat{x} + L_i \left(y_i[n] - h(\hat{x}, u[n])\right)$
14: **end if**

8.6 Attitude Estimation

This section describes the application of the EKF to estimate the roll and pitch angles of the MAV. To apply the continous-discrete extended Kalman filter derived in section 8.5 to roll and pitch estimation, we use the nonlinear propagation model

$$\dot{\phi} = p + q \sin \phi \tan \theta + r \cos \phi \tan \theta + \xi_\phi$$
$$\dot{\theta} = q \cos \phi - r \sin \phi + \xi_\theta,$$

where we have added the noise terms ξ_ϕ and ξ_θ to model the noise on p, q, and r, where $\xi_\phi \sim \mathcal{N}(0, Q_\phi)$ and $\xi_\theta \sim \mathcal{N}(0, Q_\theta)$.

We will use the accelerometers as the output equations. From equation (7.1) we have the accelerometer model

$$y_{accel} = \begin{pmatrix} \dot{u} + qw - rv + g\sin\theta \\ \dot{v} + ru - pw - g\cos\theta\sin\phi \\ \dot{w} + pv - qu - g\cos\theta\cos\phi \end{pmatrix} + \eta_{accel}. \quad (8.24)$$

However, we do not have a method for directly measuring $\dot{u}$, $\dot{v}$, $\dot{w}$, u, v, and w. We will assume that $\dot{u} = \dot{v} = \dot{w} \approx 0$. From equation (2.7) we have

$$\begin{pmatrix} u \\ v \\ w \end{pmatrix} \approx V_a \begin{pmatrix} \cos\alpha\cos\beta \\ \sin\beta \\ \sin\alpha\cos\beta \end{pmatrix}.$$

Assuming that $\alpha \approx \theta$ and $\beta \approx 0$ we obtain

$$\begin{pmatrix} u \\ v \\ w \end{pmatrix} \approx V_a \begin{pmatrix} \cos\theta \\ 0 \\ \sin\theta \end{pmatrix}.$$

Substituting into equation (8.24), we get

$$y_{accel} = \begin{pmatrix} qV_a\sin\theta + g\sin\theta \\ rV_a\cos\theta - pV_a\sin\theta - g\cos\theta\sin\phi \\ -qV_a\cos\theta - g\cos\theta\cos\phi \end{pmatrix} + \eta_{accel}.$$

Defining $x = (\phi, \theta)^\top$, $u = (p, q, r, V_a)^\top$, $\xi = (\xi_\phi, \xi_\theta)^\top$, and $\eta = (\eta_\phi, \eta_\theta)^\top$, gives

$$\dot{x} = f(x, u) + \xi$$

$$y = h(x, u) + \eta,$$

where

$$f(x, u) = \begin{pmatrix} p + q\sin\phi\tan\theta + r\cos\phi\tan\theta \\ q\cos\phi - r\sin\phi \end{pmatrix}$$

$$h(x, u) = \begin{pmatrix} qV_a\sin\theta + g\sin\theta \\ rV_a\cos\theta - pV_a\sin\theta - g\cos\theta\sin\phi \\ -qV_a\cos\theta - g\cos\theta\cos\phi \end{pmatrix}.$$

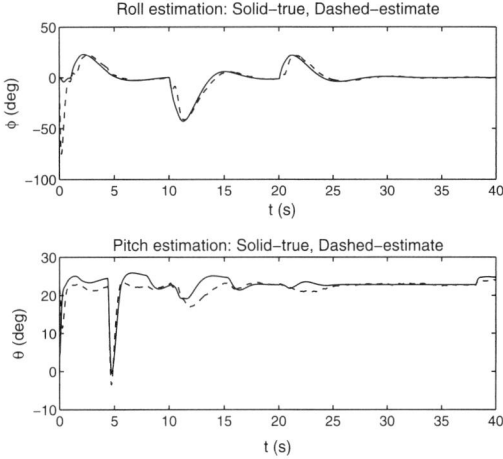

Figure 8.7 Estimation error on the roll and pitch angles using continuous-discrete extended Kalman filter.

Implementation of the Kalman filter requires the Jacobians $\frac{\partial f}{\partial x}$ and $\frac{\partial h}{\partial x}$. Accordingly we have

$$\frac{\partial f}{\partial x} = \begin{pmatrix} q\cos\phi\tan\theta - r\sin\phi\tan\theta & \frac{q\sin\phi - r\cos\phi}{\cos^2\theta} \\ -q\sin\phi - r\cos\phi & 0 \end{pmatrix}$$

$$\frac{\partial h}{\partial x} = \begin{pmatrix} 0 & qV_a\cos\theta + g\cos\theta \\ -g\cos\phi\cos\theta & -rV_a\sin\theta - pV_a\cos\theta + g\sin\phi\sin\theta \\ g\sin\phi\cos\theta & (qV_a + g\cos\phi)\sin\theta \end{pmatrix}.$$

The extended Kalman filter is implemented using algorithm 2.

For the benchmark maneuver discussed in section 8.1, the estimation error of the roll and pitch angles using algorithm 2 is shown in figure 8.7. Comparing figure 8.7 with figure 8.4 shows that the continous-discrete extended Kalman filter produces much better results during accelerated flight.

8.7 GPS Smoothing

In this section we will use GPS measurements to estimate the position, ground speed, course, wind, and heading of the MAV. If we assume that the flight path angle $\gamma = 0$, then the evolution of the position is given by

$$p_n = V_g \cos \chi$$
$$p_e = V_g \sin \chi.$$

By differentiating equation (7.21) we get that the evolution of the ground speed is given by

$$\begin{aligned}\dot{V}_g &= \frac{d}{dt}\sqrt{(V_a \cos \psi + w_n)^2 + (V_a \sin \psi + w_e)^2} \\ &= \frac{1}{V_g}\Big[(V_a \cos \psi + w_n)(\dot{V}_a \cos \psi - V_a\dot{\psi} \sin \psi + \dot{w}_n) \\ &\quad + (V_a \sin \psi + w_e)(\dot{V}_a \sin \psi + V_a\dot{\psi} \cos \psi + \dot{w}_e)\Big].\end{aligned}$$

Assuming that wind and airspeed are constant we get

$$\dot{V}_g = \frac{(V_a \cos \psi + w_n)(-V_a\dot{\psi} \sin \psi) + (V_a \sin \psi + w_e)(V_a\dot{\psi} \cos \psi)}{V_g}.$$

From equation (5.15) the evolution of χ is given by

$$\dot{\chi} = \frac{g}{V_g} \tan \phi \cos(\chi - \psi).$$

Assuming that wind is constant we have

$$\dot{w}_n = 0$$
$$\dot{w}_e = 0.$$

From equation (5.9) the evolution of ψ is given by

$$\dot{\psi} = q\frac{\sin \phi}{\cos \theta} + r\frac{\cos \phi}{\cos \theta}. \qquad (8.25)$$

Defining the state as $x = (p_n, p_e, V_g, \chi, w_n, w_e, \psi)^\top$, and the input as $u = (V_a, q, r, \phi, \theta)^\top$, the nonlinear propagation model is given by

$\dot{x} = f(x, u)$, where

$$f(x, u) \triangleq \begin{pmatrix} V_g \cos \chi \\ V_g \sin \chi \\ \frac{(V_a \cos \psi + w_n)(-V_a \dot{\psi} \sin \psi) + (V_a \sin \psi + w_e)(V_a \dot{\psi} \cos \psi)}{V_g} \\ \frac{g}{V_g} \tan \phi \cos(\chi - \psi) \\ 0 \\ 0 \\ q \frac{\sin \phi}{\cos \theta} + r \frac{\cos \phi}{\cos \theta} \end{pmatrix}.$$

The Jacobian of f is given by

$$\frac{\partial f}{\partial x} = \begin{pmatrix} 0 & 0 & \cos \chi & -V_g \sin \chi & 0 & 0 & 0 \\ 0 & 0 & \sin \chi & V_g \cos \chi & 0 & 0 & 0 \\ 0 & 0 & -\frac{\dot{V}_g}{V_g} & 0 & -\dot{\psi} V_a \sin \psi & \dot{\psi} V_a \cos \psi & \frac{\partial \dot{V}_g}{\partial \psi} \\ 0 & 0 & \frac{\partial \dot{\chi}}{\partial V_g} & \frac{\partial \dot{\chi}}{\partial \chi} & 0 & 0 & \frac{\partial \dot{\chi}}{\partial \psi} \\ 0 & 0 & 0 & 0 & 0 & 0 & 0 \\ 0 & 0 & 0 & 0 & 0 & 0 & 0 \\ 0 & 0 & 0 & 0 & 0 & 0 & 0 \end{pmatrix},$$

where

$$\frac{\partial \dot{V}_g}{\partial \psi} = \frac{-\dot{\psi} V_a (w_n \cos \psi + w_e \sin \psi)}{V_g}$$

$$\frac{\partial \dot{\chi}}{\partial V_g} = -\frac{g}{V_g^2} \tan \phi \cos(\chi - \psi)$$

$$\frac{\partial \dot{\chi}}{\partial \chi} = -\frac{g}{V_g} \tan \phi \sin(\chi - \psi)$$

$$\frac{\partial \dot{\chi}}{\partial \psi} = \frac{g}{V_g} \tan \phi \sin(\chi - \psi)$$

and $\dot{\psi}$ is given in equation (8.25).

For measurements, we will use the GPS signals for north and east position, ground speed, and course. Since the states are not independent, we will use the wind triangle relationship given in equation (2.9). Assuming that $\gamma = \gamma_a = 0$ we have that

$$V_a \cos \psi + w_n = V_g \cos \chi$$
$$V_a \sin \psi + w_e = V_g \sin \chi.$$

State Estimation

From these expressions, we define the pseudo measurements

$$y_{\text{wind},n} = V_a \cos \psi + w_n - V_g \cos \chi$$
$$y_{\text{wind},e} = V_a \sin \psi + w_e - V_g \sin \chi,$$

where the (pseudo) measurement values are equal to zero. The resulting measurement model is given by

$$y_{\text{GPS}} = h(x, u) + \eta_{\text{GPS}},$$

where $y_{\text{GPS}} = (y_{\text{GPS},n}, y_{\text{GPS},e}, y_{\text{GPS},V_g}, y_{\text{GPS},\chi}, y_{\text{GPS},n}, y_{\text{wind},e})$, $u = \hat{V}_a$, and

$$h(x, u) = \begin{pmatrix} p_n \\ p_e \\ V_g \\ \chi \\ V_a \cos \psi + w_n - V_g \cos \chi \\ V_a \sin \psi + w_e - V_g \sin \chi \end{pmatrix},$$

and where the Jacobian is given by

$$\frac{\partial h}{\partial x}(\hat{x}, u) = \begin{pmatrix} 1 & 0 & 0 & 0 & 0 & 0 & 0 \\ 0 & 1 & 0 & 0 & 0 & 0 & 0 \\ 0 & 0 & 1 & 0 & 0 & 0 & 0 \\ 0 & 0 & 0 & 1 & 0 & 0 & 0 \\ 0 & 0 & -\cos \chi & V_g \sin \chi & 1 & 0 & -V_a \sin \psi \\ 0 & 0 & -\sin \chi & -V_g \cos \chi & 0 & 1 & V_a \cos \psi \end{pmatrix}.$$

The extended Kalman filter to estimate p_n, p_e, V_g, χ, w_n, w_e, and ψ is implemented using algorithm 2.

For the benchmark maneuver discussed in section 8.1, the estimation error for the position and heading using algorithm 2 is shown in figure 8.8. Comparing figure 8.8 with figure 8.5, shows that the continous-discrete extended Kalman filter produces much better results than a simple low-pass filter.

8.8 Chapter Summary

This chapter has shown how to estimate the states that are required for the autopilot discussed in chapter 6 using the sensors described in chapter 7. We have shown that the angular rates in the body frame p, q,

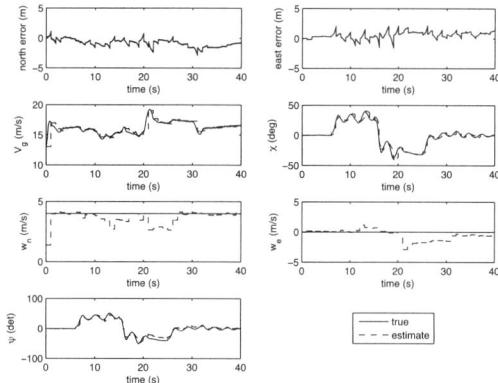

Figure 8.8 Estimation error for position, ground speed, course, wind, and heading using continuous-discrete extended Kalman filter.

and r, can be estimated by low-pass filtering the rate gyros. Similarly, the altitude h and airspeed V_a can be estimated by low-pass filtering the absolute and differential pressure sensors and inverting the sensor model. The remaining states must be estimated using extended Kalman filters. In section 8.6 we showed how a two-state EKF can be used to estimate the roll and pitch angles. In section 8.7 we showed how the position, ground speed, course, wind, and heading can be estimated using a seven-state EKF based on GPS measurements.

Notes and References

The Kalman filter was first introduced in [42]. There are many excellent texts on the Kalman filter including [43, 44, 45, 46]. Some of the results in this chapter have been discussed previously in [47, 48]. State estimation using computer vision instead of GPS has been discussed in [49, 50, 51].

8.9 Design Project

8.1. Download the simulation files for this chapter from the web. State estimation is performed in the file `estimate_states.m`.

8.2. Implement the simple schemes described in section 8.3 using low-pass filters and model inversion to estimate the states p_n, p_e, h, V_a, ϕ, θ, ψ, p, q, and r. Tune the bandwidth of the low-pass filter to observe the effect. Different states may require a different filter bandwidth.

8.3. Modify `estimate_states.m` to implement the extended Kalman filter for roll and pitch angles described in section 8.6. Tune the filter until you are satisfied with the performance.

8.4. Modify `estimate_states.m` to implement the extended Kalman filter for position, heading, and wind described in section 8.7. Tune the filter until you are satisfied with the performance.

8.5. In the Simulink model `mavsim_chap8.mdl` change the switch that directs true states to the autopilot, to direct the estimated states to the autopilot. Tune autopilot and estimator gains if necessary. By changing the bandwidth of the low-pass filter, note that the stability of the closed loop system is heavily influenced by this value.

9

Design Models for Guidance

As described in chapter 1, when the equations of motion for a system become complex, it is often necessary to develop design models that have significantly less mathematical complexity, but still capture the essential behavior of the system. If we include all the elements discussed in the previous eight chapters, including the six-degree-of-freedom model developed in chapters 3 and 4, the autopilot developed in chapter 6, the sensors developed in chapter 7, and the state-estimation scheme developed in chapter 8, the resulting model is extremely complex. This chapter approximates the performance of the closed-loop MAV system and develops reduced-order design models that are appropriate for the design of higher-level guidance strategies for MAVs. We will present several different models that are commonly used in the literature. The design models developed in this chapter will be used in chapters 10 through 13.

9.1 Autopilot Model

The guidance models developed in this chapter use a high-level representation of the autopilot loops developed in chapter 6. The airspeed-hold and roll-hold loops are represented by the first-order models

$$\dot{V}_a = b_{V_a}(V_a^c - V_a) \tag{9.1}$$

$$\dot{\phi} = b_\phi(\phi^c - \phi), \tag{9.2}$$

where b_{V_a} and b_ϕ are positive constants that depend on the implementation of the autopilot and the state estimation scheme. Drawing on the closed-loop transfer functions of chapter 6, the altitude and course-hold loops are represented by the second-order models

$$\ddot{h} = b_{\dot{h}}(\dot{h}^c - \dot{h}) + b_h(h^c - h) \tag{9.3}$$

$$\ddot{\chi} = b_{\dot{\chi}}(\dot{\chi}^c - \dot{\chi}) + b_\chi(\chi^c - \chi), \tag{9.4}$$

where $b_{\dot{h}}$, b_h, $b_{\dot{\chi}}$, and b_χ are also positive constants that depend on the implementation of the autopilot and the state estimation schemes. As explained in subsequent sections, some of the guidance models also assume autopilot-hold loops for the flight-path angle γ and the load factor n_{lf}, where load factor is defined as lift divided by weight. The first-order autopilot loops for flight-path angle and load factor are given by

$$\dot{\gamma} = b_\gamma(\gamma^c - \gamma) \tag{9.5}$$

$$\dot{n}_{lf} = b_n(n_{lf}^c - n_{lf}), \tag{9.6}$$

where b_γ and b_n are positive constants that depend on the implementation of the low-level autopilot loops.

9.2 Kinematic Model of Controlled Flight

In deriving reduced-order guidance models, the main simplification we make is to eliminate the force- and moment-balance equations of motion (those involving $\dot{u}, \dot{v}, \dot{w}, \dot{p}, \dot{q}, \dot{r}$), thus eliminating the need to calculate the complex aerodynamic forces acting on the airframe. These general equations are replaced with simpler kinematic equations derived for the specific flight conditions of a coordinated turn and an accelerating climb.

Recall from figure 2.10 that the velocity vector of the aircraft with respect to the inertial frame can be expressed in terms of the course angle and the (inertially referenced) flight-path angle as

$$\mathbf{V}_g^i = V_g \begin{pmatrix} \cos\chi \cos\gamma \\ \sin\chi \cos\gamma \\ -\sin\gamma \end{pmatrix}.$$

Therefore, the kinematics can be expressed as

$$\begin{pmatrix} \dot{p}_n \\ \dot{p}_e \\ \dot{h} \end{pmatrix} = V_g \begin{pmatrix} \cos\chi \cos\gamma \\ \sin\chi \cos\gamma \\ \sin\gamma \end{pmatrix}. \tag{9.7}$$

Because it is common to control the heading and airspeed of an aircraft, it is useful to express equation (9.7) in terms of ψ and V_a. With reference to the wind triangle expression in equation (2.9), we can write equation (9.7) as

$$\begin{pmatrix} \dot{p}_n \\ \dot{p}_e \\ \dot{h} \end{pmatrix} = V_a \begin{pmatrix} \cos\psi \cos\gamma_a \\ \sin\psi \cos\gamma_a \\ \sin\gamma_a \end{pmatrix} + \begin{pmatrix} w_n \\ w_e \\ -w_d \end{pmatrix}. \tag{9.8}$$

If we assume that the aircraft is maintained at a constant altitude and that there is no downward component of wind, then the kinematic expressions simplify as

$$\begin{pmatrix} \dot{p}_n \\ \dot{p}_e \\ \dot{h} \end{pmatrix} = V_a \begin{pmatrix} \cos\psi \\ \sin\psi \\ 0 \end{pmatrix} + \begin{pmatrix} w_n \\ w_e \\ 0 \end{pmatrix}, \tag{9.9}$$

which is a model commonly used in the UAV literature.

9.2.1 Coordinated Turn

In section 5.2 we showed that the coordinated-turn condition is described by

$$\dot{\chi} = \frac{g}{V_g} \tan\phi \cos(\chi - \psi). \tag{9.10}$$

Even though the coordinated-turn condition is not enforced by the autopilot loops described in chapter 6, the essential behavior—that the aircraft must bank to turn (as opposed to skid to turn)—is captured by this model.

The coordinated-turn condition can also be expressed in terms of the heading and the airspeed. To obtain the correct expression, we start by differentiating both sides of equation (2.9) to get

$$\begin{pmatrix} \cos\chi\cos\gamma & -V_g\sin\chi\cos\gamma & -V_g\cos\chi\sin\gamma \\ \sin\chi\cos\gamma & V_g\cos\chi\cos\gamma & -V_g\sin\chi\sin\gamma \\ -\sin\gamma & 0 & -\cos\gamma \end{pmatrix} \begin{pmatrix} \dot{V}_g \\ \dot{\chi} \\ \dot{\gamma} \end{pmatrix}$$
$$= \begin{pmatrix} \cos\psi\cos\gamma_a & -V_a\sin\psi\cos\gamma_a & -V_a\cos\psi\sin\gamma_a \\ \sin\psi\cos\gamma_a & V_a\cos\psi\cos\gamma_a & -V_a\sin\psi\sin\gamma_a \\ -\sin\gamma_a & 0 & -\cos\gamma_a \end{pmatrix} \begin{pmatrix} \dot{V}_a \\ \dot{\psi} \\ \dot{\gamma}_a \end{pmatrix}. \tag{9.11}$$

Under the condition of constant-altitude flight and no down component of wind, where γ, γ_a, $\dot{\gamma}$, $\dot{\gamma}_a$, and w_d are zero, we solve for $\dot{V}_g$ and $\dot{\psi}$ in terms of $\dot{V}_a$ and $\dot{\chi}$ to obtain

$$\dot{V}_g = \frac{\dot{V}_a}{\cos(\chi - \psi)} + V_g \dot{\chi} \tan(\chi - \psi) \tag{9.12}$$

$$\dot{\psi} = \frac{\dot{V}_a}{V_a} \tan(\chi - \psi) + \frac{V_g \dot{\chi}}{V_a \cos(\chi - \psi)}. \tag{9.13}$$

Design Models for Guidance

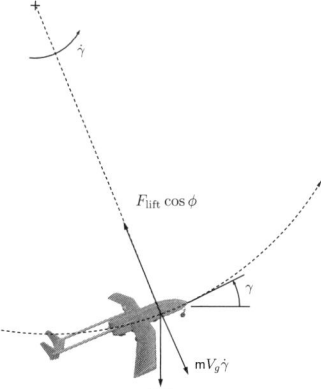

Figure 9.1 Free-body diagram for a pull-up maneuver. The MAV is at a roll angle of ϕ.

If we assume that the airspeed is constant, then from equations (9.13) and (9.10), we have

$$\dot{\psi} = \frac{g}{V_a} \tan \phi. \tag{9.14}$$

This is the familiar coordinated-turn expression. Most notable is that this equation is true in the presence of wind.

9.2.2 Accelerating Climb

To derive the dynamics for the flight-path angle, we will consider a pull-up maneuver in which the aircraft climbs along an arc. The free-body diagram of the MAV in the $\mathbf{i}^b$-$\mathbf{k}^b$ plane is shown in figure 9.1. Since the airframe is rolled at an angle of ϕ, the projection of the lift vector onto the $\mathbf{i}^b$-$\mathbf{k}^b$ plane is $F_{\text{lift}} \cos \phi$. The centripetal force due to the pull-up maneuver is $m V_g \dot{\gamma}$. Therefore, summing the forces in the $\mathbf{i}^b$-$\mathbf{k}^b$ plane gives

$$F_{\text{lift}} \cos \phi = m V_g \dot{\gamma} + mg \cos \gamma. \tag{9.15}$$

Solving for $\dot{\gamma}$ gives

$$\dot{\gamma} = \frac{g}{V_g} \left(\frac{F_{\text{lift}}}{mg} \cos \phi - \cos \gamma \right). \tag{9.16}$$

Load factor is defined as the ratio of the lift acting on the aircraft to the weight of the aircraft: $n_{lf} \triangleq F_{\text{lift}}/mg$. In wings-level, horizontal flight where the roll angle and flight-path angle are zero ($\phi = \gamma = 0$), the load factor is equal to 1. From a control perspective, it is useful to consider the load factor because it represents the force that the aircraft experiences during climbing and turning maneuvers. Although the load factor is a dimensionless number, it is often referred to by the number of "g's" that an aircraft experiences in flight. By controlling the load factor as a state, we can ensure that the aircraft is always given commands that are within its structural capability. Taking into account the definition of load factor, equation (9.16) becomes

$$\dot{\gamma} = \frac{g}{V_g}(n_{lf} \cos \phi - \cos \gamma). \qquad (9.17)$$

We note that in a constant climb, when $\dot{\gamma} = 0$, the load factor can be expressed as

$$n_{lf} = \frac{\cos \gamma}{\cos \phi}. \qquad (9.18)$$

This expression will be used in section 9.4.

9.3 Kinematic Guidance Models

In this section we summarize several different kinematic guidance models for MAVs. The guidance models that we derive will assume the presence of wind. Wind can be tricky to model correctly using a kinematic model because it introduces aerodynamic forces on the aircraft that are expressed in the dynamic model in terms of the airspeed, angle of attack, and side slip angle, as explained in chapter 4. The velocity vector can be expressed in terms of the airspeed, heading, and air-mass-referenced flight-path angle, as in equation (9.8), or in terms of the groundspeed, course, and flight-path angle, as in equation (9.7). However, we typically control the airspeed, the course angle, and the flight-path angle. Therefore, if in the simulation, we directly propagate airspeed, course angle, and flight path angle, then we will use equations (2.10) through (2.12) to solve for ground speed, heading, and air-mass-referenced flight path angle.

The first guidance model that we will consider assumes that the autopilot controls airspeed, altitude, and course angle. The corresponding equations of motion do not include the flight-path angle and are given by

$$\begin{aligned}
\dot{p}_n &= V_a \cos \psi + w_n \\
\dot{p}_e &= V_a \sin \psi + w_e \\
\ddot{\chi} &= b_{\dot{\chi}}(\dot{\chi}^c - \dot{\chi}) + b_{\chi}(\chi^c - \chi) \\
\ddot{h} &= b_{\dot{h}}(\dot{h}^c - \dot{h}) + b_h(h^c - h) \\
\dot{V}_a &= b_{V_a}(V_a^c - V_a),
\end{aligned} \qquad (9.19)$$

where the inputs are the commanded altitude h^c, the commanded airspeed V_a^c, and the commanded course χ^c, and ψ is given by equation (2.12), with $\gamma_a = 0$.

Alternatively, it is common to consider the roll angle as the input command and to control the heading through roll by using the coordinated-turn condition given in equation (9.10). In that case, the kinematic equations become

$$\begin{aligned}
\dot{p}_n &= V_a \cos \psi + w_n \\
\dot{p}_e &= V_a \sin \psi + w_e \\
\dot{\psi} &= \frac{g}{V_a} \tan \phi \\
\ddot{h} &= b_{\dot{h}}(\dot{h}^c - \dot{h}) + b_h(h^c - h) \\
\dot{V}_a &= b_{V_a}(V_a^c - V_a) \\
\dot{\phi} &= b_\phi(\phi^c - \phi),
\end{aligned} \qquad (9.20)$$

where ϕ^c is the commanded roll angle.

For the longitudinal motion, altitude is often controlled indirectly through the flight-path angle. With γ as a state, we choose to use χ as a state since both are referenced to the inertial frame. In that case, we have

$$\begin{aligned}
\dot{p}_n &= V_a \cos \psi \cos \gamma_a + w_n \\
\dot{p}_e &= V_a \sin \psi \cos \gamma_a + w_e \\
\dot{h} &= V_a \sin \gamma_a - w_d \\
\dot{\chi} &= \frac{g}{V_g} \tan \phi \cos(\chi - \psi) \\
\dot{\gamma} &= b_\gamma(\gamma^c - \gamma) \\
\dot{V}_a &= b_{V_a}(V_a^c - V_a) \\
\dot{\phi} &= b_\phi(\phi^c - \phi),
\end{aligned} \qquad (9.21)$$

where γ^c is the commanded (inertial referenced) flight-path angle, and where V_g, γ_a, and ψ are given by Equations (2.10), (2.11), and (2.12), respectively.

Some autopilots command the load factor instead of the flight-path angle. Using equation (9.17), a kinematic model that represents this

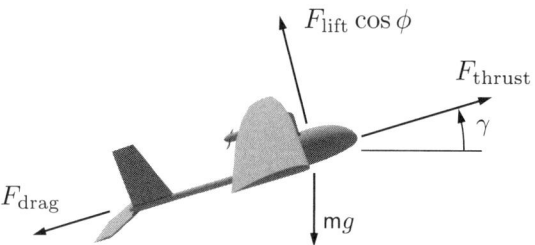

Figure 9.2 Free-body diagram indicating external forces on the UAV along the $\mathbf{i}_b$ axis. The UAV is assumed to be at a roll angle of ϕ.

situation is given by

$$\begin{aligned}
\dot{p}_n &= V_a \cos\psi \cos\gamma_a + w_n \\
\dot{p}_e &= V_a \sin\psi \cos\gamma_a + w_e \\
\dot{h} &= V_a \sin\gamma_a - w_d \\
\dot{\psi} &= \frac{g}{V_a} \tan\phi \\
\dot{\gamma} &= \frac{g}{V_g} \left(n_{lf} \cos\phi - \cos\gamma\right) \\
\dot{V}_a &= b_{V_a}(V_a^c - V_a) \\
\dot{\phi} &= b_\phi(\phi^c - \phi) \\
\dot{n}_{lf} &= b_n(n_{lf}^c - n_{lf}),
\end{aligned} \qquad (9.22)$$

where n_{lf}^c is the commanded load factor, and where V_g and γ_a are given by Equations (2.10) and (2.11) respectively.

9.4 Dynamic Guidance Model

The reduced-order guidance models derived in the previous section are based on kinematic relations between positions and velocities. Additionally, they employ first-order differential equations to model the closed-loop response of commanded states. In these equations, we took advantage of the conditions for the coordinated turn to eliminate the lift force from the equations of motion. Furthermore, we assumed that airspeed was a controlled quantity and therefore did not perform a force balance along the body-fixed $\mathbf{i}_b$ axis. In this section, we will derive an alternative set of equations of motion commonly encountered in the literature that utilize relationships drawn from free-body diagrams. Lift, drag, and thrust forces are evident in these dynamic equations.

Figure 9.2 shows a free-body diagram for a MAV climbing at flight-path angle γ and bank angle ϕ. Applying Newton's second law along the $\mathbf{i}_b$ axis and rearranging gives

$$\dot{V}_g = \frac{F_{\text{thrust}}}{m} - \frac{F_{\text{drag}}}{m} - g \sin \gamma,$$

where F_{thrust} is the thrust and F_{drag} is the drag. It is interesting to note that we arrived at this exact equation in the process of deriving transfer functions from the full nonlinear equations of motion in chapter 5, equation (5.34). The course angle can be expressed in terms of lift by combining the coordinated turn equation in (9.10) with expression (9.18) for the load factor to get

$$\dot{\chi} = \frac{g}{V_g} \tan \phi \cos(\chi - \psi) = \frac{g \sin \phi \cos(\chi - \psi)}{V_g \cos \gamma} n_{lf} = \frac{F_{\text{lift}} \sin \phi \cos(\chi - \psi)}{m V_g \cos \gamma}.$$

Similarly, equation (9.16) expresses the flight-path angle in terms of lift.

Combining these dynamic equations with the kinematic equations relating Cartesian position and velocity gives the following alternative equations of motion:

$$\begin{aligned}
\dot{p}_n &= V_g \cos \chi \cos \gamma \\
\dot{p}_e &= V_g \sin \chi \cos \gamma \\
\dot{h} &= V_g \sin \gamma \\
\dot{V}_g &= \frac{F_{\text{thrust}}}{m} - \frac{F_{\text{drag}}}{m} - g \sin \gamma \\
\dot{\chi} &= \frac{F_{\text{lift}} \sin \phi \cos(\chi - \psi)}{m V_g \cos \gamma} \\
\dot{\gamma} &= \frac{F_{\text{lift}}}{m V_g} \cos \phi - \frac{g}{V_g} \cos \gamma,
\end{aligned} \quad (9.23)$$

where ψ is given by equation (2.12). The control variables are thrust, lift coefficient, and bank angle $[F_{\text{thrust}}, C_L, \phi]^\top$. Lift and drag are given by

$$F_{\text{lift}} = \frac{1}{2} \rho V_a^2 S C_L$$

$$F_{\text{drag}} = \frac{1}{2} \rho V_a^2 S C_D,$$

with $C_D = C_{D_0} + K C_L^2$ [52]. The induced drag factor K can be determined from the aerodynamic efficiency, which is defined as

$$E_{\max} \triangleq \left(\frac{F_{\text{lift}}}{F_{\text{drag}}}\right)_{\max},$$

and the zero-lift drag coefficient C_{D_0}, using the expression

$$K = \frac{1}{4 E_{\max}^2 C_{D_0}}.$$

The popularity of this point-mass model is likely due to the fact that it models aircraft behavior in response to inputs that a pilot commonly controls: engine thrust, lift from the lifting surfaces, and bank angle as observed using the attitude indicator. In the absence of wind, $V_g = V_a$, $\gamma = \gamma_a$ and $\chi = \psi$ so that equation (9.23) can be expressed as

$$\begin{aligned}
\dot{p}_n &= V_a \cos \psi \cos \gamma \\
\dot{p}_e &= V_a \sin \psi \cos \gamma \\
\dot{h} &= V_a \sin \gamma \\
\dot{V}_a &= \frac{F_{\text{thrust}}}{m} - \frac{F_{\text{drag}}}{m} - g \sin \gamma \\
\dot{\psi} &= \frac{F_{\text{lift}}}{m V_a} \frac{\sin \phi}{\cos \gamma} \\
\dot{\gamma} &= \frac{F_{\text{lift}}}{m V_a} \cos \phi - \frac{g}{V_a} \cos \gamma.
\end{aligned} \qquad (9.24)$$

9.5 Chapter Summary

The objective of this chapter is to present high-level design models for the guidance loops. The guidance models are derived from the six-degree-of-freedom model, kinematic relations, and force balance equations. Kinematic design models are given in equations (9.19) through (9.22). A dynamic design model often found in the literature is given in equation (9.24).

Notes and References

Material supporting the development of the guidance models discussed in this chapter can be found in [2, 25, 22, 52]. The discussion of the air-mass-referenced flight-path angle is from a Boeing circular obtained at www.boeing.com. The derivation of accelerating climb in section 9.2.2 draws on the discussion in [2, p. 227–228].

Design Models for Guidance

9.6 Design Project

The objective of the assignment in this chapter is to estimate the autopilot constants b_* and to develop a reduced-order Simulink model that can be used to test and debug the guidance algorithm discussed in later chapters, prior to implementation on the full simulation model. We will focus primarily on the models given in equations (9.19) and (9.20).

9.1. Create a Simulink S-function that implements the model given in Equation (9.19) and insert it in your MAV simulator. For different inputs χ^c, h^c, and V_a^c, compare the output of the two models, and tune the autopilot coefficients b_{V_a}, b_h, $b_{\dot{h}}$, $b_{\dot{\chi}}$, and b_χ to obtain similar behavior. You may need to re-tune the autopilot gains obtained from the previous chapter. You may want to use the Simulink file `mavsim_chap9.mdl` and the Matlab function `guidance_model.m` located on the website.

9.2. Modify your autopilot function so that it uses the commanded roll angle ϕ^c as an input instead of the commanded course χ^c. Create a Simulink S-function that implements the model given in equation (9.20) and insert it in your MAV simulator. For different inputs ϕ^c, h^c, and V_a^c, compare the output of the two models, and tune the autopilot coefficients b_* to obtain similar behavior. You may need to re-tune the autopilot gains obtained from the previous chapter. Using the simulation under zero-wind conditions, find the achievable minimum turn radius $R_{\min}$ of the MAV when the commanded roll angle is $\phi^c = 30$ degrees.

10

Straight-line and Orbit Following

This chapter develops guidance laws for tracking straight-line segments and for tracking constant-altitude circular orbits. Chapter 11 will discuss techniques for combining straight-line segments and circular orbits to track more complex paths, and chapter 12 will describe techniques for path planning through obstacle fields. In the context of the architectures shown in figures 1.1 and 1.2, this chapter describes algorithms for the path following block. The primary challenge in tracking straight-line segments and circular orbits is wind, which is almost always present. For small unmanned aircraft, wind speeds are commonly 20 to 60 percent of the desired airspeed. Effective path-tracking strategies must overcome the effect of this ever-present disturbance. For most fixed-wing MAVs, the minimum turn radius is in the range of 10 to 50 m. This places a fundamental limit on the spatial frequency of paths that can be tracked. Thus, it is important that the path-tracking algorithms utilize the full capability of the MAV.

Implicit in the notion of trajectory tracking is that the vehicle is commanded to be at a particular location at a specific time and that the location typically varies in time, thus causing the vehicle to move in the desired fashion. With fixed-wing aircraft, the desired position is constantly moving (at the desired ground speed). The approach of tracking a moving point can result in significant problems for MAVs if disturbances, such as those due to wind, are not properly accounted for. If the MAV is flying into a strong wind (relative to its commanded ground speed), the progression of the trajectory point must be slowed accordingly. Similarly, if the MAV is flying downwind, the speed of the tracking point must be increased to keep it from overrunning the desired position. Given that wind disturbances vary and are often not easily predicted, trajectory tracking can be challenging in anything other than calm conditions.

Rather than using a trajectory tracking approach, this chapter focuses on path following, where the objective is to be *on the path* rather than at a certain point at a particular time. With path following, the time dependence of the problem is removed. For this chapter, we will assume that the controlled MAV is modeled by the guidance model given in equation (9.19). Our objective is to develop a method for accurate path following in the presence of wind. For a given airframe, there is

Straight-line and Orbit Following

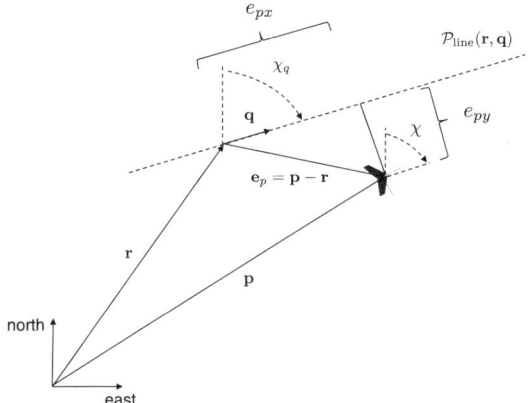

Figure 10.1 The configuration of the MAV indicated by (**p**, χ), and the straight-line path indicated by $\mathcal{P}_{\text{line}}(\mathbf{r}, \mathbf{q})$.

an optimal airspeed for which the airframe is the most aerodynamically efficient, and to conserve fuel the MAV should maintain this airspeed. Accordingly, in this chapter we will assume that the MAV is moving with a constant airspeed V_a.

10.1 Straight-line Path Following

A straight-line path is described by two vectors in $\mathbb{R}^3$, namely

$$\mathcal{P}_{\text{line}}(\mathbf{r}, \mathbf{q}) = \left\{ \mathbf{x} \in \mathbb{R}^3 : \mathbf{x} = \mathbf{r} + \lambda \mathbf{q}, \, \lambda \in \mathbb{R} \right\},$$

where $\mathbf{r} \in \mathbb{R}^3$ is the origin of the path, and $\mathbf{q} \in \mathbb{R}^3$ is a unit vector whose direction indicates the desired direction of travel. Figure 10.1 shows a top-down or lateral view of $\mathcal{P}_{\text{line}}(\mathbf{r}, \mathbf{q})$, and figure 10.2 shows a side or longitudinal view. The course angle of $\mathcal{P}_{\text{line}}(\mathbf{r}, \mathbf{q})$, as measured from north is given by

$$\chi_q \triangleq \operatorname{atan2} \frac{q_e}{q_n}, \qquad (10.1)$$

where $\mathbf{q} = (q_n \; q_e \; q_d)^\top$ expresses the north, east, and down components of the unit direction vector.

The path-following problem is most easily solved in a frame relative to the straight-line path. Selecting $\mathbf{r}$ as the center of the path frame, with the x-axis aligned with the projection of $\mathbf{q}$ onto the local north-east plane, the z-axis aligned with the inertial z-axis, and the y-axis selected

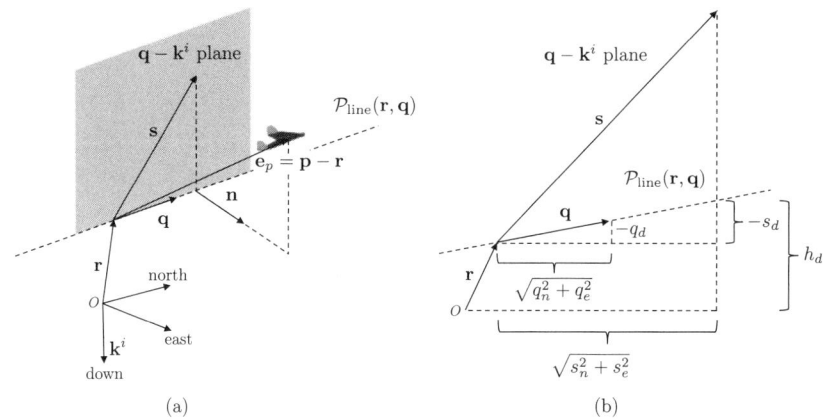

Figure 10.2 Desired altitude calculation for straight-line path following in longitudinal direction.

to create a right-handed coordinate system, then

$$\mathcal{R}_i^p \triangleq \begin{pmatrix} \cos \chi_q & \sin \chi_q & 0 \\ -\sin \chi_q & \cos \chi_q & 0 \\ 0 & 0 & 1 \end{pmatrix}$$

is the transformation from the inertial frame to the path frame, and

$$\mathbf{e}_p = \begin{pmatrix} e_{px} \\ e_{py} \\ e_{pz} \end{pmatrix} \triangleq \mathcal{R}_i^p \left(\mathbf{p}^i - \mathbf{r}^i \right)$$

is the relative path error expressed in the path frame. The relative error dynamics in the north-east inertial plane, expressed in the path frame, are given by

$$\begin{pmatrix} \dot{e}_{px} \\ \dot{e}_{py} \end{pmatrix} = \begin{pmatrix} \cos \chi_q & \sin \chi_q \\ -\sin \chi_q & \cos \chi_q \end{pmatrix} \begin{pmatrix} V_g \cos \chi \\ V_g \sin \chi \end{pmatrix}$$
$$= V_g \begin{pmatrix} \cos(\chi - \chi_q) \\ \sin(\chi - \chi_q) \end{pmatrix}. \qquad (10.2)$$

For path following, we desire to regulate the cross-track error e_{py} to zero by commanding the course angle. The relevant dynamics are therefore given by

$$\dot{e}_{py} = V_g \sin(\chi - \chi_q) \qquad (10.3)$$
$$\ddot{\chi} = b_{\dot{\chi}}(\dot{\chi}^c - \dot{\chi}) + b_\chi(\chi^c - \chi). \qquad (10.4)$$

The lateral straight-line path following problem is to select χ^c so that $e_{py} \to 0$ when χ_q is known.

The geometry for straight-line path following in the longitudinal direction is shown in figure 10.2. To calculate the desired altitude, it is necessary to project the relative path error vector onto the vertical plane containing the path direction vector $\mathbf{q}$ as shown in figure 10.2(a). We denote the projection of $\mathbf{e}_p$ as $\mathbf{s}$. Referring to the vertical plane containing the path shown in figure 10.2(b) and using similar triangles, we have the relationship

$$\frac{-s_d}{\sqrt{s_n^2 + s_e^2}} = \frac{-q_d}{\sqrt{q_n^2 + q_e^2}}.$$

The projection $\mathbf{s}$ of the relative error vector is defined as

$$\mathbf{s}^i = \begin{pmatrix} s_n \\ s_e \\ s_d \end{pmatrix}$$

$$= \mathbf{e}_p^i - (\mathbf{e}_p^i \cdot \mathbf{n})\mathbf{n},$$

where

$$\mathbf{e}_p^i = \begin{pmatrix} e_{pn} \\ e_{pe} \\ e_{pd} \end{pmatrix} \triangleq \mathbf{p}^i - \mathbf{r}^i = \begin{pmatrix} p_n - r_n \\ p_e - r_e \\ p_d - r_d \end{pmatrix}$$

and the unit vector normal to the $\mathbf{q}$-$\mathbf{k}^i$ plane is calculated as

$$\mathbf{n} = \frac{\mathbf{q} \times \mathbf{k}^i}{\|\mathbf{q} \times \mathbf{k}^i\|}.$$

From figure 10.2(b), the desired altitude for an aircraft at $\mathbf{p}$ following the straight-line path $\mathcal{P}_{\text{line}}(\mathbf{r}, \mathbf{q})$ is given by

$$h_d(\mathbf{r}, \mathbf{p}, \mathbf{q}) = -r_d + \sqrt{s_n^2 + s_e^2} \left(\frac{q_d}{\sqrt{q_n^2 + q_e^2}} \right). \tag{10.5}$$

Since the altitude dynamics are given by

$$\ddot{h} = b_{\dot{h}}(\dot{h}^c - \dot{h}) + b_h(h^c - h), \tag{10.6}$$

the longitudinal straight-line path following problem is to select h^c so that $h \to h_d(\mathbf{r}, \mathbf{p}, \mathbf{q})$.

10.1.1 Longitudinal Guidance Strategy for Straight-line Following

In this section we specify the longitudinal guidance law for tracking the altitude portion of the waypoint path. With the desired altitude specified by equation (10.5) and the dynamics modeled by equation (10.6), we will show that letting $h^c = h_d(\mathbf{r}, \mathbf{p}, \mathbf{q})$ and utilizing the altitude state machine of figure 6.20, good path-following performance will result, with zero steady-state error in altitude for straight-line paths.

With respect to the altitude state machine, we will assume that the control laws in the climb and descend zones will cause the MAV to climb or descend into the altitude-hold zone. In the altitude-hold zone, pitch attitude is used to control the altitude of the MAV, as shown in figure 6.16. Assuming that successive loop closure has been properly implemented, figure 6.17 shows a simplified representation of the outer-loop dynamics, which has the transfer function

$$\frac{h}{h^c} = \frac{b_{h_s} s + b_h}{s^2 + b_{h_s} s + b_h}.$$

Defining the altitude error as

$$e_h \triangleq h - h_d(\mathbf{r}, \mathbf{p}, \mathbf{q}) = h - h^c,$$

we get that

$$\frac{e_h}{h^c} = 1 - \frac{h}{h^c}$$
$$= \frac{s^2}{s^2 + b_{h_s} s + b_h}.$$

By applying the final value theorem, we find that

$$e_{h,ss} = \lim_{s \to 0} s \frac{s^2}{s^2 + b_{h_s} s + b_h} h^c$$
$$= 0, \quad \text{for } h^c = \frac{H_0}{s}, \frac{H_0}{s^2}.$$

The analysis in chapter 6 also shows that constant disturbances are rejected. Thus, utilizing the altitude state machine, we can track constant-altitude and inclined straight-line paths with zero steady-state altitude error provided we do not exceed the physical capabilities of the MAV and disturbances (such as vertical components of wind) are zero or constant in magnitude.

Straight-line and Orbit Following

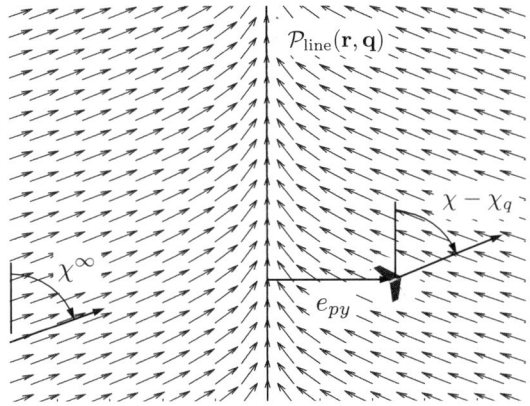

Figure 10.3 Vector field for straight-line path following. Far away from the waypoint path, the vector field is directed with an angle χ^∞ from the perpendicular to the path.

10.1.2 Lateral Guidance Strategy for Straight-line Following

The objective in this section is to select the commanded course angle χ^c in equation (10.4) so that e_{py} in equation (10.3) is driven to zero asymptotically. The strategy in this section will be to construct a desired course angle at every spatial point relative to the straight-line path that results in the MAV moving toward the path. The set of desired course angles at every point will be called a vector field because the desired course angle specifies a vector (relative to the straight line) with a magnitude of unity. Figure 10.3 depicts an example vector field for straight-line path following. The objective is to construct the vector field so that when e_{py} is large, the MAV is directed to approach the path with course angle $\chi^\infty \in (0, \frac{\pi}{2}]$, and so that as e_{py} approaches zero, the desired course also approaches zero. Toward that end, we define the desired course of the MAV as

$$\chi_d(e_{py}) = -\chi^\infty \frac{2}{\pi} \tan^{-1}(k_{\text{path}} e_{py}), \qquad (10.7)$$

where k_{path} is a positive constant that influences the rate of the transition from χ^∞ to zero. Figure 10.4 shows how the choice of k_{path} affects the rate of transition. Large values of k_{path} result in short, abrupt transitions, while small values of k_{path} cause long, smooth transitions in the desired course.

If χ^∞ is restricted to be in the range $\chi^\infty \in (0, \frac{\pi}{2}]$, then clearly

$$-\frac{\pi}{2} < \chi^\infty \frac{2}{\pi} \tan^{-1}(k_{\text{path}} e_{py}) < \frac{\pi}{2}$$

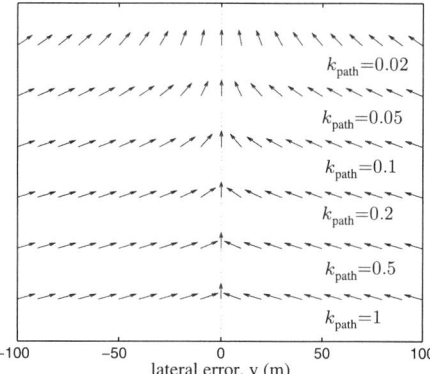

Figure 10.4 Vector fields for various values of k_{path}. Large values of k_{path} yield abrupt transitions from χ^∞ to zero, while small values of k_{path} give smooth transitions.

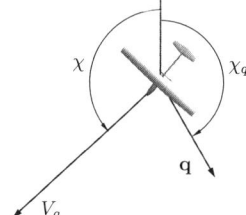

Figure 10.5 The calculation of χ_q needs to account for the current course angle of the MAV. In this scenario, the MAV should turn left to align with the waypoint path, but if χ_q is computed with `atan2`, the angle will be a positive number slightly smaller than $+\pi$, which will cause the MAV to turn right to align with the waypoint path.

for all values of e_{py}. Therefore, since $\tan^{-1}(\cdot)$ is an odd function and $\sin(\cdot)$ is odd over $(-\frac{\pi}{2}, \frac{\pi}{2})$, we can use the Lyapunov function $W(e_{py}) = \frac{1}{2}e_{py}^2$ to argue that if $\chi = \chi_q + \chi^d(e_{py})$, then $e_{py} \to 0$ asymptotically, since

$$\dot{W} = -V_g e_{py} \sin\left(\chi^\infty \frac{2}{\pi} \tan^{-1}(k_{\text{path}} e_{py})\right)$$

is less than zero for $e_{py} \neq 0$. The command for lateral path following is therefore given by

$$\chi^c(t) = \chi_q - \chi^\infty \frac{2}{\pi} \tan^{-1}(k_{\text{path}} e_{py}(t)). \tag{10.8}$$

Before moving to orbit following, we note that using equation (10.8) may result in undesirable behavior if χ_q is computed directly from equation (10.1), where `atan2` returns an angle between $\pm\pi$. As an example, consider the scenario shown in figure 10.5, where χ_q is a positive number slightly smaller than $+\pi$. Since the current course is negative, equation (10.8) will cause the MAV to turn right to align with the waypoint path. As an alternative, if χ_q is expressed as a negative angle slightly less than $-\pi$, then the MAV will turn left to align with

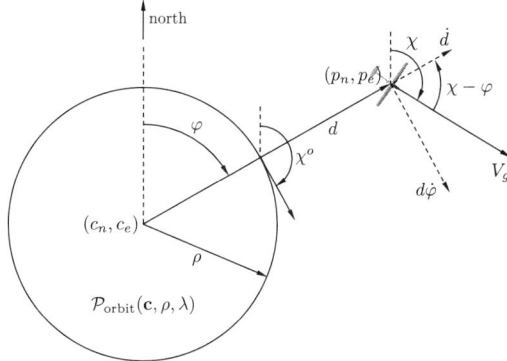

Figure 10.6 Orbital path with center (c_n, c_e), and radius ρ. The distance from the orbit center to the MAV is d, and the angular position of the MAV relative to the orbit is φ.

the waypoint path. To alleviate this problem, χ_q should be computed as

$$\chi_q = \texttt{atan2}(q_e, q_n) + 2\pi m,$$

where $m \in \mathcal{N}$ is selected so that $-\pi \leq \chi_q - \chi \leq \pi$, and $\texttt{atan2}$ is a four-quadrant $\tan^{-1}$ function.

10.2 Orbit Following

An orbit path is described by a center $\mathbf{c} \in \mathbb{R}^3$, a radius $\rho \in \mathbb{R}$, and a direction $\lambda \in \{-1, 1\}$, as

$$\mathcal{P}_{\text{orbit}}(\mathbf{c}, \rho, \lambda) = \left\{ \mathbf{r} \in \mathbb{R}^3 : \mathbf{r} = \mathbf{c} + \lambda \rho \left(\cos \varphi, \sin \varphi\ 0\right)^\top, \varphi \in [0, 2\pi) \right\},$$

where $\lambda = 1$ signifies a clockwise orbit and $\lambda = -1$ signifies a counter-clockwise orbit. We assume that the center of the orbit is expressed in inertial coordinates so that $\mathbf{c} = (c_n, c_e, c_d)^\top$, where $-c_d$ represents the desired altitude of the orbit and to maintain altitude we let $h^c = -c_d$

Figure 10.6 shows a top-down view of an orbital path. The guidance strategy for orbit following is best derived in polar coordinates. Let d be the radial distance from the desired center of the orbit to the MAV, and let φ be the phase angle of the relative position, as shown in figure 10.6. The constant-altitude MAV dynamics in polar coordinates can be derived by rotating the differential equations that describe the motion of the MAV in the north and east directions as

$$\begin{pmatrix} \dot{p}_n \\ \dot{p}_e \end{pmatrix} = \begin{pmatrix} V_g \cos \chi \\ V_g \sin \chi \end{pmatrix},$$

through the phase angle φ so that the equations of motion represent the MAV motion in the normal and tangential directions to the orbit as

$$\begin{pmatrix} \dot{d} \\ d\dot{\varphi} \end{pmatrix} = \begin{pmatrix} \cos \varphi & \sin \varphi \\ -\sin \varphi & \cos \varphi \end{pmatrix} \begin{pmatrix} \dot{p}_n \\ \dot{p}_e \end{pmatrix}$$

$$= \begin{pmatrix} \cos \varphi & \sin \varphi \\ -\sin \varphi & \cos \varphi \end{pmatrix} \begin{pmatrix} V_g \cos \chi \\ V_g \sin \chi \end{pmatrix}$$

$$= \begin{pmatrix} V_g \cos(\chi - \varphi) \\ V_g \sin(\chi - \varphi) \end{pmatrix}.$$

These expressions can also be derived from the geometry illustrated in figure 10.6. The MAV dynamics in polar coordinates are therefore given by

$$\dot{d} = V_g \cos(\chi - \varphi) \tag{10.9}$$

$$\dot{\varphi} = \frac{V_g}{d} \sin(\chi - \varphi) \tag{10.10}$$

$$\ddot{\chi} = -b_{\dot{\chi}} \dot{\chi} + b_\chi (\chi^c - \chi). \tag{10.11}$$

As shown in figure 10.6, for a clockwise orbit, the desired course angle when the MAV is located on the orbit is given by $\chi^o = \varphi + \pi/2$. Similarly, for a counterclockwise orbit, the desired angle is given by $\chi^o = \varphi - \pi/2$. Therefore, in general we have

$$\chi^o = \varphi + \lambda \frac{\pi}{2}.$$

The control objective is to drive $d(t)$ to the orbit radius ρ and to drive the course angle $\chi(t)$ to χ^o in the presence of wind.

Our approach to orbit following is similar to the ideas developed in section 10.1.2. The strategy is to construct a desired course field that moves the MAV onto the orbit $\mathcal{P}_{\text{orbit}}(\mathbf{c}, \rho, \lambda)$. When the distance between the MAV and the center of the orbit is large, it is desirable for the MAV to fly toward the orbit center. In other words, when $d \gg \rho$, the desired course is

$$\chi_d \approx \chi^o + \lambda \frac{\pi}{2},$$

and when $d = \rho$, the desired course is $\chi_d = \chi^o$. Therefore, a candidate course field is given by

$$\chi_d(d - \rho, \lambda) = \chi^o + \lambda \tan^{-1}\left(k_{\text{orbit}}\left(\frac{d - \rho}{\rho}\right)\right), \tag{10.12}$$

where $k_{\text{orbit}} > 0$ is a constant that specifies the rate of transition from $\lambda\pi/2$ to zero. This expression for χ_d is valid for all values of $d \geq 0$.

We can again use the Lyapunov function $W = \frac{1}{2}(d - \rho)^2$ to argue that if $\chi = \chi_d$, then the tracking objective is satisfied. Differentiating W along the system trajectory gives

$$\dot{W} = -V_g(d - \rho)\sin\left(\tan^{-1}\left(k_{\text{orbit}}\left(\frac{d - \rho}{\rho}\right)\right)\right),$$

which is negative definite since the argument of sin is in the set $(-\pi/2, \pi/2)$ for all $d > 0$, implying that $d \to \rho$ asymptotically. The course command for orbit following is therefore given by

$$\chi^c(t) = \varphi + \lambda\left[\frac{\pi}{2} + \tan^{-1}\left(k_{\text{orbit}}\left(\frac{d - \rho}{\rho}\right)\right)\right]. \tag{10.13}$$

Similar to the computation of the path angle χ_q, if the angular position in the orbit φ is computed to be between $\pm\pi$, then there will be a sudden jump of 2π in the commanded course as the MAV transitions from $\varphi = \pi$ to $\varphi = -\pi$. To alleviate this problem, φ should be computed as

$$\varphi = \text{atan2}(p_e - c_e, p_n - c_n) + 2\pi m,$$

where $m \in \mathcal{N}$ is selected so that $-\pi \leq \varphi - \chi \leq \pi$.

10.3 Chapter Summary

This chapter introduced algorithms for following straight-line paths and circular orbits in the presence of wind. The idea is to construct a heading field that directs the MAV onto the path and is therefore distinctly different from trajectory tracking, where the vehicle would be commanded to follow a time-varying location. The algorithms developed in this chapter are summarized in algorithms 3 and 4.

Algorithm 3 Straight-line Following: $[h^c, \chi^c]$ = followStraightLine $(\mathbf{r}, \mathbf{q}, \mathbf{p}, \chi)$

Input: Path definition $\mathbf{r} = (r_n, r_e, r_d)^\top$ and $\mathbf{q} = (q_n, q_e, q_d)^\top$, MAV position $\mathbf{p} = (p_n, p_e, p_d)^\top$, course χ, gains χ_∞, k_{path}, sample rate T_s.
1: Compute commanded altitude using equation (10.5).
2: $\chi_q \leftarrow \text{atan2}(q_e, q_n)$
3: **while** $\chi_q - \chi < -\pi$ **do**
4: $\quad \chi_q \leftarrow \chi_q + 2\pi$
5: **end while**
6: **while** $\chi_q - \chi > \pi$ **do**
7: $\quad \chi_q \leftarrow \chi_q - 2\pi$
8: **end while**
9: $e_{py} \leftarrow -\sin\chi_q(p_n - r_n) + \cos\chi_q(p_e - r_e)$
10: Compute commanded course angle using equation (10.8).
11: **return** h^c, χ^c

Algorithm 4 Circular Orbit Following: $[h^c, \chi^c]$ = followOrbit $(\mathbf{c}, \rho, \lambda, \mathbf{p}, \chi)$

Input: Orbit center $\mathbf{c} = (c_n, c_e, c_d)^\top$, radius ρ, and direction λ, MAV position $\mathbf{p} = (p_n, p_e, p_d)^\top$, course χ, gains k_{orbit}, sample rate T_s.
1: $h^c \leftarrow -c_d$
2: $d \leftarrow \sqrt{(p_n - c_n)^2 + (p_e - c_e)^2}$
3: $\varphi \leftarrow \text{atan2}(p_e - c_e, p_n - c_n)$
4: **while** $\varphi - \chi < -\pi$ **do**
5: $\quad \varphi \leftarrow \varphi + 2\pi$
6: **end while**
7: **while** $\varphi - \chi > \pi$ **do**
8: $\quad \varphi \leftarrow \varphi - 2\pi$
9: **end while**
10: Compute commanded course angle using equation (10.13).
11: **return** h^c, χ^c

Notes and References

The methods described in sections 10.1 and 10.2 are variations on those described in [29, 53, 54] and are based on the notion of a vector field, which calculates a desired heading based on the distance from the path. A nice extension of [53] is given in [55], which derives general stability conditions for vector-field based methods. The focus is entirely on orbits, but elongated oval orbits and elliptical orbits can be produced. The method in [55], which is based on Lyapunov techniques, could be extended to straight lines.

The notion of vector fields is similar to that of potential fields, which have been widely used as a tool for path planning in the robotics community (see, e.g., [56]). It has also been suggested in [57] that potential fields can be used in UAV navigation for obstacle and collision avoidance applications. The method of [57] provides a way for groups of UAVs to use the gradient of a potential field to navigate through heavily populated areas safely while still aggressively approaching their targets. Vector fields are different from potential fields in that they do not necessarily represent the gradient of a potential. Rather, the vector field simply indicates a desired direction of travel.

Several approaches have been proposed for UAV trajectory tracking. An approach for tight tracking of curved trajectories is presented in [58]. For straight-line paths, the approach approximates PD control. For curved paths, an additional anticipatory control element that improves the tracking capability is implemented. The approach accommodates the addition of an adaptive element to account for disturbances such as wind. This approach is validated with flight experiments.

Reference [59] describes an integrated approach for developing guidance and control algorithms for autonomous vehicle trajectory tracking. Their approach builds upon the theory of gain scheduling and produces controllers for tracking trajectories that are defined in an inertial reference frame. The approach is illustrated through simulations of a small UAV. Reference [26] presents a path-following method for UAVs that provides a constant line of sight between the UAV and an observation target.

10.4 Design Project

The objective of this assignment is to implement Algorithms 3 and 4. Download the sample code for this chapter from the book website and note the addition of two blocks labeled PathManager and

`PathFollower`. The output of the path manager is

$$y_{\text{manager}} = \begin{pmatrix} \texttt{flag} \\ V_g^d \\ \mathbf{r} \\ \mathbf{q} \\ \mathbf{c} \\ \rho \\ \lambda \end{pmatrix},$$

where `flag=1` indicates that $\mathcal{P}_{\text{line}}(\mathbf{r}, \mathbf{q})$ should be followed and `flag=2` indicates that $\mathcal{P}_{\text{orbit}}(\mathbf{c}, \rho, \lambda)$ should be followed, and where V_g^d is the desired airspeed.

10.1 Modify `path_follow.m` to implement algorithms 3 and 4. By modifying `path_manager_chap10.m` test both the straight-line and orbit-following algorithms on the guidance model given in equation (9.19). An example Simulink diagram is given in `mavsim_chap10_model.mdl`. Test your design with significant constant winds (e.g., $w_n = 3$, $w_e = -3$). Tune the gains to get acceptable performance.

10.2 Implement the path following algorithms on the full six-DOF simulation of the MAV. An example Simulink diagram is given in `mavsim_chap10.mdl`. Test your design with significant constant winds (e.g., $w_n = 3$, $w_e = -3$). If necessary, tune the gains to get acceptable performance.

11

Path Manager

In Chapter 10 we developed guidance strategies for following straight-line paths and circular orbits. The objective of this chapter is to describe two simple strategies that combine straight-line paths and orbits to synthesize general classes of paths that are useful for autonomous operation of MAVs. In section 11.1, we show how the straight-line and orbit guidance strategies can be used to follow a series of waypoints. In section 11.2, the straight-line and orbit guidance strategies are used to synthesize Dubins paths, which for constant-altitude, constant-velocity vehicles with turning constraints, are time-optimal paths between two configurations. In reference to the architectures shown in figures 1.1 and 1.2, this chapter describes the path manager.

11.1 Transitions Between Waypoints

Define a waypoint path as an ordered sequence of waypoints

$$\mathcal{W} = \{\mathbf{w}_1, \mathbf{w}_2, \ldots, \mathbf{w}_N\}, \tag{11.1}$$

where $\mathbf{w}_i = (w_{n,i}, w_{e,i}, w_{d,i})^\top \in \mathbb{R}^3$. In this section, we address the problem of switching from one waypoint segment to another. Consider the scenario shown in figure 11.1 that depicts a MAV tracking the straight line segment $\overline{\mathbf{w}_{i-1}\mathbf{w}_i}$. Intuitively, when the MAV reaches $\mathbf{w}_i$, we desire to switch the guidance algorithm so that it will track the straight-line segment $\overline{\mathbf{w}_i\mathbf{w}_{i+1}}$. What is the best method for determining whether the MAV has reached $\mathbf{w}_i$? One possible strategy is to switch when the MAV enters a ball around $\mathbf{w}_i$. In other words, the guidance algorithm would switch at the first time instant when

$$\|\mathbf{p}(t) - \mathbf{w}_i)\| \leq b,$$

where b is the size of the ball and $\mathbf{p}(t)$ is the location of the MAV. However, if there are disturbances like wind or if b is chosen too small or if the segment from $\mathbf{w}_{i-1}$ to $\mathbf{w}_i$ is short and the tracking algorithm has not had time to converge, then the MAV may never enter the b-ball around $\mathbf{w}_i$.

A better approach, one that is not sensitive to tracking error, is to use a half-plane switching criteria. Given a point $\mathbf{r} \in \mathbb{R}^3$ and a normal vector

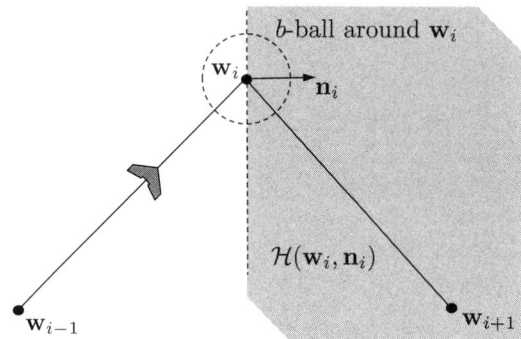

Figure 11.1 When transitioning from one straight-line segment to another, a criterium is needed to indicate when the MAV has completed the first straight-line segment. A possible option is to switch when the MAV enters a *b*-ball around the transition waypoint. A better option is to switch when the MAV enters the half-plane $\mathcal{H}(\mathbf{w}_i, \mathbf{n}_i)$.

$\mathbf{n} \in \mathbb{R}^3$, define the half plane

$$\mathcal{H}(\mathbf{r}, \mathbf{n}) \triangleq \{\mathbf{p} \in \mathbb{R}^3 : (\mathbf{p} - \mathbf{r})^\top \mathbf{n} \geq 0\}.$$

Referring to figure 11.1, we can define the unit vector pointing in the direction of the line $\overline{\mathbf{w}_i \mathbf{w}_{i+1}}$ as

$$\mathbf{q}_i \triangleq \frac{\mathbf{w}_{i+1} - \mathbf{w}_i}{\|\mathbf{w}_{i+1} - \mathbf{w}_i\|}, \tag{11.2}$$

and accordingly, the unit normal to the 3-D half plane that separates the line $\overline{\mathbf{w}_{i-1} \mathbf{w}_i}$ from the line $\overline{\mathbf{w}_i \mathbf{w}_{i+1}}$ is given by

$$\mathbf{n}_i \triangleq \frac{\mathbf{q}_{i-1} + \mathbf{q}_i}{\|\mathbf{q}_{i-1} + \mathbf{q}_i\|}.$$

The MAV tracks the straight-line path from $\mathbf{w}_{i-1}$ to $\mathbf{w}_i$ until it enters $\mathcal{H}(\mathbf{w}_i, \mathbf{n}_i)$, at which point it will track the straight-line path from $\mathbf{w}_i$ to $\mathbf{w}_{i+1}$.

A simple algorithm for following the sequence of waypoints in equation (11.1) is given in algorithm 5. The first time that the algorithm is executed, the waypoint pointer is initialized to $i = 2$ in line 2. The MAV will be commanded to follow the straight-line segment $\overline{\mathbf{w}_{i-1} \mathbf{w}_i}$. The index i is a static variable and retains its value from one execution of the algorithm to the next. Lines 4 and 5 define $\mathbf{r}$ and $\mathbf{q}$ for the current waypoint segment. Line 6 defines the unit vector along the next waypoint path, and line 7 is a vector that is perpendicular to the half plane that separates $\overline{\mathbf{w}_{i-1}\mathbf{w}_i}$ from $\overline{\mathbf{w}_i\mathbf{w}_{i+1}}$. Line 8 checks to see if the half plane defining the next waypoint segment has been reached by the MAV. If it has, then line 9 will increment the pointer until reaching the last waypoint segment.

Algorithm 5 will produce paths like that shown in figure 11.2. The advantages of algorithm 5 are that it is extremely simple and that the MAV reaches the waypoint before transitioning to the next straight-line path. However, the paths shown in figure 11.2 provide neither a smooth nor balanced transition between the straight-line segments. An alternative is to smoothly transition between waypoints by inserting a fillet as shown in figure 11.3. The disadvantage with the path shown in figure 11.3 is that the MAV does not directly pass through waypoint $\mathbf{w}_i$, which may sometimes be desired.

Algorithm 5 Follow Waypoints: $(\mathbf{r}, \mathbf{q}) = \text{followWpp}(\mathcal{W}, \mathbf{p})$

Input: Waypoint path $\mathcal{W} = \{\mathbf{w}_1, \ldots, \mathbf{w}_N\}$, MAV position $\mathbf{p} = (p_n, p_e, p_d)^\top$.
Require: $N \geq 3$
1: **if** New waypoint path $\mathcal{W}$ is received **then**
2:     Initialize waypoint index: $i \leftarrow 2$
3: **end if**
4: $\mathbf{r} \leftarrow \mathbf{w}_{i-1}$
5: $\mathbf{q}_{i-1} \leftarrow \frac{\mathbf{w}_i - \mathbf{w}_{i-1}}{\|\mathbf{w}_i - \mathbf{w}_{i-1}\|}$
6: $\mathbf{q}_i \leftarrow \frac{\mathbf{w}_{i+1} - \mathbf{w}_i}{\|\mathbf{w}_{i+1} - \mathbf{w}_i\|}$
7: $\mathbf{n}_i \leftarrow \frac{\mathbf{q}_{i-1} + \mathbf{q}_i}{\|\mathbf{q}_{i-1} + \mathbf{q}_i\|}$
8: **if** $\mathbf{p} \in \mathcal{H}(\mathbf{w}_i, \mathbf{n}_i)$ **then**
9:     Increment $i \leftarrow (i + 1)$ until $i = N - 1$
10: **end if**
11: **return** $\mathbf{r}, \mathbf{q} = \mathbf{q}_{i-1}$ at each time step

---

In the remainder of this section we will focus on smoothed paths like those shown in figure 11.3. The geometry near the transition is shown in figure 11.4. With the unit vector $\mathbf{q}_i$ aligned with the line between waypoints $\mathbf{w}_i$ and $\mathbf{w}_{i+1}$ defined as in equation (11.2), the angle between $\overline{\mathbf{w}_{i-1}\mathbf{w}_i}$ and $\overline{\mathbf{w}_i\mathbf{w}_{i+1}}$ is given by

$$\varrho \triangleq \cos^{-1}\left(-\mathbf{q}_{i-1}^\top \mathbf{q}_i\right). \tag{11.3}$$

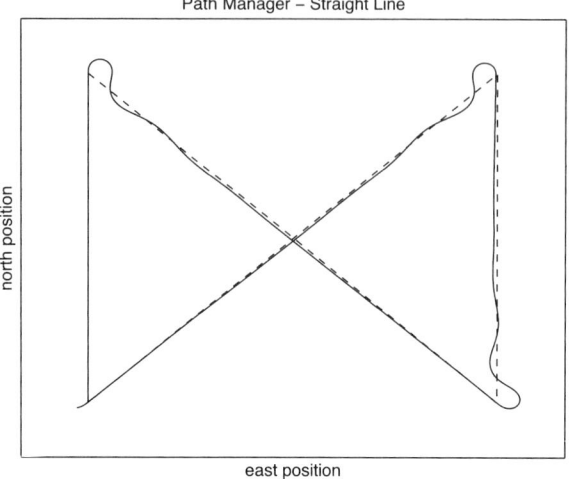

Figure 11.2 Path generated using the path-following approach given in algorithm 5. The MAV follows the straight-line path until reaching the waypoint, and then maneuvers onto the next straight-line section.

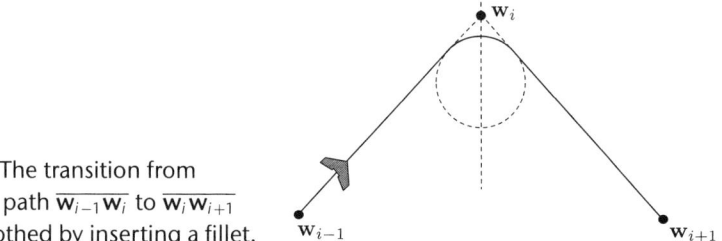

Figure 11.3 The transition from straight-line path $\overline{\mathbf{w}_{i-1}\mathbf{w}_i}$ to $\overline{\mathbf{w}_i\mathbf{w}_{i+1}}$ can be smoothed by inserting a fillet.

If the radius of the fillet is R, as shown in figure 11.4, then the distance between the waypoint $\mathbf{w}_i$ and the location where the fillet intersects the line $\overline{\mathbf{w}_i\mathbf{w}_{i+1}}$ is $R/\tan\frac{\varrho}{2}$, and the distance between $\mathbf{w}_i$ and the center of the fillet circle is $R/\sin\frac{\varrho}{2}$. Therefore, the distance between $\mathbf{w}_i$ and the edge of the fillet circle along the bisector of ϱ is given by $R/\sin\frac{\varrho}{2} - R$.

To implement the fillet maneuver using the path-following algorithms described in chapter 10, we will follow the straight-line segment $\overline{\mathbf{w}_{i-1}\mathbf{w}_i}$ until entering the half plane $\mathcal{H}_1$ shown in figure 11.5. The right-handed orbit of radius R is then followed until entering the half plane $\mathcal{H}_2$ shown in figure 11.5, at which point the straight-line segment $\overline{\mathbf{w}_i\mathbf{w}_{i+1}}$ is followed.

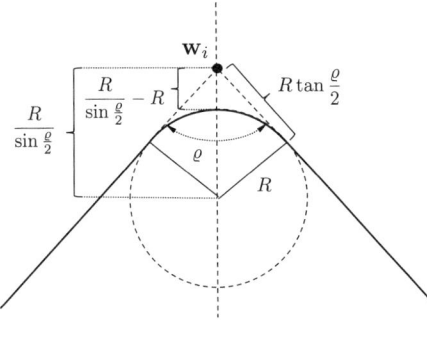

Figure 11.4 The geometry associated with inserting a fillet between waypoint segments.

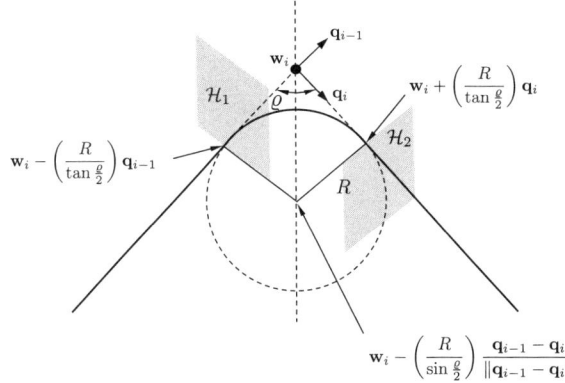

Figure 11.5 Definitions of the half planes associated with following a fillet inserted between waypoint segments.

As shown in figure 11.5, the center of the fillet is given by

$$\mathbf{c} = \mathbf{w}_i - \left(\frac{R}{\sin \frac{\varrho}{2}}\right) \frac{\mathbf{q}_{i-1} - \mathbf{q}_i}{\|\mathbf{q}_{i-1} - \mathbf{q}_i\|}.$$

Similarly, the half plane $\mathcal{H}_1$ is defined by the location

$$\mathbf{r}_1 = \mathbf{w}_i - \left(\frac{R}{\tan \frac{\varrho}{2}}\right) \mathbf{q}_{i-1},$$

and the normal vector $\mathbf{q}_{i-1}$. The half plane $\mathcal{H}_2$ is defined by the location

$$\mathbf{r}_2 = \mathbf{w}_i + \left(\frac{R}{\tan \frac{\varrho}{2}}\right) \mathbf{q}_i$$

and the normal vector $\mathbf{q}_i$.

The algorithm for maneuvering along the waypoint path $\mathcal{W}$ using fillets to smooth between the straight-line segments is given by algorithm 6. The If statement in line 1 tests to see if a new waypoint

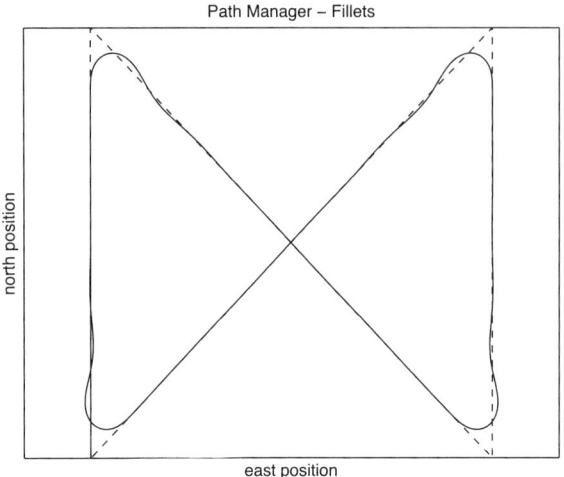

Figure 11.6 An example of the types of flight paths produced by algorithm 6.

path has been received, including when the algorithm is instantiated. If a new waypoint path has been received by the path manager, then the waypoint pointer and the state machine are initialized in line 2. The unit vectors $\mathbf{q}_{i-1}$ and $\mathbf{q}_i$ and the angle ϱ are computed in lines 4–6.

When the state machine is in state state=1, the MAV is commanded to follow the straight-line path along $\overline{\mathbf{w}_{i-1}\mathbf{w}_i}$, which is parameterized by $\mathbf{r} = \mathbf{w}_{i-1}$, and $\mathbf{q} = \mathbf{q}_{i-1}$, and are assigned in lines 8–10. Lines 11–14 test to see if the MAV has transitioned into the half plane shown as $\mathcal{H}_1$ in figure 11.5. If the MAV has transitioned into $\mathcal{H}_2$, then the state machine is updated to state=2.

When the state machine is in state=2, the MAV is commanded to follow the orbit that defines the fillet. The center, radius, and direction of the orbit are assigned in lines 17–19. In Line 19, $q_{i-1,n}$ and $q_{i-1,e}$ denote the north and east components of $\mathbf{q}_{i-1}$. Lines 21–24 test to see if the MAV has transitioned into the half plane shown as $\mathcal{H}_2$ in figure 11.5. If the MAV has transitioned into $\mathcal{H}_2$, then the waypoint pointer is incremented, and the state machine is switched back to state=1 to follow the segment $\overline{\mathbf{w}_i\mathbf{w}_{i+1}}$. Algorithm 6 produces paths like that shown in figure 11.6.

One of the disadvantages of the fillet method as given in algorithm 6 is that the path length is changed when fillets are inserted. For certain applications, like the cooperative timing problems discussed in [60], it is important to have a high quality estimate of the path length, or the time required to traverse a certain waypoint path. We will conclude this section by deriving an expression for the path length of $\mathcal{W}$ after fillets have been inserted.

Algorithm 6 Follow Waypoints with Fillets: (flag, r, q, c, ρ, λ) = followWppFillet($\mathcal{W}$, p, R)

Input: Waypoint path $\mathcal{W} = \{\mathbf{w}_1, \ldots, \mathbf{w}_N\}$, MAV position $\mathbf{p} = (p_n, p_e, p_d)^\top$, fillet radius R.
Require: $N \geq 3$

1: **if** New waypoint path $\mathcal{W}$ is received **then**
2:     Initialize waypoint index: $i \leftarrow 2$, and state machine: state $\leftarrow 1$.
3: **end if**
4: $\mathbf{q}_{i-1} \leftarrow \frac{\mathbf{w}_i - \mathbf{w}_{i-1}}{\|\mathbf{w}_i - \mathbf{w}_{i-1}\|}$
5: $\mathbf{q}_i \leftarrow \frac{\mathbf{w}_{i+1} - \mathbf{w}_i}{\|\mathbf{w}_{i+1} - \mathbf{w}_i\|}$
6: $\varrho \leftarrow \cos^{-1}(-\mathbf{q}_{i-1}^\top \mathbf{q}_i)$
7: **if** state = 1 **then**
8:     flag $\leftarrow 1$
9:     $\mathbf{r} \leftarrow \mathbf{w}_{i-1}$
10:     $\mathbf{q} \leftarrow \mathbf{q}_{i-1}$
11:     $\mathbf{z} \leftarrow \mathbf{w}_i - \left(\frac{R}{\tan(\varrho/2)}\right) \mathbf{q}_{i-1}$
12:     **if** $\mathbf{p} \in \mathcal{H}(\mathbf{z}, \mathbf{q}_{i-1})$ **then**
13:         state $\leftarrow 2$
14:     **end if**
15: **else if** state = 2 **then**
16:     flag $\leftarrow 2$
17:     $\mathbf{c} \leftarrow \mathbf{w}_i - \left(\frac{R}{\sin(\varrho/2)}\right) \frac{\mathbf{q}_{i-1} - \mathbf{q}_i}{\|\mathbf{q}_{i-1} - \mathbf{q}_i\|}$
18:     $\rho \leftarrow R$
19:     $\lambda \leftarrow \text{sign}(q_{i-1,n} q_{i,e} - q_{i-1,e} q_{i,n})$
20:     $\mathbf{z} \leftarrow \mathbf{w}_i + \left(\frac{R}{\tan(\varrho/2)}\right) \mathbf{q}_i$
21:     **if** $\mathbf{p} \in \mathcal{H}(\mathbf{z}, \mathbf{q}_i)$ **then**
22:         $i \leftarrow (i+1)$ until $i = N - 1$.
23:         state $\leftarrow 1$
24:     **end if**
25: **end if**
26: **return** flag, r, q, c, ρ, λ.

To be precise, let

$$|\mathcal{W}| \triangleq \sum_{i=2}^{N} \|\mathbf{w}_i - \mathbf{w}_{i-1}\|$$

be defined as the length of the waypoint path $\mathcal{W}$. Define $|\mathcal{W}|_F$ as the path length of the fillet-corrected waypoint path that will be obtained using algorithm 6. From figure 11.4 we see that the length of the fillet traversed by the corrected path is $R(\varrho_i)$. In addition, it is clear that the length of the straight-line segment removed from $|\mathcal{W}|$ by traversing the fillet is $2R \tan \frac{\varrho_i}{2}$. Therefore,

$$|\mathcal{W}|_F = |\mathcal{W}| + \sum_{i=2}^{N} \left(R(\varrho_i) - \frac{2R}{\tan \frac{\varrho_i}{2}} \right), \qquad (11.4)$$

where ϱ_i is given in equation (11.3).

11.2 Dubins Paths

11.2.1 Definition of Dubins Path

This section focuses on so-called Dubins paths, where, rather than following a waypoint path, the objective is to transition from one configuration (position and course) to another. It was shown in [61] that for a vehicle with kinematics given by

$$\dot{p}_n = V \cos \vartheta$$
$$\dot{p}_e = V \sin \vartheta$$
$$\dot{\vartheta} = u,$$

where V is constant and $u \in [-\bar{u}, \bar{u}]$, the time-optimal path between two different configurations consists of a circular arc, followed by a straight line, and concluding with another circular arc to the final configuration, where the radius of the circular arcs is $V/\bar{u}$. These turn-straight-turn paths are one of several classes of Dubins paths defined for optimal transitions between configurations. In the context of unmanned aircraft, we will restrict our attention to constant-altitude, constant-groundspeed scenarios.

The radius of the circular arcs that define a Dubins path will be denoted by R, where we assume that R is at least as large as the minimum turn radius of the UAV. Throughout this section, a MAV configuration is defined as $(\mathbf{p}, \chi)$, where $\mathbf{p}$ is inertial position and χ is course angle.

Given a start configuration denoted as $(\mathbf{p}_s, \chi_s)$ and an end configuration denoted as $(\mathbf{p}_e, \chi_e)$, a Dubins path consists of an arc of radius R that starts at the initial configuration, followed by a straight line, and concluded by another arc of radius R that ends at the end configuration. As shown in figure 11.7, for any given start and end configurations, there are four possible paths consisting of an arc, followed by a straight line, followed by an arc. Case I (R-S-R) is a right-handed arc

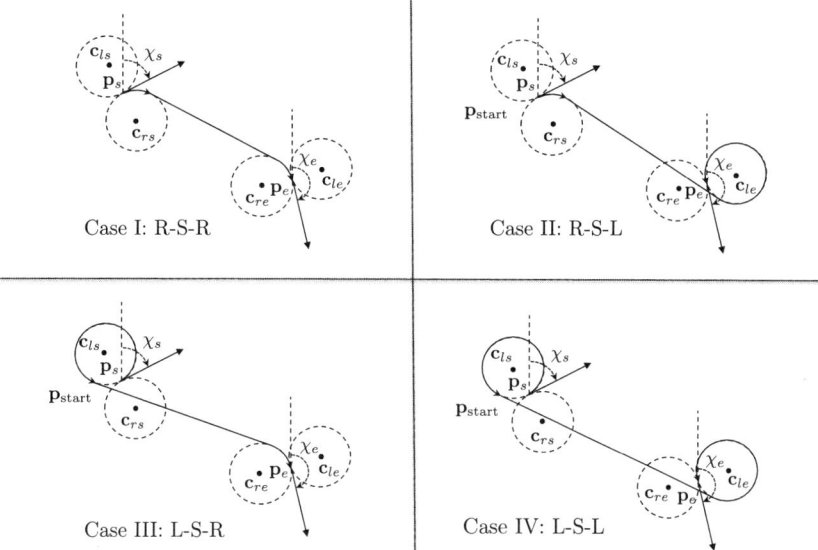

Figure 11.7 Given a start configuration ($\mathbf{p}_s$, χ_s), an end configuration ($\mathbf{p}_e$, χ_e), and a radius R, there are four possible paths consisting of an arc, a straight line, and an arc. The Dubins path is defined as the case that results in the shortest path length, which for this scenario is Case I.

followed by a straight line followed by another right-handed arc. Case II (R-S-L) is a right-handed arc followed by a straight line followed by a left-handed arc. Case III (L-S-R) is a left-handed arc followed by a straight line followed by a right-handed arc. Case IV (L-S-L) is a left-handed arc followed by a straight line followed by another left-handed arc. The Dubins path is defined as the case with the shortest path length.

11.2.2 Path Length Computation

To determine the Dubins path, it is necessary to compute the path length for the four cases shown in figure 11.7. In this section, we will derive explicit formulas for the path length for each case. Given the position $\mathbf{p}$, the course χ, and the radius R, the centers of the right and left turning circles are given by

$$\mathbf{c}_r = \mathbf{p} + R \left(\cos(\chi + \tfrac{\pi}{2}), \sin(\chi + \tfrac{\pi}{2}), 0\right)^\top \tag{11.5}$$

$$\mathbf{c}_l = \mathbf{p} + R \left(\cos(\chi - \tfrac{\pi}{2}), \sin(\chi - \tfrac{\pi}{2}), 0\right)^\top. \tag{11.6}$$

To compute the path length of the different trajectories, we need a general equation for angular distances on a circle. Figure 11.8 shows the geometry for both clockwise (CW) and counter clockwise (CCW) circles.

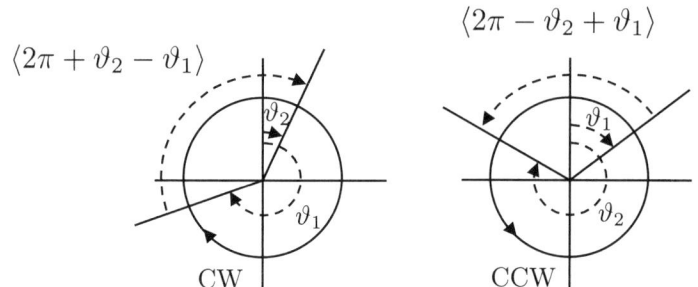

Figure 11.8 The angular distance between angles ϑ_1 and ϑ_2 for clockwise (CW) and counter clockwise (CCW) circles.

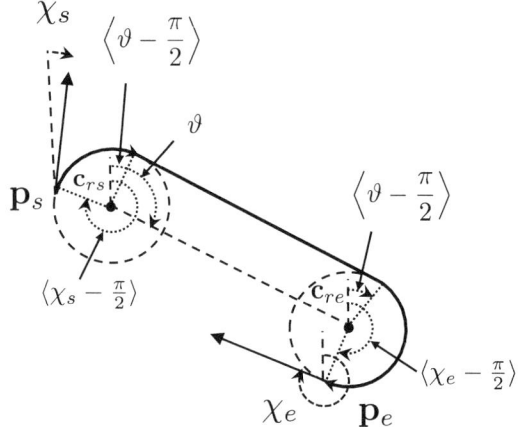

Figure 11.9 Dubins path, case I.

We will assume that both ϑ_1 and ϑ_2 are between 0 and 2π. For clockwise circles, the angular distance between ϑ_1 and ϑ_2 is given by

$$|\vartheta_2 - \vartheta_1|_{CW} \triangleq \langle 2\pi + \vartheta_2 - \vartheta_1 \rangle, \tag{11.7}$$

where

$$\langle \varphi \rangle \triangleq \varphi \mod 2\pi.$$

Similarly, for counter clockwise circles, we get

$$|\vartheta_2 - \vartheta_1|_{CCW} \triangleq \langle 2\pi - \vartheta_2 + \vartheta_1 \rangle. \tag{11.8}$$

Case I: R-S-R

The geometry for case I is shown in figure 11.9, where ϑ is the angle formed by the line between $\mathbf{c}_{rs}$ and $\mathbf{c}_{re}$. Using equation (11.7), the

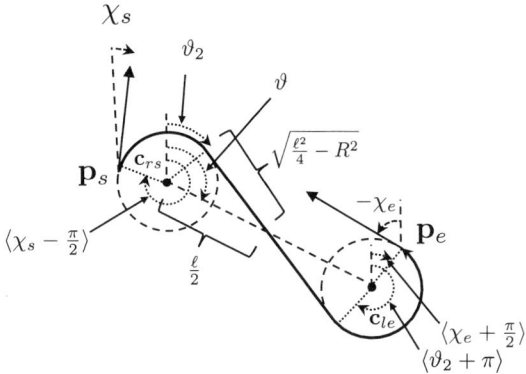

Figure 11.10 Dubins path, case II.

distance traveled along c_{rs} is given by

$$R \left\langle 2\pi + \left\langle \vartheta - \frac{\pi}{2} \right\rangle - \left\langle \chi_s - \frac{\pi}{2} \right\rangle \right\rangle.$$

Similarly, using equation (11.7), the distance traveled along c_{re} is given by

$$R \left\langle 2\pi + \left\langle \chi_e - \frac{\pi}{2} \right\rangle - \left\langle \vartheta - \frac{\pi}{2} \right\rangle \right\rangle.$$

The total path length for case I is therefore given by

$$L_1 = \|c_{rs} - c_{re}\| + R \left\langle 2\pi + \left\langle \vartheta - \frac{\pi}{2} \right\rangle - \left\langle \chi_s - \frac{\pi}{2} \right\rangle \right\rangle \\ + R \left\langle 2\pi + \left\langle \chi_e - \frac{\pi}{2} \right\rangle - \left\langle \vartheta - \frac{\pi}{2} \right\rangle \right\rangle. \tag{11.9}$$

Case II: R-S-L
The geometry for case II is shown in figure 11.10, where ϑ is the angle formed by the line between c_{rs} and c_{le}, $\ell = \|c_{le} - c_{rs}\|$, and

$$\vartheta_2 = \vartheta - \frac{\pi}{2} + \sin^{-1}\left(\frac{2R}{\ell}\right).$$

Using equation (11.7), the distance traveled along c_{rs} is given by

$$R \left\langle 2\pi + \langle \vartheta_2 \rangle - \left\langle \chi_s - \frac{\pi}{2} \right\rangle \right\rangle.$$

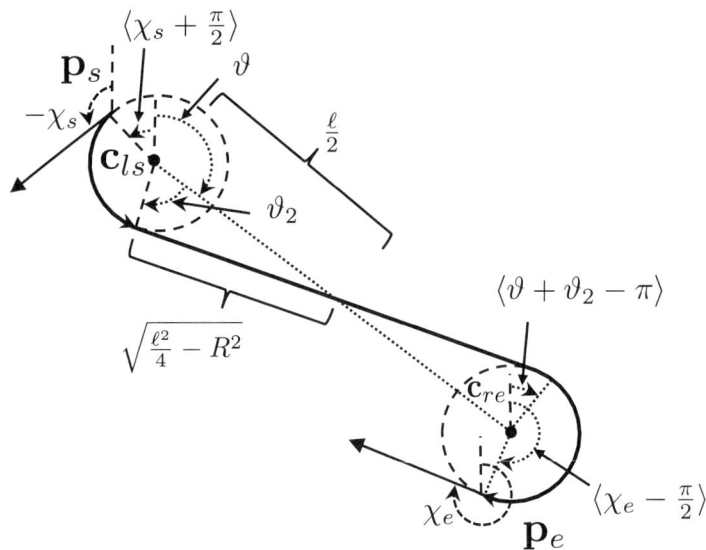

Figure 11.11 Dubins path, case III.

Similarly, using equation (11.8), the distance traveled along c_{le} is given by

$$R\left\langle 2\pi + \langle \vartheta_2 + \pi \rangle - \left\langle \chi_e + \frac{\pi}{2} \right\rangle \right\rangle.$$

The total path length for case II is therefore given by

$$L_2 = \sqrt{\ell^2 - 4R^2} + R\left\langle 2\pi + \langle \vartheta_2 \rangle - \left\langle \chi_s - \frac{\pi}{2} \right\rangle \right\rangle$$
$$+ R\left\langle 2\pi + \langle \vartheta_2 + \pi \rangle - \left\langle \chi_e + \frac{\pi}{2} \right\rangle \right\rangle. \tag{11.10}$$

Case III: L-S-R

The geometry for case III is shown in figure 11.11, where ϑ is the angle formed by the line between c_{ls} and c_{re}, $\ell = \|c_{re} - c_{ls}\|$, and

$$\vartheta_2 = \cos^{-1}\frac{2R}{\ell}.$$

Using equation (11.8), the distance traveled along c_{ls} is given by

$$R\left\langle 2\pi + \left\langle \chi_s + \frac{\pi}{2} \right\rangle - \langle \vartheta + \vartheta_2 \rangle \right\rangle.$$

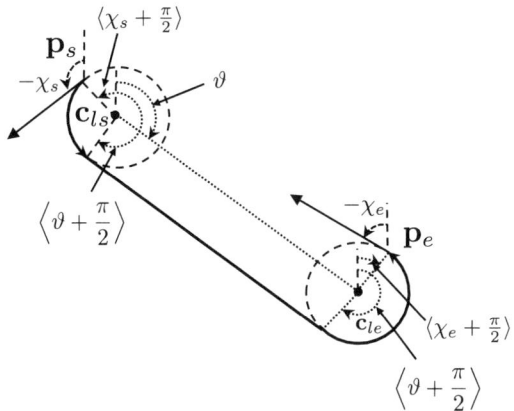

Figure 11.12 Dubins path, case IV.

Similarly, using equation (11.7), the distance traveled along c_{re} is given by

$$R \left\langle 2\pi + \left\langle \chi_e - \frac{\pi}{2} \right\rangle - \langle \vartheta + \vartheta_2 - \pi \rangle \right\rangle.$$

The total path length for case III is therefore given by

$$L_3 = \sqrt{\ell^2 - 4R^2} + R \left\langle 2\pi + \left\langle \chi_s + \frac{\pi}{2} \right\rangle - \langle \vartheta + \vartheta_2 \rangle \right\rangle$$
$$+ R \left\langle 2\pi + \left\langle \chi_e - \frac{\pi}{2} \right\rangle - \langle \vartheta + \vartheta_2 - \pi \rangle \right\rangle. \quad (11.11)$$

Case IV: L-S-L
The geometry for case IV is shown in figure 11.12, where ϑ is the angle formed by the line between c_{ls} and c_{le}. Using equation (11.8), the distance traveled along c_{ls} is given by

$$R \left\langle 2\pi + \left\langle \chi_s + \frac{\pi}{2} \right\rangle - \left\langle \vartheta + \frac{\pi}{2} \right\rangle \right\rangle.$$

Similarly, using equation (11.8), the distance traveled along c_{le} is given by

$$R \left\langle 2\pi + \left\langle \vartheta + \frac{\pi}{2} \right\rangle - \left\langle \chi_e + \frac{\pi}{2} \right\rangle \right\rangle.$$

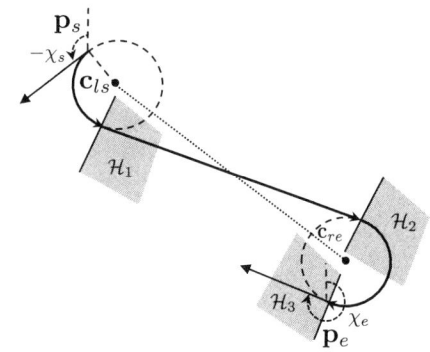

Figure 11.13 Definition of switching half planes for Dubins paths. The algorithm begins in a circular orbit and switches to straight-line tracking when $\mathcal{H}_1$ is entered. Orbit tracking is again initialized upon entering $\mathcal{H}_2$. The half plane $\mathcal{H}_3$ defines the end of the Dubins path.

The total path length for case IV is therefore given by

$$L_4 = \|\mathbf{c}_{ls} - \mathbf{c}_{le}\| + R \left\langle 2\pi + \left\langle \chi_s + \frac{\pi}{2} \right\rangle - \left\langle \vartheta + \frac{\pi}{2} \right\rangle \right\rangle$$
$$+ R \left\langle 2\pi + \left\langle \vartheta + \frac{\pi}{2} \right\rangle - \left\langle \chi_e + \frac{\pi}{2} \right\rangle \right\rangle. \qquad (11.12)$$

11.2.3 Algorithm for Tracking Dubins Paths

The guidance algorithm for tracking a Dubins path is shown graphically in figure 11.13 for case III. The algorithm is initialized in a left-handed orbit about $\mathbf{c}_{ls}$ and continues in that orbit until the MAV enters the half plane denoted as $\mathcal{H}_1$. After entering $\mathcal{H}_1$, a straight-line guidance strategy is used until the MAV enters the half plane denoted as $\mathcal{H}_2$. A right-handed orbit around $\mathbf{c}_{re}$ is then followed until the MAV enters the half plane denoted as $\mathcal{H}_3$, which defines the completion of the Dubins path.

It follows that a Dubins path can be parameterized by the start circle $\mathbf{c}_s$, the direction of the start circle λ_s, the end circle $\mathbf{c}_e$, the direction of the end circle λ_e, the parameters of the half plane $\mathcal{H}_1$ denoted as $\mathbf{z}_1$ and $\mathbf{q}_1$, the parameters of the half plane $\mathcal{H}_2$ denoted as $\mathbf{z}_2$ and $\mathbf{q}_2 = \mathbf{q}_1$, and the parameters of the half plane $\mathcal{H}_3$ denoted as $\mathbf{z}_3$ and $\mathbf{q}_3$. The parameters of the Dubins path associated with the start configuration $(\mathbf{p}_s, \chi_s)$, the end configuration $(\mathbf{p}_e, \chi_e)$, and the radius R are computed in algorithm 7. The length of the Dubins path L is also computed. The notation $\mathcal{R}_z(\vartheta)$ denotes the rotation matrix for a right-handed rotation of ϑ about the z-axis and $\vec{\mathbf{e}}_1 = (1, 0, 0)^\top$.

If we define the sequence of configurations

$$\mathcal{P} = \{(\mathbf{w}_1, \chi_1), (\mathbf{w}_2, \chi_2), \cdots, (\mathbf{w}_N, \chi_N)\}, \qquad (11.13)$$

then a guidance algorithm that follows Dubins paths between the configurations is given in algorithm 8. In line 4 the Dubins parameters

Algorithm 7 Find Dubins Parameters:
$(L, \mathbf{c}_s, \lambda_s, \mathbf{c}_e, \lambda_e, \mathbf{z}_1, \mathbf{q}_1, \mathbf{z}_2, \mathbf{z}_3, \mathbf{q}_3) =$
findDubinsParameters$(\mathbf{p}_s, \chi_s, \mathbf{p}_e, \chi_e, R)$

Input: Start configuration $(\mathbf{p}_s, \chi_s)$, End configuration $(\mathbf{p}_e, \chi_e)$, Radius R.
Require: $\|\mathbf{p}_s - \mathbf{p}_e\| \geq 3R$
Require: R is larger than minimum turn radius of MAV

1: $\mathbf{c}_{rs} \leftarrow \mathbf{p}_s + R\mathcal{R}_z\left(\frac{\pi}{2}\right)(\cos \chi_s, \sin \chi_s, 0)^\top$
2: $\mathbf{c}_{ls} \leftarrow \mathbf{p}_s + R\mathcal{R}_z\left(\frac{-\pi}{2}\right)(\cos \chi_s, \sin \chi_s, 0)^\top$
3: $\mathbf{c}_{re} \leftarrow \mathbf{p}_e + R\mathcal{R}_z\left(\frac{\pi}{2}\right)(\cos \chi_e, \sin \chi_e, 0)^\top$
4: $\mathbf{c}_{le} \leftarrow \mathbf{p}_e + R\mathcal{R}_z\left(-\frac{\pi}{2}\right)(\cos \chi_e, \sin \chi_e, 0)^\top$
5: Compute $L_1, L_2, L_3,$ and L_4 using equations (11.9) through (11.12).
6: $L \leftarrow \min\{L_1, L_2, L_3, L_4\}$
7: **if** $\arg\min\{L_1, L_2, L_3, L_4\} = 1$ **then**
8: $\quad \mathbf{c}_s \leftarrow \mathbf{c}_{rs}, \quad \lambda_s \leftarrow +1, \quad \mathbf{c}_e \leftarrow \mathbf{c}_{re}, \quad \lambda_e \leftarrow +1$
9: $\quad \mathbf{q}_1 \leftarrow \frac{\mathbf{c}_e - \mathbf{c}_s}{\|\mathbf{c}_e - \mathbf{c}_s\|}$
10: $\quad \mathbf{z}_1 \leftarrow \mathbf{c}_s + R\mathcal{R}_z\left(-\frac{\pi}{2}\right)\mathbf{q}_1$
11: $\quad \mathbf{z}_2 \leftarrow \mathbf{c}_e + R\mathcal{R}_z\left(-\frac{\pi}{2}\right)\mathbf{q}_1$
12: **else if** $\arg\min\{L_1, L_2, L_3, L_4\} = 2$ **then**
13: $\quad \mathbf{c}_s \leftarrow \mathbf{c}_{rs}, \quad \lambda_s \leftarrow +1, \quad \mathbf{c}_e \leftarrow \mathbf{c}_{le}, \quad \lambda_e \leftarrow -1$
14: $\quad \ell \leftarrow \|\mathbf{c}_e - \mathbf{c}_s\|$
15: $\quad \vartheta \leftarrow \text{angle}(\mathbf{c}_e - \mathbf{c}_s)$
16: $\quad \vartheta_2 \leftarrow \vartheta - \frac{\pi}{2} + \sin^{-1}\frac{2R}{\ell}$
17: $\quad \mathbf{q}_1 \leftarrow \mathcal{R}_z\left(\vartheta_2 + \frac{\pi}{2}\right)\mathbf{e}_1$
18: $\quad \mathbf{z}_1 \leftarrow \mathbf{c}_s + R\mathcal{R}_z(\vartheta_2)\mathbf{e}_1$
19: $\quad \mathbf{z}_2 \leftarrow \mathbf{c}_e + R\mathcal{R}_z(\vartheta_2 + \pi)\mathbf{e}_1$
20: **else if** $\arg\min\{L_1, L_2, L_3, L_4\} = 3$ **then**
21: $\quad \mathbf{c}_s \leftarrow \mathbf{c}_{ls}, \quad \lambda_s \leftarrow -1, \quad \mathbf{c}_e \leftarrow \mathbf{c}_{re}, \quad \lambda_e \leftarrow +1$
22: $\quad \ell \leftarrow \|\mathbf{c}_e - \mathbf{c}_s\|,$
23: $\quad \vartheta \leftarrow \text{angle}(\mathbf{c}_e - \mathbf{c}_s),$
24: $\quad \vartheta_2 \leftarrow \cos^{-1}\frac{2R}{\ell}$
25: $\quad \mathbf{q}_1 \leftarrow \mathcal{R}_z\left(\vartheta + \vartheta_2 - \frac{\pi}{2}\right)\mathbf{e}_1,$
26: $\quad \mathbf{z}_1 \leftarrow \mathbf{c}_s + R\mathcal{R}_z(\vartheta + \vartheta_2)\mathbf{e}_1,$
27: $\quad \mathbf{z}_2 \leftarrow \mathbf{c}_e + R\mathcal{R}_z(\vartheta + \vartheta_2 - \pi)\mathbf{e}_1$
28: **else if** $\arg\min\{L_1, L_2, L_3, L_4\} = 4$ **then**
29: $\quad \mathbf{c}_s \leftarrow \mathbf{c}_{ls}, \quad \lambda_s \leftarrow -1, \quad \mathbf{c}_e \leftarrow \mathbf{c}_{le}, \quad \lambda_e \leftarrow -1$
30: $\quad \mathbf{q}_1 \leftarrow \frac{\mathbf{c}_e - \mathbf{c}_s}{\|\mathbf{c}_e - \mathbf{c}_s\|},$
31: $\quad \mathbf{z}_1 \leftarrow \mathbf{c}_s + R\mathcal{R}_z\left(\frac{\pi}{2}\right)\mathbf{q}_1,$
32: $\quad \mathbf{z}_2 \leftarrow \mathbf{c}_e + R\mathcal{R}_z\left(\frac{\pi}{2}\right)\mathbf{q}_2$
33: **end if**
34: $\mathbf{z}_3 \leftarrow \mathbf{p}_e$
35: $\mathbf{q}_3 \leftarrow \mathcal{R}_z(\chi_e)\mathbf{e}_1$

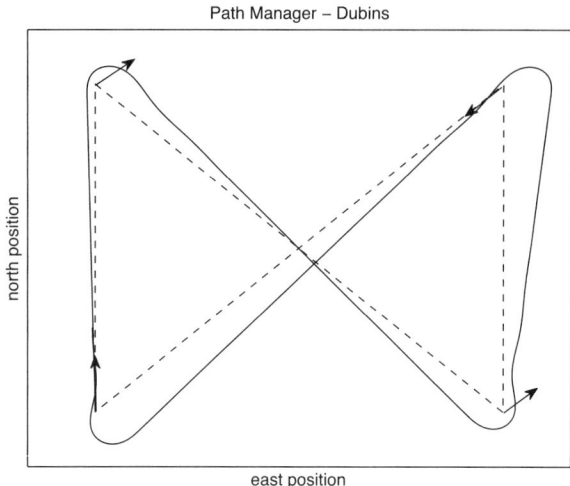

Figure 11.14 An example of the types of flight paths produced by algorithm 8.

are found for the current waypoint segment using algorithm 7. Since the initial configuration may be in the far side of the circle that is already in $\mathcal{H}_1$, the start circle is followed in state=1 until crossing into the part of circle opposite $\mathcal{H}_1$, as shown in lines 7-9. The start circle is then followed in state=2 until the MAV has crossed half plane $\mathcal{H}_1$, as shown in lines 10-13. After crossing into $\mathcal{H}_1$, the straight-line segment of the Dubins path is followed in state=3 as shown in lines 14-18. Line 16 tests to see if the MAV has crossed half plane $\mathcal{H}_2$. When it has, the end circle is followed in state=4 and state=5. Two states are again needed since the MAV may already be in $\mathcal{H}_3$ at the instant that it enters $\mathcal{H}_2$. If so, it follows the end circle until entering $\mathcal{H}_3$ as shown in lines 19-23. After the MAV has entered into $\mathcal{H}_3$, as detected in line 25, the waypoints are cycled and new Dubins parameters are found in lines 27-28. Figure 11.14 shows an example of a path generated using algorithm 8.

11.3 Chapter Summary

This chapter introduced several schemes for transitioning between waypoint configurations using the straight-line and orbit-following algorithms described in chapter 10. Section 11.1 discussed transitioning between waypoint segments using a half plane and by inserting a fillet between the waypoint segments. Section 11.2 introduced Dubins paths and showed how to construct Dubins paths between waypoint config-

Algorithm 8 Follow Waypoints with Dubins: (flag, $\mathbf{r}$, $\mathbf{q}$, $\mathbf{c}$, ρ, λ) = followWppDubins($\mathcal{P}$, $\mathbf{p}$, R)

Input: Configuration path $\mathcal{P} = \{(\mathbf{w}_1, \chi_1), \ldots, (\mathbf{w}_N, \chi_N)\}$, MAV position $\mathbf{p} = (p_n, p_e, p_d)^\top$, fillet radius R.
Require: $N \geq 3$
1: **if** New configuration path $\mathcal{P}$ is received **then**
2:     Initialize waypoint pointer: $i \leftarrow 2$, and state machine: state $\leftarrow 1$.
3: **end if**
4: $(L, \mathbf{c}_s, \lambda_s, \mathbf{c}_e, \lambda_e, \mathbf{z}_1, \mathbf{q}_1, \mathbf{z}_2, \mathbf{z}_3, \mathbf{q}_3) \leftarrow$ findDubinsParameters($\mathbf{w}_{i-1}, \chi_{i-1}, \mathbf{w}_i, \chi_i, R$)
5: **if** state $= 1$ **then**
6:     flag $\leftarrow 2, \mathbf{c} \leftarrow \mathbf{c}_s, \rho \leftarrow R, \lambda \leftarrow \lambda_s$
7:     **if** $\mathbf{p} \in \mathcal{H}(\mathbf{z}_1, -\mathbf{q}_1)$ **then**
8:         state $\leftarrow 2$
9:     **end if**
10: **else if** state $= 2$ **then**
11:     **if** $\mathbf{p} \in \mathcal{H}(\mathbf{z}_1, \mathbf{q}_1)$ **then**
12:         state $\leftarrow 3$
13:     **end if**
14: **else if** state $= 3$ **then**
15:     flag $\leftarrow 1, \mathbf{r} \leftarrow \mathbf{z}_1, \mathbf{q} \leftarrow \mathbf{q}_1$
16:     **if** $\mathbf{p} \in \mathcal{H}(\mathbf{z}_2, \mathbf{q}_1)$ **then**
17:         state $\leftarrow 4$
18:     **end if**
19: **else if** state $= 4$ **then**
20:     flag $\leftarrow 2, \mathbf{c} \leftarrow \mathbf{c}_e, \rho \leftarrow R, \lambda \leftarrow \lambda_e$
21:     **if** $\mathbf{p} \in \mathcal{H}(\mathbf{z}_3, -\mathbf{q}_3)$ **then**
22:         state $\leftarrow 5$
23:     **end if**
24: **else if** state $= 5$ **then**
25:     **if** $\mathbf{p} \in \mathcal{H}(\mathbf{z}_3, \mathbf{q}_3)$ **then**
26:         state $\leftarrow 1$
27:         $i \leftarrow (i + 1)$ until $i = N$.
28:         $(L, \mathbf{c}_s, \lambda_s, \mathbf{c}_e, \lambda_e, \mathbf{z}_1, \mathbf{q}_1, \mathbf{z}_2, \mathbf{z}_3, \mathbf{q}_3) \leftarrow$ findDubinsParameters($\mathbf{w}_{i-1}, \chi_{i-1}, \mathbf{w}_i, \chi_i, R$)
29:     **end if**
30: **end if**
31: **return** flag, $\mathbf{r}, \mathbf{q}, \mathbf{c}, \rho, \lambda$.

urations. In the next chapter, we will describe several path-planning algorithms that find waypoint paths and waypoint configurations in order to maneuver through an obstacle field.

Notes and References

Section 11.1 is based largely on [62]. Dubins paths were introduced in [61]. In certain degenerate cases, the Dubins path may not contain one of the three elements. For example, if the start and end configurations are on a straight line, then the beginning and end arcs will not be necessary. Or if the start and end configurations lie on a circle of radius R, then the straight line and end arc will not be necessary. In this chapter, we have ignored these degenerate cases. Reference [63] builds upon Dubins's ideas to generate feasible trajectories for UAVs given kinematic and path constraints by algorithmically finding the optimal location of Dubins circles and straight-line paths. In [64], Dubins circles are superimposed as fillets at the junction of straight-line waypoint paths produced from a Voronoi diagram. In some applications, like the cooperative-timing problem described in [60], it may be desirable to transition between waypoints in a way that preserves the path length. A path manager for this scenario is described in [62].

11.4 Design Project

The objective of this assignment is to implement algorithms 5 and 6 for following a set of waypoints denoted as $\mathcal{W}$, and algorithm 8 for following a set of configurations denoted as $\mathcal{P}$. The input to the path manager is either $\mathcal{W}$ or $\mathcal{P}$, and the output is the path definition

$$y_{\text{manager}} = \begin{pmatrix} \texttt{flag} \\ V_a^d \\ \mathbf{r} \\ \bar{\mathbf{q}} \\ \mathbf{c} \\ \rho \\ \lambda \end{pmatrix}.$$

Skeleton code for this chapter is given on the website.

11.1. Modify `path_manager_line.m` to implement algorithm 5 to follow the waypoint path defined in `path_planner_chap11.m`. Test and debug the algorithm on the guidance model given in equation (9.19). When the algorithm is working well on the guidance model, verify that it performs adequately for the full six-DOF model.

11.2. Modify `path_manager_fillet.m` and implement algorithm 6 to follow the waypoint path defined in `path_planner_chap11.m`. Test and debug the algorithm on the guidance model given in equation (9.19). When the algorithm is working well on the guidance model, verify that it performs adequately for the full six-DOF model.

11.3. Modify `path_manager_dubins.m` and implement algorithm 8 to follow the path configuration defined in `path_planner_chap11.m`. Test and debug the algorithm on the guidance model given in equation (9.19). When the algorithm is working well on the guidance model, verify that it performs adequately for the full six-DOF model.

12

Path Planning

In the robotics literature, there are essentially two different approaches to motion planning: *deliberative* motion planning, where explicit paths and trajectories are computed based on global world knowledge [65, 66, 67], and *reactive* motion planning, which uses behavioral methods to react to local sensor information [68, 69]. In general, deliberative motion planning is useful when the environment is known a priori, but can become computationally intensive in highly dynamic environments. Reactive motion planning, on the other hand, is well suited for dynamic environments, particularly collision avoidance, where information is incomplete and uncertain, but it lacks the ability to specify and direct motion plans.

This chapter focuses on deliberative path planning techniques that we have found to be effective and efficient for miniature air vehicles. In deliberative approaches, the MAV's trajectories are planned explicitly. The drawback of deliberative approaches is that they are strongly dependent upon the models used to describe the state of the world and the motion of the vehicle. Unfortunately, precise modeling of the atmosphere and the vehicle dynamics is not possible. To compensate for this inherent uncertainty, the path planning algorithms need to be executed on a regular basis in an outer feedback loop. It is essential, therefore, that the path planning algorithms be computationally efficient. To reduce the computational demand, we will use simple low-order navigation models for the vehicle and constant-wind models for the atmosphere. We assume that a terrain elevation map is available to the path planning algorithms. Obstacles that are known a priori are represented on the elevation map.

This chapter describes several simple and efficient path planning algorithms that are suitable for miniature air vehicles. The methods that we present are by no means exhaustive and might not be the best possible methods. However, we feel that they provide an accessible introduction to path planning. In reference to the architecture shown in figure 1.1, this chapter describes the design of the path planner. We will describe path planning algorithms for two types of problems. In section 12.1 we will address point-to-point problems, where the objective is to plan a waypoint path from one point to another through an obstacle field. In section 12.2 we will address coverage problems, where

Path Planning

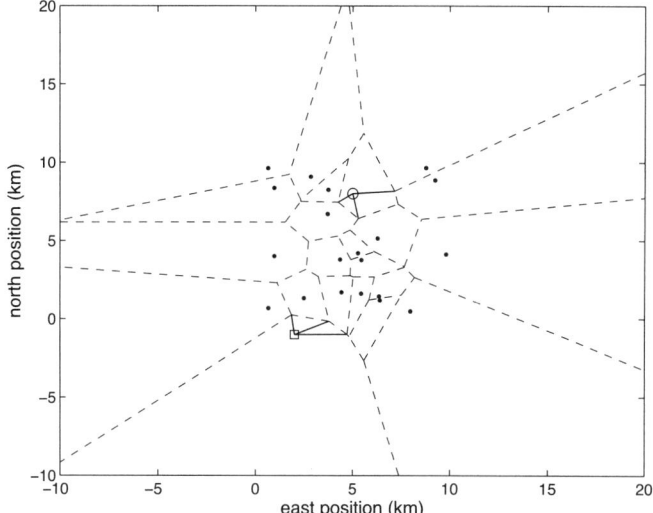

Figure 12.1 An example of a Voronoi graph with $Q = 20$ point obstacles.

the objective is to plan a waypoint path so that the MAV covers all of the area in a certain region. The output of the path planning algorithms developed in this chapter will be a sequence of either waypoints or configurations (waypoint plus orientation) and will therefore interface with the path management algorithms developed in chapter 11.

12.1 Point-to-Point Algorithms

12.1.1 Voronoi Graphs

The Voronoi graph is particularly well suited to applications that require the MAV to maneuver through a congested airspace with obstacles that are small relative to the turning radius of the vehicle. The relative size allows the obstacles to be modeled as points with zero area. The Voronoi method is essentially restricted to 2.5-D (or constant predefined altitude) path planning, where the altitude at each node is fixed in the map.

Given a finite set Q of points in $\mathbb{R}^2$, the Voronoi graph divides $\mathbb{R}^2$ into Q convex cells, each containing exactly one point in Q. The Voronoi graph is constructed so that the interior of each convex cell is closer to its associated point than to any other point in Q. An example of a Voronoi graph is shown in figure 12.1.

The key feature of the Voronoi graph that makes it useful for MAV path planning is that the edges of the graph are perpendicular bisectors between the points in Q. Therefore, following the edges of the Voronoi

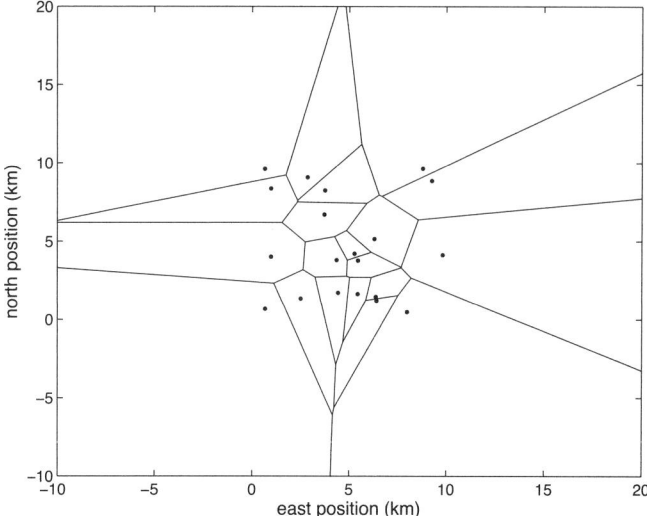

Figure 12.2 The Voronoi graph of $\mathcal{Q}$ is augmented with start and end nodes and with edges that connect the start and end notes to $\mathcal{Q}$.

graph potentially produces paths that avoid the points in $\mathcal{Q}$. However, figure 12.1 illustrates several potential pitfalls of using the Voronoi graph. First, graph edges that extend to infinity are obviously not good potential waypoint paths. Second, even for Voronoi cells with finite area, following the edges of the Voronoi graph may lead to unnecessarily long excursions. Finally, note that for the two points in the lower right-hand corner of figure 12.1, the Voronoi graph produces an edge between the two points; however, since the edge is so close to the points, the corresponding waypoint path may not be desirable.

There are well-established and widely available algorithms for generating Voronoi graphs. For example, Matlab has a built-in Voronoi function, and C++ implementations are publicly available on the Internet. Given the availability of Voronoi code, we will not discuss implementation of the algorithm. For additional discussions, see [70, 71, 72].

To use the Voronoi graph for point-to-point path planning, let $G = (V, E)$ be a graph produced by implementing the Voronoi algorithm on the set $\mathcal{Q}$. The node set V is augmented with the desired start and end locations as

$$V^+ = V \cup \{\mathbf{p}_s, \mathbf{p}_e\},$$

where $\mathbf{p}_s$ is the start position and $\mathbf{p}_e$ is the end position. The edge set E is then augmented with edges that connect the start and end nodes to the three closest nodes in V. The associated graph is shown in figure 12.2.

Path Planning

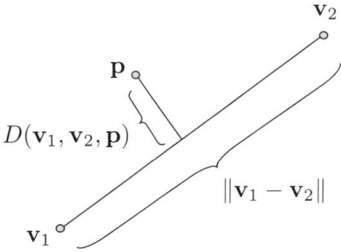

Figure 12.3 The cost penalty assigned to each edge of the Voronoi graph is proportional to the path length $\|\mathbf{v}_1 - \mathbf{v}_2\|$ and the reciprocal of the minimum distance from the path to a point in $\mathcal{Q}$.

The next step is to assign a cost to each edge in the Voronoi graph. Edge costs can be assigned in a variety of ways. For illustrative purposes, we will assume that the cost of traversing each path is a function of the path length and the distance from the path to points in $\mathcal{Q}$. The geometry for deriving the metric is shown in figure 12.3.

Let the nodes of the graph edge be denoted by $\mathbf{v}_1$ and $\mathbf{v}_2$. The length of the edge is given by $\|\mathbf{v}_1 - \mathbf{v}_2\|$. Any point on the line segment can be written as

$$\mathbf{w}(\sigma) = (1-\sigma)\mathbf{v}_1 + \sigma\mathbf{v}_2,$$

where $\sigma \in [0, 1]$. The minimum distance between $\mathbf{p}$ and the graph edge can be expressed as

$$D(\mathbf{v}_1, \mathbf{v}_2, \mathbf{p}) \triangleq \min_{\sigma \in [0,1]} \|\mathbf{p} - \mathbf{w}(\sigma)\|$$

$$= \min_{\sigma \in [0,1]} \sqrt{(\mathbf{p} - \mathbf{w}(\sigma))^\top (\mathbf{p} - \mathbf{w}(\sigma))}$$

$$= \min_{\sigma \in [0,1]} \sqrt{\begin{array}{l} \mathbf{p}^\top \mathbf{p} - 2(1-\sigma)\sigma \mathbf{p}^\top \mathbf{v}_1 - \sigma \mathbf{p}^\top \mathbf{v}_2 \\ + (1-\sigma)^2 \mathbf{v}_1^\top \mathbf{v}_1 + 2(1-\sigma)\sigma \mathbf{v}_1^\top \mathbf{v}_2 + \sigma^2 \mathbf{v}_2^\top \mathbf{v}_2 \end{array}}$$

$$= \min_{\sigma \in [0,1]} \sqrt{\begin{array}{l} \|\mathbf{p} - \mathbf{v}_1\|^2 + 2\sigma(\mathbf{p} - \mathbf{v}_1)^\top (\mathbf{v}_1 - \mathbf{v}_2) \\ + \sigma^2 \|\mathbf{v}_1 - \mathbf{v}_2\|^2 \end{array}}.$$

If σ is unconstrained, then its optimizing value is

$$\sigma^* = \frac{(\mathbf{v}_1 - \mathbf{p})^\top (\mathbf{v}_1 - \mathbf{v}_2)}{\|\mathbf{v}_1 - \mathbf{v}_2\|^2},$$

and

$$w(\sigma^*) = \sqrt{\|\mathbf{p}-\mathbf{v}_1\|^2 - \frac{((\mathbf{v}_1-\mathbf{p})^\top(\mathbf{v}_1-\mathbf{v}_2))^2}{\|\mathbf{v}_1-\mathbf{v}_2\|^2}}.$$

If we define

$$D'(\mathbf{v}_1, \mathbf{v}_2, \mathbf{p}) \triangleq \begin{cases} w(\sigma^*) & \text{if } \sigma^* \in [0, 1] \\ \|\mathbf{p}-\mathbf{v}_1\| & \text{if } \sigma^* < 0 \\ \|\mathbf{p}-\mathbf{v}_2\| & \text{if } \sigma^* > 1, \end{cases}$$

then the distance between the point set $\mathcal{Q}$ and the line segment $\overline{\mathbf{v}_1 \mathbf{v}_2}$ is given by

$$D(\mathbf{v}_1, \mathbf{v}_2, \mathcal{Q}) = \min_{\mathbf{p} \in \mathcal{Q}} D'(\mathbf{v}_1, \mathbf{v}_2, \mathbf{p}).$$

The cost for the edge defined by $(\mathbf{v}_1, \mathbf{v}_2)$ is assigned as

$$J(\mathbf{v}_1, \mathbf{v}_2) = k_1 \|\mathbf{v}_1 - \mathbf{v}_2\| + \frac{k_2}{D(\mathbf{v}_1, \mathbf{v}_2, \mathcal{Q})}, \tag{12.1}$$

where k_1 and k_2 are positive weights. The first term in equation (12.1) is the length of the edge, and the second term is the reciprocal of the distance from the edge to the closest point in $\mathcal{Q}$.

The final step is to search the Voronoi graph to determine the lowest cost path from the start node to the end node. There are numerous existing graph search techniques that might be appropriate to accomplish this task [72]. A well-known algorithm with readily available code is Dijkstra's algorithm [73], which has a computational complexity equal to $O(|V|)$. An example of a path found by Dijkstra's algorithm with $k_1 = 0.1$ and $k_2 = 0.9$ is shown in figure 12.4.

Pseudo-code for the Voronoi path planning method is listed in algorithm 9. If there are not a sufficient number of obstacle points in $\mathcal{Q}$, the resulting Voronoi graph will be sparse and could potentially have many edges extending to infinity. To avoid that situation, algorithm 9 requires that $\mathcal{Q}$ has at least 10 points. That number is, of course, arbitrary. In line 1 the Voronoi graph is constructed using a standard algorithm. In line 2 the start and end points are added to the Voronoi graph, and the edges between the start and end points and the closest nodes in $\mathcal{Q}$ are added in lines 3–4. Edge costs are assigned in lines 5–7 according to equation (12.1), and the waypoint path is determined via a Dijkstra search in line 8.

Path Planning

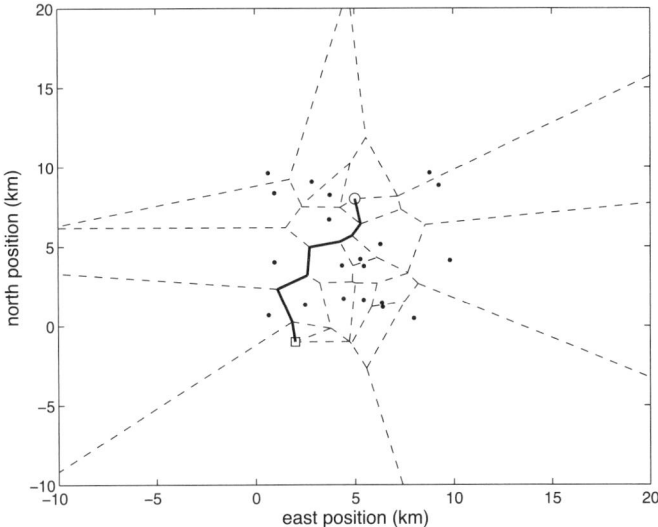

Figure 12.4 Optimal path through the Voronoi graph.

Algorithm 9 Plan Voronoi Path: $\mathcal{W} = \text{planVoronoi}(\mathcal{Q}, \mathbf{p}_s, \mathbf{p}_e)$

Input: Obstacle points $\mathcal{Q}$, start position $\mathbf{p}_s$, end position $\mathbf{p}_e$
Require: $|\mathcal{Q}| \geq 10$ Randomly add points if necessary.
1: $(V, E) = \text{constructVoronoiGraph}(\mathcal{Q})$
2: $V^+ = V \bigcup \{\mathbf{p}_s\} \bigcup \{\mathbf{p}_e\}$
3: Find $\{\mathbf{v}_{1s}, \mathbf{v}_{2s}, \mathbf{v}_{3s}\}$, the three closest points in V to $\mathbf{p}_s$, and $\{\mathbf{v}_{1e}, \mathbf{v}_{2e}, \mathbf{v}_{3e}\}$, the three closest points in V to $\mathbf{p}_e$
4: $E^+ = E \bigcup_{i=1,2,3}(\mathbf{v}_{is}, \mathbf{p}_s) \bigcup_{i=1,2,3}(\mathbf{v}_{ie}, \mathbf{p}_e)$
5: **for** Each element $(\mathbf{v}_a, \mathbf{v}_b) \in E$ **do**
6:    Assign edge cost $\mathbf{J}_{ab} = J(\mathbf{v}_a, \mathbf{v}_b)$ according to equation (12.1).
7: **end for**
8: $\mathcal{W} = \text{DijkstraSearch}(V^+, E^+, \mathbf{J})$
9: **return** $\mathcal{W}$

One of the disadvantages of the Voronoi method described in algorithm 9 is that it is limited to point obstacles. However, there are straightforward modifications for non-point obstacles. For example, consider the obstacle field shown in figure 12.5(a). A Voronoi graph can be constructed by first adding points around the perimeter of the obstacles that exceed a certain size, as shown in figure 12.5(b). The associated Voronoi graph, including connections to start and end nodes, is shown in figure 12.5(c). However, it is obvious from figure 12.5(c) that the Voronoi graph includes many infeasible links that are contained inside

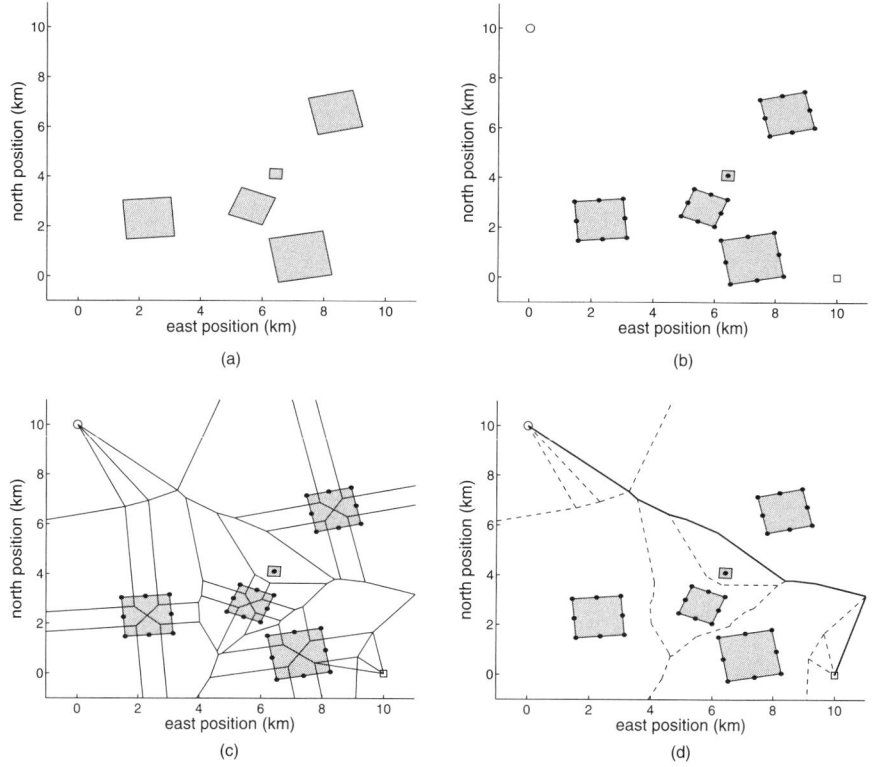

Figure 12.5 (a) An obstacle field with non-point obstacles. (b) The first step in using the Voronoi method to construct a path through the obstacle field is to insert points around the perimeter of the obstacles. (c) The resulting Voronoi graph includes many infeasible links contained inside the obstacles or that terminate on the obstacles. (d) When infeasible links are removed, the resulting graph can be used to plan paths through the obstacle field.

an obstacle or terminate on the obstacle. The final step is to remove the infeasible links, as shown in figure 12.5(d), which also displays the resulting optimal path.

12.1.2 Rapidly Exploring Random Trees

Another method for planning paths through an obstacle field from a start node to an end node is the Rapidly Exploring Random Tree (RRT) method. The RRT scheme is a random exploration algorithm that uniformly, but randomly, explores the search space. It has the advantage that it can be extended to vehicles with complicated non-linear dynamics. We assume throughout this section that obstacles are represented in a terrain map that can be queried to detect possible collisions.

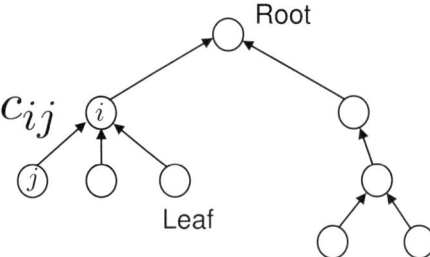

Figure 12.6 A tree is a special graph where every node, except the root, has exactly one parent.

The RRT algorithm is implemented using a data structure called a *tree*. A tree is a special case of a directed graph. Figure 12.6 is a graphical depiction of a tree. Edges in trees are directed from a child node to its parent. In a tree, every node has exactly one parent, except the root, which does not have any parents. In the RRT framework, the nodes represent physical states, or configurations, and the edges represent feasible paths between the states. The cost associated with each edge, c_{ij}, is the cost associated with traversing the feasible path between states represented by the nodes.

The basic idea of the RRT algorithm is to build a tree that uniformly explores the search space. The uniformity is achieved by randomly sampling from a uniform probability distribution. To illustrate the basic idea, let the nodes represent north-east locations at a constant altitude, and let the cost c_{ij} of the edges between nodes be the length of the straight-line path between the nodes.

Figure 12.7 depicts the basic RRT algorithm. As shown in figure 12.7(a), the input to the RRT algorithm is a start configuration $\mathbf{p}_s$, an end configuration $\mathbf{p}_e$, and the terrain map. The first step of the algorithm is to randomly select a configuration $\mathbf{p}$ in the workspace. As shown in figure 12.7(b), a new configuration $\mathbf{v}_1$ is selected a fixed distance D from $\mathbf{p}_s$ along the line $\overline{\mathbf{pp}_s}$, and inserted into the tree. At each subsequent step, a random configuration $\mathbf{p}$ is generated in the workspace, and the tree is searched to find the node that is closest to $\mathbf{p}$. As shown in figure 12.7(c), a new configuration is generated that is a distance D from the closest node in the tree, along the line connecting $\mathbf{p}$ to the closest node. Before a path segment is added to the tree, it needs to be checked for collisions with the terrain. If a collision is detected, as shown in figure 12.7(d), then the segment is deleted and the process is repeated. When a new node is added, its distance from the end node $\mathbf{p}_e$ is checked. If it is less than D, then a path segment from $\mathbf{p}_e$ is added to the tree, as shown in figure 12.7(f), indicating that a complete path through the terrain has been found.

Let $\mathcal{T}$ be the terrain map, and let $\mathbf{p}_s$ and $\mathbf{p}_e$ be the start and end configurations in the map. Algorithm 10 gives the basic RRT algorithm.

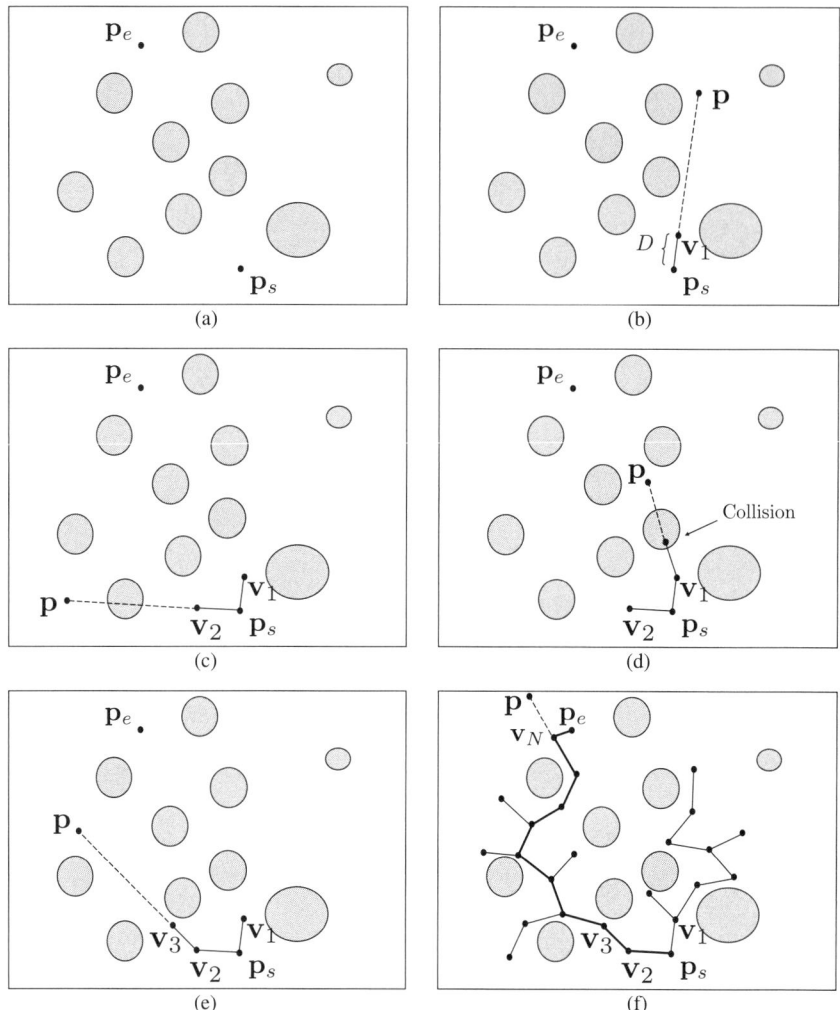

Figure 12.7 (a) The RRT algorithm is initialized with a terrain map and a start node and an end node. (b) and (c) The RRT graph is extended by randomly generating a point **p** in the terrain and planning a path of length D in the direction of **p**. (d) If the resulting configuration is not feasible, then it is not added to the RRT graph, and the process continues as shown in (e). (f) The RRT algorithm completes when the end node is added to the the RRT graph.

In line 1, the RRT graph G is initialized to contain only the start node. The while loop in lines 2–14 adds nodes to the RRT graph until the end node is included in the graph, indicating that a path from $\mathbf{p}_s$ to $\mathbf{p}_e$ has been found. In line 3 a random configuration is drawn from the terrain according to a uniform distribution over $\mathcal{T}$. Line 4 finds the closest node $\mathbf{v}^* \in G$ to the randomly selected point **p**. Since the distance between **p** and $\mathbf{v}^*$ may be large, line 5 plans a path of fixed length D from $\mathbf{v}^*$ in the direction of $\mathbf{p}$. The resulting configuration is denoted as $\mathbf{v}^+$. If the resulting path is feasible, as checked in line 6, then $\mathbf{v}^+$ is added to G in line 7 and the cost matrix is updated in line 8. The *if* statement in line 10 checks to see if the new node $\mathbf{v}^+$ can be connected directly to the end node $\mathbf{p}_e$. If so, $\mathbf{p}_e$ is added to G in line 11–12, and the algorithm ends in line 15 by returning the shortest waypoint path in G.

Algorithm 10 Plan RRT Path: $\mathcal{W} = \text{planRRT}(\mathcal{T}, \mathbf{p}_s, \mathbf{p}_e)$

Input: Terrain map $\mathcal{T}$, start configuration $\mathbf{p}_s$, end configuration $\mathbf{p}_e$
1: Initialize RRT graph $G = (V, E)$ as $V = \{\mathbf{p}_s\}, E = \emptyset$
2: **while** The end node $\mathbf{p}_e$ is not connected to G, i.e., $\mathbf{p}_e \notin V$ **do**
3:     $\mathbf{p} \leftarrow \text{generateRandomConfiguration}(\mathcal{T})$
4:     $\mathbf{v}^* \leftarrow \text{findClosestConfiguration}(\mathbf{p}, V)$
5:     $\mathbf{v}^+ \leftarrow \text{planPath}(\mathbf{v}^*, \mathbf{p}, D)$
6:     **if** existFeasiblePath$(\mathcal{T}, \mathbf{v}^*, \mathbf{v}^+)$ **then**
7:         Update graph $G = (V, E)$ as $V \leftarrow V \bigcup \{\mathbf{v}^+\}, E \leftarrow E \bigcup \{(\mathbf{v}^*, \mathbf{v}^+)\}$
8:         Update edge costs as $C[(\mathbf{v}^*, \mathbf{v}^+)] \leftarrow \text{pathLength}(\mathbf{v}^*, \mathbf{v}^+)$
9:     **end if**
10:    **if** existFeasiblePath$(\mathcal{T}, \mathbf{v}^+, \mathbf{p}_e)$ **then**
11:        Update graph $G = (V, E)$ as $V \leftarrow V \bigcup \{\mathbf{p}_e\}\ E \leftarrow E \bigcup \{(\mathbf{v}^*, \mathbf{p}_e)\}$
12:        Update edge costs as $C[(\mathbf{v}^*, \mathbf{p}_e)] \leftarrow \text{pathLength}(\mathbf{v}^*, \mathbf{p}_e)$
13:    **end if**
14: **end while**
15: $\mathcal{W} = \text{findShortestPath}(G, C)$.
16: **return** $\mathcal{W}$

The result of implementing algorithm 10 for four different randomly generated obstacle fields and randomly generated start and end nodes is displayed with a dashed line in figure 12.8. Note that the paths generated by Algorithm 10 sometimes wander needlessly and that eliminating some nodes may result in a more efficient path. Algorithm 11 gives a simple scheme for smoothing the paths generated by algorithm 10. The basic idea is to remove intermediate nodes if a feasible path still exists. The result of applying algorithm 11 is shown with a solid line in figure 12.8.

There are numerous extensions to the basic RRT algorithm. A common extension, which is discussed in [74], is to extend the tree from both the start and the end nodes and, at the end of each extension, to attempt to connect the two trees. In the next two subsections, we

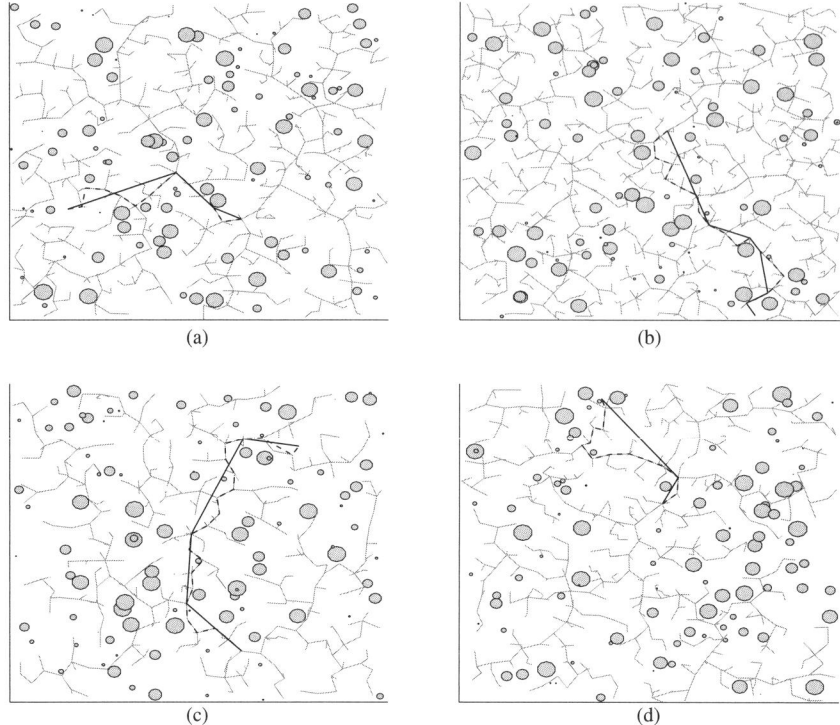

Figure 12.8 The results of algorithm 10 for four randomly generated obstacle fields and randomly generated start and end nodes are indicated by dashed lines. The smoothed paths generated by algorithm 11 are indicated by the solid lines.

will give two simple extensions that are useful for MAV applications: waypoint planning over 3-D terrain and using Dubins paths to plan kinematically feasible paths in complex 2-D terrain.

RRT Waypoint Planning over 3-D Terrain
In this section we will consider the extension of the basic RRT algorithm to planning waypoint paths over 3-D terrain. We will assume that the terrain $\mathcal{T}$ can be queried for the altitude of the terrain at any north-east position. The primary question that must be answered to extend the basic RRT algorithm to 3-D is how to generate the altitude at random nodes. For example, one option is to randomly select the altitude as a uniform distribution of height-above-ground, up to a maximum limit. Another option is to pre-select several altitude levels and then randomly select one of these levels.

In this section, we will select the altitude as a fixed height-above-ground h_{AGL}. Therefore, the (unsmoothed) RRT graph will, in essence, be a 2-D graph that follows the contour of the terrain. The output of

> **Algorithm 11** Smooth RRT Path: $(\mathcal{W}_s, C_s) = \text{smoothRRT}(\mathcal{T}, \mathcal{W}, C)$
>
> **Input:** Terrain map $\mathcal{T}$, waypoint path $\mathcal{W} = \{\mathbf{w}_1, \ldots, \mathbf{w}_N\}$, cost matrix C
>
> 1: Initialized smoothed path $\mathcal{W}_s \leftarrow \{\mathbf{w}_1\}$
> 2: Initialize pointer to current node in $\mathcal{W}_s$: $i \leftarrow 1$
> 3: Initialize pointer to next node in $\mathcal{W}$: $j \leftarrow 2$
> 4: **while** $j < N$ **do**
> 5: $\mathbf{w}_s \leftarrow \text{getNode}(\mathcal{W}_s, i)$
> 6: $\mathbf{w}^+ \leftarrow \text{getNode}(\mathcal{W}, j+1)$
> 7:   **if** existFeasiblePath$(\mathcal{T}, \mathbf{w}_s, \mathbf{w}^+) = \text{FALSE}$ **then**
> 8:   Get last node: $\mathbf{w} \leftarrow \text{getNode}(\mathcal{W}, j)$
> 9:   Add deconflicted node to smoothed path: $\mathcal{W}_s \leftarrow \mathcal{W}_s \bigcup \{\mathbf{w}\}$
> 10:   Update smoothed cost: $C_s[(\mathbf{w}_s, \mathbf{w})] \leftarrow \text{pathLength}(\mathbf{w}_s, \mathbf{w})$
> 11: $i \leftarrow i + 1$
> 12:   **end if**
> 13: $j \leftarrow j + 1$
> 14: **end while**
> 15: Add last node from $\mathcal{W}$: $\mathcal{W}_s \leftarrow \mathcal{W}_s \bigcup \{\mathbf{w}_N\}$
> 16: Update smoothed cost: $C_s[(\mathbf{w}_i, \mathbf{w}_N)] \leftarrow \text{pathLength}(\mathbf{w}_i, \mathbf{w}_N)$
> 17: **return** $\mathcal{W}_s$

algorithm 10 will be a path that follows the terrain at a fixed altitude h_{AGL}. However, the smoothing step represented by algorithm 11 will result in paths with much less altitude variation. For 3-D terrain, the climb rate and descent rates of the MAV are usually constrained to be within certain limits. The function existFeasiblePath in algorithms 10 and 11 can be modified to ensure that the climb and descent rates are satisfied and that terrain collisions are avoided.

The results of the 3-D RRT algorithm are shown in figures 12.9 and 12.10, where the thin lines represent the RRT tree, the thick dashed line is the RRT path generated by algorithm 10, and the thick solid line is the smoothed path generated by algorithm 11.

RRT Dubins Path Planning in a 2-D Obstacle Field
In this section, we will consider the extension of the basic RRT algorithm to planning paths subject to turning constraints. Assuming that the vehicle moves at constant velocity, optimal paths between configurations are given by Dubins paths, as discussed in section 11.2. Dubins paths are planned between two different configurations, where a configuration is given by three numbers representing the north and east positions and the course angle at that position. To apply the RRT algorithm to this scenario, we need to have a technique for generating random configurations.

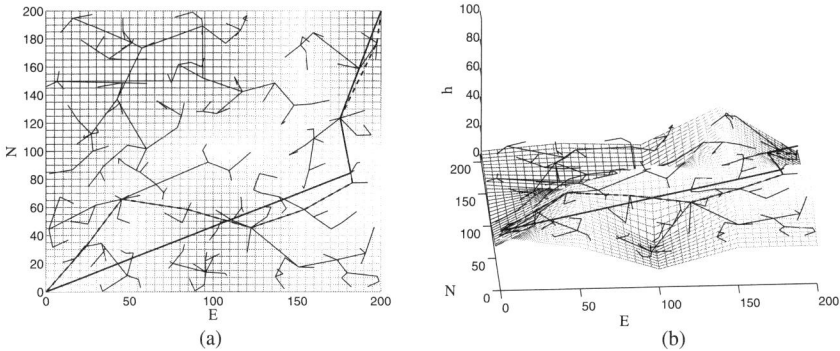

Figure 12.9 (a) Overhead view of the results of the 3-D RRT waypoint path planning algorithm. (b) Side view of the results of the 3-D RRT waypoint path planning algorithm. The thin lines are the RRT graph, the thick dotted line is the RRT path returned by algorithm 10, and the thick solid line is the smoothed path. The RRT graph is generated at a fixed height above the terrain.

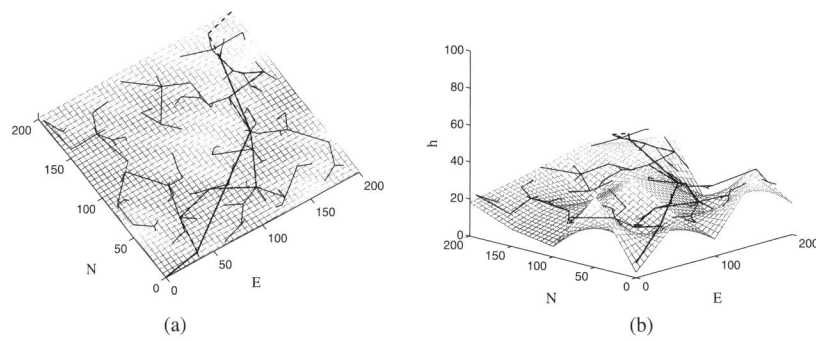

Figure 12.10 (a) Overhead view of the results of the 3-D RRT waypoint path planning algorithm. (b) Side view of the results of the 3-D RRT waypoint path planning algorithm. The thin lines are the RRT graph, the thick dashed line is the RRT path returned by algorithm 10, and the thick solid line is the smoothed path. The RRT graph is generated at a fixed height above the terrain.

We will generate a random configuration as follows:

1. Generate a random north-east position in the environment.
2. Find the closest node in the RRT graph to the new point.
3. Select a position of distance L from the closest RRT node, and use that position as the north-east coordinates of the new configuration.
4. Select the course angle for the configuration as the angle of the line that connects the new configuration to the RRT tree.

Path Planning

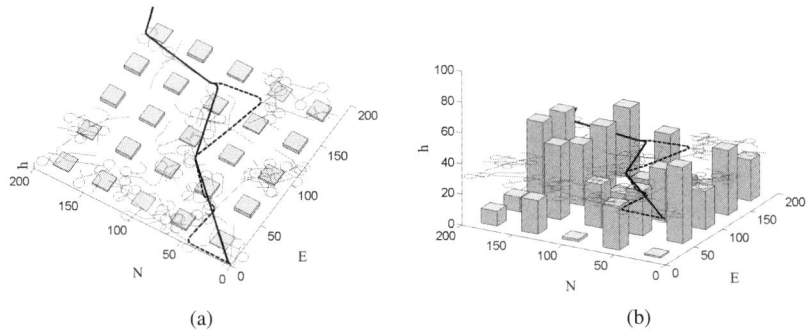

Figure 12.11 (a) Overhead view of the results of the RRT Dubins path planning algorithm. (b) Side view of the results of the RRT Dubins path planning algorithm. The thin lines are the RRT graph, the thick dashed line is the RRT path returned by algorithm 10, and the thick solid line is the smoothed path. The RRT graph is generated at a fixed height above the terrain.

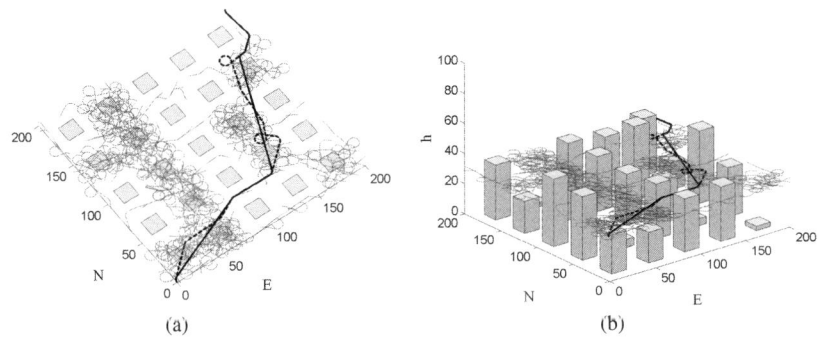

Figure 12.12 (a) Overhead view of the results of the 3-D RRT waypoint path planning algorithm. (b) Side view of the results of the 3-D RRT waypoint path planning algorithm. The thin lines are the RRT graph, the thick dashed line is the RRT path returned by algorithm 10, and the thick solid line is the smoothed path. The RRT graph is generated at a fixed height above the terrain.

The RRT algorithm is then implemented as described in algorithm 10, where the function pathLength returns the length of the Dubins path between configurations.

The results of the RRT Dubins algorithm are shown in figures 12.11 and 12.12, where the thin lines represent the RRT tree, the thick dashed line is the RRT path generated by algorithm 10, and the thick solid line is the smoothed path generated by algorithm 11.

12.2 Coverage Algorithms

In this section we will briefly discuss coverage algorithms, where the objective is not to transition from a start configuration to an end configuration, but rather to cover as much area as possible. Coverage algorithms are used, for example, in search problems, where the air vehicle is searching for objects of interest within a given region. Since the location of the object may be unknown, the region must be searched as uniformly as possible. The particular algorithm that we present in this section allows for prior information about possible locations of objects to be included.

The basic idea is to maintain two maps in memory: the terrain map and the return map. The terrain map is used to detect possible collisions with the environment. We will model the benefit of being at a particular location in the terrain by a return value Υ_i, where i indexes the location in the terrain. To ensure uniform coverage of an area, the return map is initialized so that all locations in the terrain have the same initial return value. As locations are visited, the return value is decremented by a fixed amount according to

$$\Upsilon_i[k] = \Upsilon_i[k-1] - c, \tag{12.2}$$

where c is a positive constant. The path planner searches for paths that provide the largest possible return over a finite look-ahead window.

The basic coverage algorithm is listed in algorithm 12. At each step of the algorithm, a look-ahead tree is generated from the current MAV configuration and used to search for regions of the terrain with a large return value. In line 1 the look-ahead tree is initialized to the start configuration $\mathbf{p}_s$, and in line 8 the look-ahead tree is reset to the current configuration. More advanced algorithms could be designed to retain portions of the look-ahead tree that have already been explored. The return map is initialized in line 2. We initialize the return map to be a large constant number plus additive noise. The additive noise facilitates choices in the initial stages of the algorithm, where all regions of the terrain need to be searched and, therefore, produce equal return values. After the MAV has moved into a new region of the terrain, the return map Υ is updated in line 9 according to equation 12.2. In line 5 the look-ahead tree is generated, starting from the current configuration. The look-ahead tree is generated in a way that avoids obstacles in the terrain. In line 6 the look-ahead tree G is searched for the the path that produces the largest return value.

There are several techniques that can be used to generate the look-ahead tree in line 5 of algorithm 12; we will briefly describe two possible

Path Planning

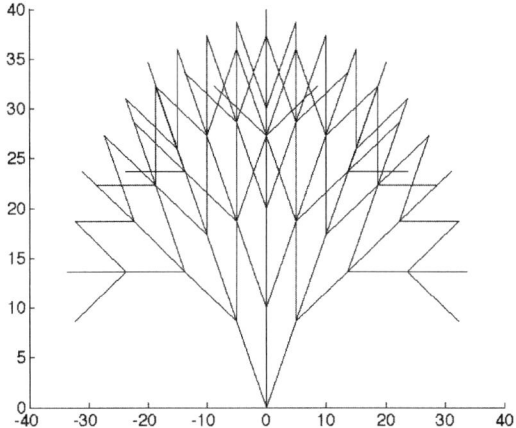

Figure 12.13 A look-ahead tree of depth three is generated from the position $(0, 0)$, where $L = 10$, and $\vartheta = \pi/6$.

methods. In the first method, the path is given by waypoints, and the look-ahead tree is generated by taking finite steps of length L and, at the end of each step, by allowing the MAV to move straight or change heading by $\pm\vartheta$ at each configuration. A three-step look-ahead tree where $\vartheta = \pi/6$ is shown in figure 12.13.

Algorithm 12 Plan Cover Path: planCover($\mathcal{T}$, Υ, **p**)

Input: Terrain map $\mathcal{T}$, return map Υ, initial configuration $\mathbf{p}_s$
1: Initialize look-ahead tree $G = (V, E)$ as $V = \{\mathbf{p}_s\}$, $E = \emptyset$
2: Initialize return map $\Upsilon = \{\Upsilon_i : i$ indexes the terrain$\}$
3: $\mathbf{p} = \mathbf{p}_s$
4: **for** Each planning cycle **do**
5: $\quad G = $ generateTree($\mathbf{p}$, $\mathcal{T}$, Υ)
6: $\quad W = $ highestReturnPath(G)
7: $\quad$ Update **p** by moving along the first segment of W
8: $\quad$ Reset $G = (V, E)$ as $V = \{\mathbf{p}\}$, $E = \emptyset$
9: $\quad \Upsilon = $ updateReturnMap(Υ, **p**)
10: **end for**

The results of using a waypoint look-ahead tree in algorithm 12 are shown in figure 12.14. figure 12.14(a) shows paths through an obstacle field where the look-ahead length is $L = 5$, the allowed heading change

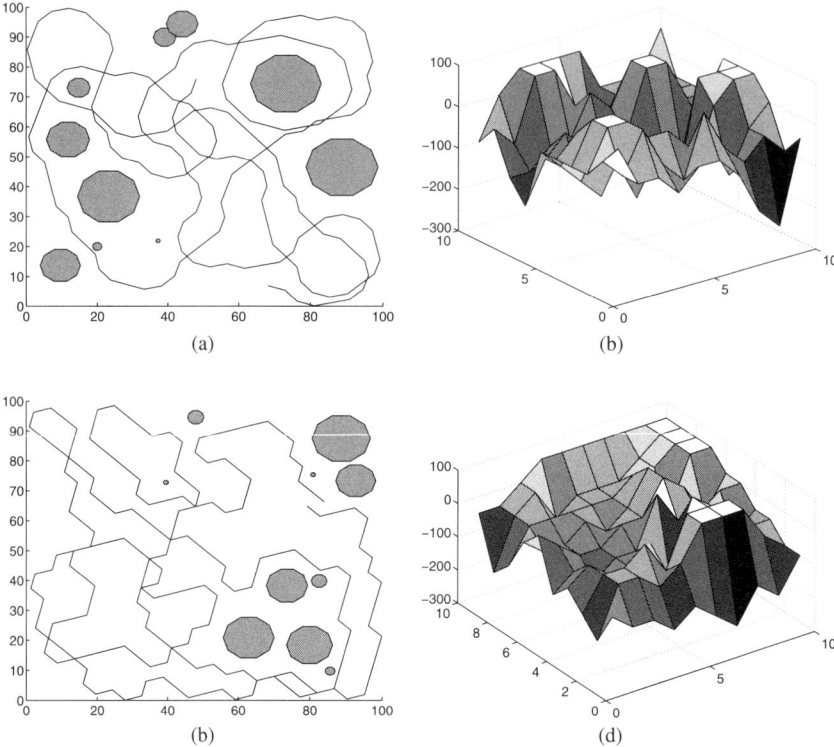

Figure 12.14 Plots (a) and (c) provide an overhead view of the results of the coverage algorithm using a three-angle expansion tree. The associated return maps after 200 planning cycles are shown in plots (b) and (d). In (a) and (b), the allowed heading change is $\vartheta = \pi/6$, and in (c) and (d), the allowed heading change is $\vartheta = \pi/3$.

at each step is $\vartheta = \pi/6$, and the depth of the look-ahead tree is three. The associated return map after 200 iterations of the algorithm is shown in figure 12.14(a). figure 12.14(b) shows paths through an obstacle field where the look-ahead length is $L = 5$, the allowed heading change at each step is $\vartheta = \pi/3$, and the depth of the look-ahead tree is three. The associated return map after 200 iterations of the algorithm is shown in figure 12.14(c). Note that the area is approximately uniformly covered, but that paths through regions are often repeated, especially through tight regions.

Another method that can be used to generate the look-ahead tree in line 5 of algorithm 12 is an RRT algorithm using Dubins paths. Given the current configuration, N steps of the RRT tree expansion algorithm are used to generate the look-ahead tree. Results using this algorithm in a 3-D urban environment are shown in figure 12.15. Figures 12.15(a)

Path Planning 223

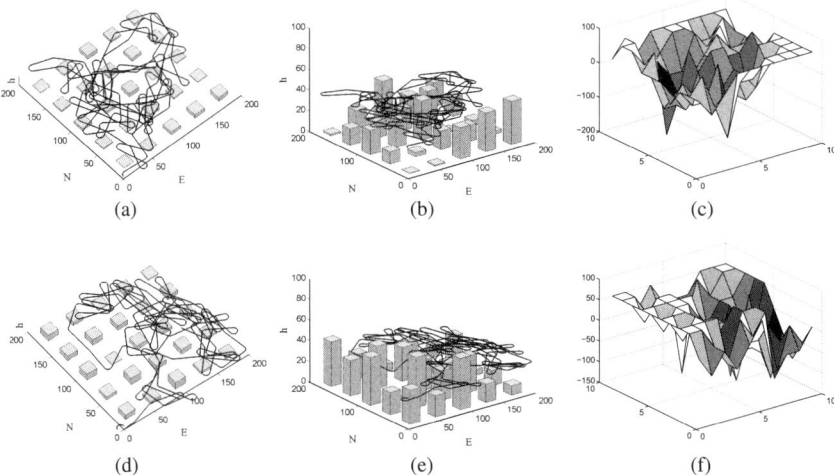

Figure 12.15 Plots (a) and (d) provide an overhead view of the results of the coverage algorithm using the RRT planner to find new regions using Dubins paths between configurations. Plots (b) and (e) show side views of the results. Plots (c) and (f) show the return map after 100 iterations of the search algorithm.

and 12.15(b) show overhead views of two different instances of the algorithm. The altitude of the MAV is fixed at a certain height, so buildings prevent certain regions from being searched. A side view of the results are shown in figures 12.15(b) and 12.15(e) to provide the 3-D perspective. The associated return maps are shown in figures 12.15(c) and 12.15(f). The results again show that the area is covered fairly uniformly, but it also highlights that the coverage algorithm is not particularly efficient.

12.3 Chapter Summary

This chapter has provided a brief introduction to path planning methods for small unmanned aircraft. The algorithms presented in this chapter are not intended to be complete or even the best algorithms for MAVs. Rather we have selected algorithms that are easy to understand and implement and that provide a good springboard for further research. We have presented two classes of algorithms: point-to-point algorithms for planing paths between two configurations, and coverage algorithms for planning paths that uniformly cover a region, given the constraints of the obstacle field. Our primary focus has been the use of the rapidly exploring random tree (RRT) algorithm using Dubins paths between nodes.

Notes and References

There is extensive literature on path planning methods, including the textbooks by Latombe [75], Choset, et al. [76], and LaValle [77], which contain thorough reviews of related path planning research. An introduction to the Voronoi graph is contained in [71] and early applications of Voronoi techniques to UAV path planning are described in [64, 78, 79, 80]. Incremental construction of Voronoi graphs based on sensor measurements is described in [81, 82]. An effective search technique for Voronoi diagrams providing multiple path options is the Eppstein's *k*-shortest paths algorithm [83].

The RRT algorithm was first introduced in [84] and applied to non-holonomic robotic vehicles in [85, 74]. There are numerous applications of RRTs reported in the literature, as well as extensions to the basic algorithms [86]. A recent extension to the RRT algorithm that converges with probability one to the optimal path is described in [87]. The RRT algorithm is closely related to the probabilistic roadmap technique described in [88], which is applied to UAVs in [89].

There are several coverage algorithms discussed in the literature. Reference [90] describes a coverage algorithm that plans paths such that the robot passes over all points in its free space and also includes a nice survey of other coverage algorithms reported in the literature. A coverage algorithm in the presence of moving obstacles is described in [91]. Multiple vehicle coverage algorithms are discussed in [92], and coverage algorithms in the context of mobile sensor networks are described in [93, 94].

12.4 Design Project

The objective of this assignment is to implement several of the path planning algorithms described in this chapter. Skeleton code for this chapter is given on the website. The file `createWorld.m` creates a map similar to those shown in figures 12.11 and 12.12. The file `drawEnvironment.m` draws the map, the waypoint path, and the current straight-line or orbit that is being followed. The file `path_planner.m` contains a switch statement for manually choosing between different path planning algorithms. The sample code contains the file `planRRT.m` for planning straight line paths through a synthetic urban terrain.

> 12.1. Using `planRRT.m` as a template, create `planRRTDubins.m` and modify `path_planner.m` so that it calls `planRRTDubins.m` to plan Dubins paths through the map. Modify `planRRTDubins.m` to implement the RRT

algorithm based on Dubins paths and the associated smoothing algorithm. Test and debug the algorithm on the guidance model given in equation (9.19). When the algorithm is working well on the guidance model, verify that it performs adequately for the full six-DOF model.

12.2. Using `planCover.m` as a template, create `planCoverRRTDubins.m` and modify `path_planner.m` so that it calls the function `planRRTCoverDubins.m`. Modify `planCoverRRTDubins.m` to implement the coverage algorithm described in algorithm 12 where the motion of the vehicle is based on Dubins paths. Test and debug the algorithm on the guidance model given in equation (9.19). When the algorithm is working well on the guidance model, verify that it performs adequately for the full six-DOF model.

13

Vision-guided Navigation

One of the primary reasons for the current interest in small unmanned aircraft is that they offer an inexpensive platform to carry electro-optical (EO) and infrared (IR) cameras. Almost all small and miniature air vehicles that are currently deployed carry either an EO or IR camera. While the camera's primary use is to relay information to a user, it makes sense to also attempt to use the camera for the purpose of navigation, guidance, and control. Further motivation comes from the fact that birds and flying insects use vision as their primary guidance sensor [95].

This chapter briefly introduces some of the issues that arise in vision-based guidance and control of MAVs. In section 13.1 we revisit coordinate frame geometry and expand upon the discussion in chapter 2 by introducing the gimbal and camera frames. We also discuss the image plane and the projective geometry that relates the position of 3-D objects to their 2-D projection on the image plane. In section 13.2 we give a simple algorithm for pointing a pan-tilt gimbal at a known world coordinate. In section 13.3 we describe a geolocation algorithm that estimates the position of a ground-based target based on the location and motion of the target in the video sequence. In this chapter we will assume that an algorithm exists for tracking the features of a target in the video sequence. The motion of the target on the image plane is influenced by both the target motion and by the translational and rotational motion of the aircraft. In section 13.4 we describe a method that compensates for the apparent target motion that is induced by gimbal movement and angular rates of the air platform. As a final application of vision-based guidance, section 13.6 describes an algorithm that uses vision to land accurately at a user-specified location on the ground.

13.1 Gimbal and Camera Frames and Projective Geometry

In this section we will assume that the origins of the gimbal and camera frames are located at the center of mass of the vehicle. For more general geometry, see [96]. Figure 13.1 shows the relationship between the vehicle and body frames of the MAV and the gimbal and camera frames. There are three frames of interest: the gimbal-1 frame denoted by $\mathcal{F}^{g1} = (\mathbf{i}^{g1}, \mathbf{j}^{g1}, \mathbf{k}^{g1})$, the gimbal frame denoted by $\mathcal{F}^{g} = (\mathbf{i}^{g}, \mathbf{j}^{g}, \mathbf{k}^{g})$,

Vision-guided Navigation

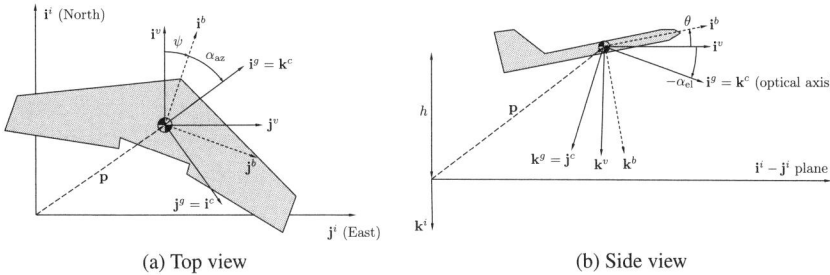

(a) Top view   (b) Side view

Figure 13.1 A graphic showing the relationship between the gimbal and camera frames and the vehicle and body frames.

and the camera frame denoted by $\mathcal{F}^c = (\mathbf{i}^c, \mathbf{j}^c, \mathbf{k}^c)$. The gimbal-1 frame is obtained by rotating the body frame about the $\mathbf{k}^b$ axis by an angle of α_{az}, which is called the gimbal azimuth angle. The rotation from the body to the gimbal-1 frame is given by

$$\mathcal{R}_b^{g1}(\alpha_{az}) \triangleq \begin{pmatrix} \cos\alpha_{az} & \sin\alpha_{az} & 0 \\ -\sin\alpha_{az} & \cos\alpha_{az} & 0 \\ 0 & 0 & 1 \end{pmatrix}. \quad (13.1)$$

The gimbal frame is obtained by rotating the gimbal-1 frame about the $\mathbf{j}^{g1}$ axis by an angle of α_{el}, which is called the gimbal elevation angle. Note that a negative elevation angle points the camera toward the ground. The rotation from the gimbal-1 frame to the gimbal frame is given by

$$\mathcal{R}_{g1}^{g}(\alpha_{el}) \triangleq \begin{pmatrix} \cos\alpha_{el} & 0 & -\sin\alpha_{el} \\ 0 & 1 & 0 \\ \sin\alpha_{el} & 0 & \cos\alpha_{el} \end{pmatrix}. \quad (13.2)$$

The rotation from the body to the gimbal frame is, therefore, given by

$$\mathcal{R}_b^g = \mathcal{R}_{g1}^g \mathcal{R}_b^{g1} = \begin{pmatrix} \cos\alpha_{el}\cos\alpha_{az} & \cos\alpha_{el}\sin\alpha_{az} & -\sin\alpha_{el} \\ -\sin\alpha_{az} & \cos\alpha_{az} & 0 \\ \sin\alpha_{el}\cos\alpha_{az} & \sin\alpha_{el}\sin\alpha_{az} & \cos\alpha_{el} \end{pmatrix}. \quad (13.3)$$

The literature in computer vision and image processing traditionally aligns the coordinate axis of the camera such that $\mathbf{i}^c$ points to the right in the image, $\mathbf{j}^c$ points down in the image, and $\mathbf{k}^c$ points along the optical axis. It follows that the transformation from the gimbal frame to the camera frame is given by

$$\mathcal{R}_g^c = \begin{pmatrix} 0 & 1 & 0 \\ 0 & 0 & 1 \\ 1 & 0 & 0 \end{pmatrix}. \quad (13.4)$$

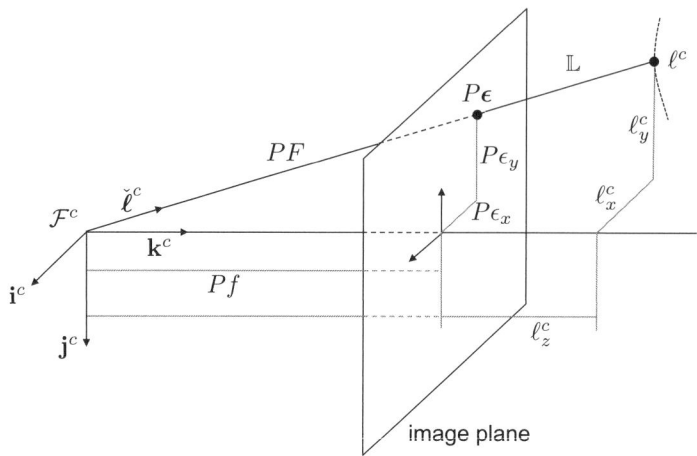

Figure 13.2 The camera frame. The target in the camera frame is represented by ℓ^c. The projection of the target onto the image plane is represented by ϵ. The pixel location (0,0) corresponds to the center of the image, which is assumed to be aligned with the optical axis. The distance to the target is given by $\mathbb{L}$, ϵ and f are in units of pixels, ℓ is in units of meters.

13.1.1 Camera Model

The geometry in the camera frame is shown in figure 13.2, where f is the focal length in units of pixels and P converts pixels to meters. To simplify the discussion, we will assume that the pixels and the pixel array are square. If the width of the square pixel array in units of pixels is M and the field-of-view of the camera v is known, then the focal length f can be expressed as

$$f = \frac{M}{2\tan\left(\frac{v}{2}\right)}. \tag{13.5}$$

The location of the projection of the object is expressed in the camera frame as $(P\epsilon_x, P\epsilon_y, Pf)$, where ϵ_x and ϵ_y are the pixel location (in units of pixels) of the object. The distance from the origin of the camera frame to the pixel location (ϵ_x, ϵ_y), as shown in figure 13.2, is PF where

$$F = \sqrt{f^2 + \epsilon_x^2 + \epsilon_y^2}. \tag{13.6}$$

Using similar triangles in figure 13.2, we get

$$\frac{\ell_x^c}{\mathbb{L}} = \frac{P\epsilon_x}{PF} = \frac{\epsilon_x}{F}. \tag{13.7}$$

Similarly, we get that $\ell_y^c/\mathbb{L} = \epsilon_y/F$ and $\ell_z^c/\mathbb{L} = f/F$. Combining, we have that

$$\ell^c = \frac{\mathbb{L}}{F}\begin{pmatrix} \epsilon_x \\ \epsilon_y \\ f \end{pmatrix}, \tag{13.8}$$

where ℓ is the vector to the object of interest and $\mathbb{L} = \|\ell\|$.

Note that ℓ^c cannot be determined strictly from camera data since $\mathbb{L}$ is unknown. However, we can determine the unit direction vector to the target as

$$\frac{\ell^c}{\mathbb{L}} = \frac{1}{F}\begin{pmatrix} \epsilon_x \\ \epsilon_y \\ f \end{pmatrix} = \frac{1}{\sqrt{\epsilon_x^2 + \epsilon_y^2 + f^2}}\begin{pmatrix} \epsilon_x \\ \epsilon_y \\ f \end{pmatrix}. \tag{13.9}$$

Since the unit vector $\ell^c/\mathbb{L}$ plays a major role throughout this chapter, we will use the notation

$$\check{\ell} \triangleq \begin{pmatrix} \check{\ell}_x \\ \check{\ell}_y \\ \check{\ell}_z \end{pmatrix} \triangleq \frac{\ell}{\mathbb{L}}$$

to denote the normalized version of ℓ.

13.2 Gimbal Pointing

Small and miniature air vehicles are used primarily for intelligence, surveillance, and reconnaissance (ISR) tasks. If the MAV is equipped with a gimbal, this involves maneuvering the gimbal so that the camera points at certain objects. The objective of this section is to describe a simple gimbal-pointing algorithm. We assume a pan-tilt gimbal and that the equations of motion for the gimbal are given by

$$\dot{\alpha}_{az} = u_{az}$$
$$\dot{\alpha}_{el} = u_{el},$$

where u_{az} and u_{el} are control variables for the gimbal's azimuth and elevation angles, respectively.

We will consider two pointing scenarios. In the first scenario, the objective is to point the gimbal at a given world coordinate. In the second scenario, the objective is to point the gimbal so that the optical axis aligns with a certain point in the image plane. For the second scenario, we envision a user watching a video stream from the MAV and using a mouse to click on a location in the image plane. The gimbal is then maneuvered to push that location to the center of the image plane.

For the first scenario, let $\mathbf{p}^i_{\text{obj}}$ be the known location of an object in the inertial frame. The objective is to align the optical axis of the camera with the desired relative position vector

$$\boldsymbol{\ell}^i_d \triangleq \mathbf{p}^i_{\text{obj}} - \mathbf{p}^i_{\text{MAV}},$$

where $\mathbf{p}^i_{\text{MAV}} = (p_n, p_e, p_d)^\top$ is the inertial position of the MAV and where the subscript d indicates a desired quantity. The body-frame unit vector that points in the desired direction of the object is given by

$$\check{\boldsymbol{\ell}}^b_d = \frac{1}{\|\boldsymbol{\ell}^i_d\|} \mathcal{R}^b_i \, \boldsymbol{\ell}^i_d.$$

For the second scenario, suppose that we desire to maneuver the gimbal so that the pixel location ϵ is pushed to the center of the image. Using equation (13.9), the desired direction of the optical axis in the camera frame is given by

$$\check{\boldsymbol{\ell}}^c_d = \frac{1}{\sqrt{f^2 + \epsilon_x^2 + \epsilon_y^2}} \begin{pmatrix} \epsilon_x \\ \epsilon_y \\ f \end{pmatrix}.$$

In the body frame, the desired direction of the optical axis is

$$\check{\boldsymbol{\ell}}^b_d = \mathcal{R}^b_g \mathcal{R}^g_c \, \check{\boldsymbol{\ell}}^c_d.$$

The next step is to determine the desired azimuth and elevation angles that will align the optical axis with $\check{\boldsymbol{\ell}}^b_d$. In the camera frame, the optical axis is given by $(0, 0, 1)^c$. Therefore, the objective is to select the commanded gimbal angles α^c_{az} and α^c_{el} so that

$$\check{\boldsymbol{\ell}}^b_d \triangleq \begin{pmatrix} \check{\ell}^b_{xd} \\ \check{\ell}^b_{yd} \\ \check{\ell}^b_{zd} \end{pmatrix} = \mathcal{R}^b_g(\alpha^c_{\text{az}}, \alpha^c_{\text{el}}) \mathcal{R}^g_c \begin{pmatrix} 0 \\ 0 \\ 1 \end{pmatrix} \quad (13.10)$$

$$= \begin{pmatrix} \cos\alpha^c_{\text{el}}\cos\alpha^c_{\text{az}} & -\sin\alpha^c_{\text{el}} & -\sin\alpha^c_{\text{el}}\cos\alpha^c_{\text{az}} \\ \cos\alpha^c_{\text{el}}\sin\alpha^c_{\text{az}} & \cos\alpha^c_{\text{az}} & -\sin\alpha^c_{\text{el}}\sin\alpha^c_{\text{az}} \\ \sin\alpha^c_{\text{el}} & 0 & \cos\alpha^c_{\text{el}} \end{pmatrix} \begin{pmatrix} 0 & 0 & 1 \\ 1 & 0 & 0 \\ 0 & 1 & 0 \end{pmatrix} \begin{pmatrix} 0 \\ 0 \\ 1 \end{pmatrix} \quad (13.11)$$

$$= \begin{pmatrix} \cos\alpha_{el}^c \cos\alpha_{az}^c \\ \cos\alpha_{el}^c \sin\alpha_{az}^c \\ \sin\alpha_{el}^c \end{pmatrix}. \tag{13.12}$$

Solving for α_{el}^c and α_{az}^c gives the desired azimuth and elevation angles as

$$\alpha_{az}^c = \tan^{-1}\left(\frac{\check{\ell}_{yd}^b}{\check{\ell}_{xd}^b}\right) \tag{13.13}$$

$$\alpha_{el}^c = \sin^{-1}\left(\check{\ell}_{zd}^b\right). \tag{13.14}$$

The gimbal servo commands can be selected as

$$u_{az} = k_{az}(\alpha_{az}^c - \alpha_{az}) \tag{13.15}$$
$$u_{el} = k_{el}(\alpha_{el}^c - \alpha_{el}),$$

where k_{az} and k_{el} are positive control gains.

13.3 Geolocation

This section presents a method for determining the location of objects in world/inertial coordinates using a gimbaled EO/IR camera on board a fixed-wing MAV. We assume that the MAV can measure its own world coordinates using, for example, a GPS receiver, and that other MAV state variables are also available.

Following section 13.1, let $\ell = \mathbf{p}_{obj} - \mathbf{p}_{MAV}$ be the relative position vector between the target of interest and the MAV, and define $\mathbb{L} = \|\ell\|$ and $\check{\ell} = \ell/\mathbb{L}$. From geometry, we have the relationship

$$\begin{aligned} \mathbf{p}_{obj}^i &= \mathbf{p}_{MAV}^i + \mathcal{R}_b^i \mathcal{R}_g^b \mathcal{R}_c^g \ell^c \\ &= \mathbf{p}_{MAV}^i + \mathbb{L}\left(\mathcal{R}_b^i \mathcal{R}_g^b \mathcal{R}_c^g \check{\ell}^c\right), \end{aligned} \tag{13.16}$$

where $\mathbf{p}_{MAV}^i = (p_n, p_e, p_d)^\top$, $\mathcal{R}_b^i = \mathcal{R}_b^i(\phi, \theta, \psi)$, and $\mathcal{R}_g^b = \mathcal{R}_g^b(\alpha_{az}, \alpha_{el})$. The only element on the right-hand side of equation (13.16) that is unknown is $\mathbb{L}$. Therefore, solving the geolocation problem reduces to the problem of estimating the range to the target $\mathbb{L}$.

13.3.1 Range to Target Using the Flat-earth Model

If the MAV is able to measure height-above-ground, then a simple strategy for estimating $\mathbb{L}$ is to assume a flat-earth model [96]. Figure 13.3

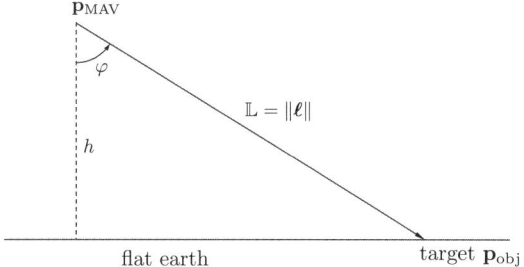

Figure 13.3 Range estimation using the flat-earth assumption.

shows the geometry of the situation, where $h = -p_d$ is the height-above-ground, and φ is the angle between ℓ and the $\mathbf{k}^i$ axis. It is clear from figure 13.3 that

$$\mathbb{L} = \frac{h}{\cos \varphi},$$

where

$$\cos \varphi = \mathbf{k}^i \cdot \check{\boldsymbol{\ell}}^i$$
$$= \mathbf{k}^i \cdot \mathcal{R}_b^i \mathcal{R}_g^b \mathcal{R}_c^g \check{\boldsymbol{\ell}}^c.$$

Therefore, the range estimate using the flat-earth model is given by

$$\mathbb{L} = \frac{h}{\mathbf{k}^i \cdot \mathcal{R}_b^i \mathcal{R}_g^b \mathcal{R}_c^g \check{\boldsymbol{\ell}}^c}. \tag{13.17}$$

The geolocation estimate is given by combining equations (13.16) and (13.17) to obtain

$$\mathbf{p}_{\text{obj}}^i = \begin{pmatrix} p_n \\ p_e \\ p_d \end{pmatrix} + h \frac{\mathcal{R}_b^i \mathcal{R}_g^b \mathcal{R}_c^g \check{\boldsymbol{\ell}}^c}{\mathbf{k}^i \cdot \mathcal{R}_b^i \mathcal{R}_g^b \mathcal{R}_c^g \check{\boldsymbol{\ell}}^c}. \tag{13.18}$$

13.3.2 Geolocation Using an Extended Kalman Filter

The geolocation estimate in equation (13.18) provides a one-shot estimate of the target location. Unfortunately, this equation is highly sensitive to measurement errors, especially attitude estimation errors of the airframe. In this section we will describe the use of the extended Kalman filter (EKF) to solve the geolocation problem.

Rearranging equation (13.16), we get

$$\mathbf{p}_{\text{MAV}}^i = \mathbf{p}_{\text{obj}}^i - \mathbb{L}\left(\mathcal{R}_b^i \mathcal{R}_g^b \mathcal{R}_c^g \check{\boldsymbol{\ell}}^c\right), \tag{13.19}$$

Vision-guided Navigation

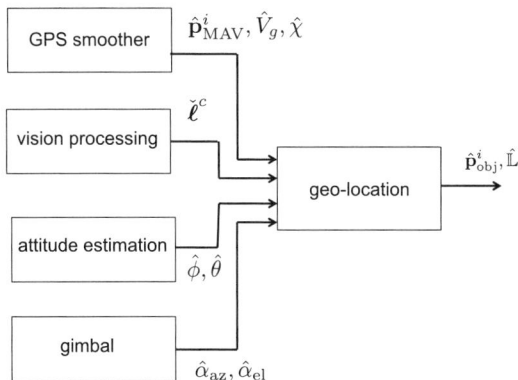

Figure 13.4 The geolocation algorithm uses the output of the GPS smoother, the normalized line-of-sight vector from the vision algorithm, and the attitude to estimate the position of the object in the inertial frame and the distance to the object.

which, since $\mathbf{p}_{\text{MAV}}^i$ is measured by GPS, will be used as the measurement equation, assuming that GPS noise is zero-mean Gaussian. However, since GPS measurement error contains a constant bias, the geolocation error will also contain a bias. If we assume that the object is stationary, we have

$$\dot{\mathbf{p}}_{\text{obj}}^i = 0.$$

Since $\mathbb{L} = \|\mathbf{p}_{\text{obj}}^i - \mathbf{p}_{\text{MAV}}^i\|$, we have

$$\dot{\mathbb{L}} = \frac{d}{dt}\sqrt{(\mathbf{p}_{\text{obj}}^i - \mathbf{p}_{\text{MAV}}^i)^\top (\mathbf{p}_{\text{obj}}^i - \mathbf{p}_{\text{MAV}}^i)}$$
$$= \frac{(\mathbf{p}_{\text{obj}}^i - \mathbf{p}_{\text{MAV}}^i)^\top (\dot{\mathbf{p}}_{\text{obj}}^i - \dot{\mathbf{p}}_{\text{MAV}}^i)}{\mathbb{L}}$$
$$= -\frac{(\mathbf{p}_{\text{obj}}^i - \mathbf{p}_{\text{MAV}}^i)^\top \dot{\mathbf{p}}_{\text{MAV}}^i}{\mathbb{L}},$$

where for constant-altitude flight, $\dot{\mathbf{p}}_{\text{MAV}}^i$ can be approximated as

$$\dot{\mathbf{p}}_{\text{MAV}}^i = \begin{pmatrix} \hat{V}_g \cos \hat{\chi} \\ \hat{V}_g \sin \hat{\chi} \\ 0 \end{pmatrix},$$

and where $\hat{V}_g$ and $\hat{\chi}$ are estimated using the EKF discussed in section 8.7.

A block diagram of the geolocation algorithm is shown in figure 13.4. The input to the geolocation algorithm is the position and the velocity of the MAV in the inertial frame as estimated by the GPS smoother described in section 8.7, the estimate of the normalized line-of-sight vector as given in equation (13.9), and the attitude as estimated by the scheme described in section 8.6.

The geolocation algorithm is an extended Kalman filter with state $\hat{x} = (\hat{\mathbf{p}}_{obj}^{iT}, \hat{L})^\top$ and prediction equations given by

$$\begin{pmatrix} \dot{\hat{\mathbf{p}}}_{obj}^i \\ \dot{\hat{L}} \end{pmatrix} = \begin{pmatrix} 0 \\ -\dfrac{(\hat{\mathbf{p}}_{obj}^i - \hat{\mathbf{p}}_{MAV}^i)^\top \dot{\hat{\mathbf{p}}}_{MAV}^i}{\hat{L}} \end{pmatrix}.$$

The Jacobian is therefore given by

$$\frac{\partial f}{\partial x} = \begin{pmatrix} 0 & 0 \\ -\dfrac{\dot{\hat{\mathbf{p}}}_{MAV}^{iT}}{\hat{L}} & \dfrac{(\hat{\mathbf{p}}_{obj}^i - \hat{\mathbf{p}}_{MAV}^i)^\top \dot{\hat{\mathbf{p}}}_{MAV}^i}{\hat{L}^2} \end{pmatrix}.$$

The output equation is given by equation (13.19), where the Jacobian of the output equation is

$$\frac{\partial h}{\partial x} = \begin{pmatrix} I & \mathcal{R}_b^i \mathcal{R}_g^b \mathcal{R}_c^g \check{\ell}^c \end{pmatrix}.$$

13.4 Estimating Target Motion in the Image Plane

We assume in this chapter that a computer vision algorithm is used to track the pixel location of the target. Since the video stream often contains noise and the tracking algorithms are imperfect, the pixel location returned by these algorithms is noisy. The guidance algorithm described in the next section needs both the pixel location of the target and the pixel velocity of the target. In section 13.4.1 we show how to construct a simple low-pass filter that returns a filtered version of both the pixel location and the pixel velocity.

The pixel velocity is influenced by both the relative (translational) motion between the target and the MAV, and the rotational motion of the MAV-gimbal combination. In section 13.4.2 we derive an explicit expression for pixel velocity and show how to compensate for the apparent motion induced by the rotational rates of the MAV and gimbal.

13.4.1 Digital Low-pass Filter and Differentiation

Let $\bar{\epsilon} = (\bar{\epsilon}_x, \bar{\epsilon}_y)^\top$ denote the raw pixel measurements, $\epsilon = (\epsilon_x, \epsilon_y)^\top$ denote the filtered pixel location, and $\dot{\epsilon} = (\dot{\epsilon}_x, \dot{\epsilon}_y)^\top$ denote the filtered pixel velocity. The basic idea is to low-pass filter the raw pixel measurements as

$$\epsilon(s) = \frac{1}{\tau s + 1}\bar{\epsilon}, \qquad (13.20)$$

and to differentiate the raw pixel measurements as

$$\dot{\epsilon}(s) = \frac{s}{\tau s + 1}\bar{\epsilon}. \qquad (13.21)$$

Using the Tustin approximation [28],

$$s \mapsto \frac{2}{T_s}\frac{z-1}{z+1},$$

to convert to the z-domain gives

$$\epsilon[z] = \frac{1}{\frac{2\tau}{T_s}\frac{z-1}{z+1} + 1}\bar{\epsilon} = \frac{\frac{T_s}{2\tau+T_s}(z+1)}{z - \frac{2\tau-T_s}{2\tau+T_s}}\bar{\epsilon}$$

$$\dot{\epsilon}[z] = \frac{\frac{2}{T_s}\frac{z-1}{z+1}}{\frac{2\tau}{T_s}\frac{z-1}{z+1} + 1}\bar{\epsilon} = \frac{\frac{2}{2\tau+T_s}(z-1)}{z - \frac{2\tau-T_s}{2\tau+T_s}}\bar{\epsilon}.$$

Taking the inverse z-transform gives the difference equations

$$\epsilon[n] = \left(\frac{2\tau - T_s}{2\tau + T_s}\right)\epsilon[n-1] + \left(\frac{T_s}{2\tau + T_s}\right)(\bar{\epsilon}[n] + \bar{\epsilon}[n-1])$$

$$\dot{\epsilon}[n] = \left(\frac{2\tau - T_s}{2\tau + T_s}\right)\dot{\epsilon}[n-1] + \left(\frac{2}{2\tau + T_s}\right)(\bar{\epsilon}[n] - \bar{\epsilon}[n-1]),$$

where $\epsilon[0] = \bar{\epsilon}[0]$ and $\dot{\epsilon}[0] = 0$.

13.4.2 Apparent Motion Due to Rotation

Motion of the target on the image plane is induced by both relative translational motion of the target with respect to the MAV, as well as rotational motion of the MAV and gimbal platform. For most guidance tasks, we are primarily interested in the relative translational motion and desire to remove the apparent motion due to the rotation of the MAV and gimbal platforms. Following the notation introduced in section 13.1, let $\check{\ell} \triangleq \ell/\mathbb{L} = (\mathbf{p}_{\text{obj}} - \mathbf{p}_{\text{MAV}})/\|\mathbf{p}_{\text{obj}} - \mathbf{p}_{\text{MAV}}\|$ be the normalized relative position vector between the target and the MAV. Using the Coriolis formula in equation (2.17), we get

$$\frac{d\check{\ell}}{dt_i} = \frac{d\check{\ell}}{dt_c} + \boldsymbol{\omega}_{c/i} \times \check{\ell}. \qquad (13.22)$$

The expression on the left-hand side of equation (13.22) is the true relative translational motion between the target and the MAV. The first expression on the right-hand side of equation (13.22) is the motion of the target on the image plane, which can be computed from camera information. The second expression on the right-hand side of equation (13.22) is the apparent motion due to the rotation of the MAV and gimbal platform. Equation (13.22) can be expressed in the camera frame as

$$\frac{d\check{\ell}^c}{dt_i} = \frac{d\check{\ell}^c}{dt_c} + \omega^c_{c/i} \times \check{\ell}^c. \tag{13.23}$$

The first expression on the right-hand side of equation (13.23) can be computed as

$$\frac{d\check{\ell}^c}{dt_c} = \frac{d}{dt_c} \frac{\begin{pmatrix} \epsilon_x \\ \epsilon_y \\ f \end{pmatrix}}{F} = \frac{F \begin{pmatrix} \dot{\epsilon}_x \\ \dot{\epsilon}_y \\ 0 \end{pmatrix} - \dot{F} \begin{pmatrix} \epsilon_x \\ \epsilon_y \\ f \end{pmatrix}}{F^2} = \frac{F \begin{pmatrix} \dot{\epsilon}_x \\ \dot{\epsilon}_y \\ 0 \end{pmatrix} - \frac{\epsilon_x \dot{\epsilon}_x + \epsilon_y \dot{\epsilon}_y}{F} \begin{pmatrix} \epsilon_x \\ \epsilon_y \\ f \end{pmatrix}}{F^2}$$

$$= \frac{1}{F^3} \begin{pmatrix} F^2 - \epsilon_x^2 & -\epsilon_x \epsilon_y \\ -\epsilon_x \epsilon_y & F^2 - \epsilon_y^2 \\ -\epsilon_x f & -\epsilon_y f \end{pmatrix} \begin{pmatrix} \dot{\epsilon}_x \\ \dot{\epsilon}_y \end{pmatrix} = \frac{1}{F^3} \begin{pmatrix} \epsilon_y^2 + f^2 & -\epsilon_x \epsilon_y \\ -\epsilon_x \epsilon_y & \epsilon_x^2 + f^2 \\ -\epsilon_x f & -\epsilon_y f \end{pmatrix} \dot{\epsilon}$$

$$= Z(\epsilon)\dot{\epsilon}, \tag{13.24}$$

where

$$Z(\epsilon) \triangleq \frac{1}{F^3} \begin{pmatrix} \epsilon_y^2 + f^2 & -\epsilon_x \epsilon_y \\ -\epsilon_x \epsilon_y & \epsilon_x^2 + f^2 \\ -\epsilon_x f & -\epsilon_y f \end{pmatrix}.$$

To compute the second term on the right-hand side of equation (13.23), we need an expression for $\omega^c_{c/i}$, which can be decomposed as

$$\omega^c_{c/i} = \omega^c_{c/g} + \omega^c_{g/b} + \omega^c_{b/i}. \tag{13.25}$$

Since the camera is fixed in the gimbal frame, we have that $\omega^c_{c/g} = 0$. Letting p, q, and r denote the angular body rates of the platform, as measured by onboard rate gyros that are aligned with the body frame axes, gives $\omega_{b/i}^b = (p, q, r)^\top$. Expressing $\omega_{b/i}$ in the camera frame gives

$$\omega_{b/i}^c = \mathcal{R}_g^c \mathcal{R}_b^g \omega_{b/i}^b = \mathcal{R}_g^c \mathcal{R}_b^g \begin{pmatrix} p \\ q \\ r \end{pmatrix}. \quad (13.26)$$

To derive an expression for $\omega_{g/b}$ in terms of the measured gimbal angle rates $\dot{\alpha}_{el}$ and $\dot{\alpha}_{az}$, recall that the azimuth angle α_{az} is defined with respect to the body frame, and that the elevation angle α_{el} is defined with respect to the gimbal-1 frame. The gimbal frame is obtained by rotating the gimbal-1 frame about its y-axis by α_{el}. Therefore, $\dot{\alpha}_{az}$ is defined with respect to the gimbal-1 frame, and $\dot{\alpha}_{el}$ is defined with respect to the gimbal frame. This implies that

$$\omega_{g/b}^b = \mathcal{R}_{g1}^b(\alpha_{az}) \mathcal{R}_g^{g1}(\alpha_{el}) \begin{pmatrix} 0 \\ \dot{\alpha}_{el} \\ 0 \end{pmatrix}^g + \mathcal{R}_{g1}^b(\alpha_{az}) \begin{pmatrix} 0 \\ 0 \\ \dot{\alpha}_{az} \end{pmatrix}^{g1}.$$

Noting that $\mathcal{R}_g^{g1}$ is a y-axis rotation, we get

$$\omega_{g/b}^b = \mathcal{R}_{g1}^b(\alpha_{az}) \begin{pmatrix} 0 \\ \dot{\alpha}_{el} \\ \dot{\alpha}_{az} \end{pmatrix}$$

$$= \begin{pmatrix} \cos\alpha_{az} & -\sin\alpha_{az} & 0 \\ \sin\alpha_{az} & \cos\alpha_{az} & 0 \\ 0 & 0 & 1 \end{pmatrix} \begin{pmatrix} 0 \\ \dot{\alpha}_{el} \\ \dot{\alpha}_{az} \end{pmatrix} = \begin{pmatrix} -\sin(\alpha_{az})\dot{\alpha}_{el} \\ \cos(\alpha_{az})\dot{\alpha}_{el} \\ \dot{\alpha}_{az} \end{pmatrix},$$

and it follows that

$$\omega_{g/b}^c = \mathcal{R}_g^c \mathcal{R}_b^g \omega_{g/b}^b = \mathcal{R}_g^c \mathcal{R}_b^g \begin{pmatrix} -\sin(\alpha_{az})\dot{\alpha}_{el} \\ \cos(\alpha_{az})\dot{\alpha}_{el} \\ \dot{\alpha}_{az} \end{pmatrix}. \quad (13.27)$$

Drawing on equations (13.24) and (13.27), equation (13.23) can be expressed as

$$\frac{d\check{\ell}^c}{dt_i} = Z(\epsilon)\dot{\epsilon} + \check{\ell}_{app}^c, \quad (13.28)$$

where

$$\check{\ell}_{app}^c \triangleq \frac{1}{F} \left[\mathcal{R}_g^c \mathcal{R}_b^g \begin{pmatrix} p - \sin(\alpha_{az})\dot{\alpha}_{el} \\ q + \cos(\alpha_{az})\dot{\alpha}_{el} \\ r + \dot{\alpha}_{az} \end{pmatrix} \right] \times \begin{pmatrix} \epsilon_x \\ \epsilon_y \\ f \end{pmatrix} \quad (13.29)$$

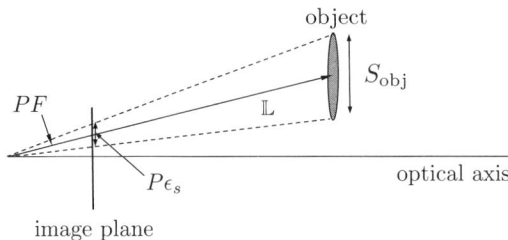

Figure 13.5 The size and growth of the target in the image frame can be used to estimate the time to collision.

is the apparent motion of the normalized line-of-sight vector in the camera frame due to the rotation of the gimbal and the aircraft.

13.5 Time to Collision

For collision avoidance algorithms and for the precision landing algorithm described in section 13.6, it is important to estimate the time to collision for objects in the camera field of view. If $\mathbb{L}$ is the length of the line-of-sight vector between the MAV and an object, then the time to collision is given by

$$t_c \triangleq \frac{\mathbb{L}}{\dot{\mathbb{L}}}.$$

It is not possible to accurately calculate time to collision using only a monocular camera because of scale ambiguity. However, if additional side information is known, then t_c can be estimated. In section 13.5.1 we assume that the target size in the image plane can be computed and then use that information to estimate t_c. Alternatively, in section 13.5.2 we assume that the target is stationary on a flat earth and then use that information to estimate t_c.

13.5.1 Computing Time to Collision from Target Size

In this section we assume that the computer vision algorithm can estimate the size of the target in the image frame. Consider the geometry shown in figure 13.5. Using similar triangles, we obtain the relationship

$$\frac{S_{\text{obj}}}{\mathbb{L}} = \frac{P\epsilon_s}{PF} = \frac{\epsilon_s}{F}, \quad (13.30)$$

where the size of the target in meters is S_{obj}, and the size of the target in pixels is given by ϵ_s. We assume that the size of the target S_{obj} is not changing in time. Differentiating equation (13.30) and solving for $\dot{\mathbb{L}}/\mathbb{L}$, we obtain

$$\frac{\dot{\mathbb{L}}}{\mathbb{L}} = \frac{\mathbb{L}}{S_{\text{obj}}} \left[\frac{\epsilon_s \dot{F}}{F\,F} - \frac{\dot{\epsilon}_s}{F} \right]$$

$$= \frac{F}{\epsilon_s} \left[\frac{\epsilon_s \dot{F}}{F\,F} - \frac{\dot{\epsilon}_s}{F} \right]$$

$$= \frac{\dot{F}}{F} - \frac{\dot{\epsilon}_s}{\epsilon_s}$$

$$= \frac{\epsilon_x \dot{\epsilon}_x + \epsilon_y \dot{\epsilon}_y}{F} - \frac{\dot{\epsilon}_s}{\epsilon_s}, \tag{13.31}$$

the inverse of which is the time to collision t_c.

13.5.2 Computing Time to Collision from the Flat-earth Model

A popular computer vision algorithm for target tracking is to track features on the target [97]. If a feature tracking algorithm is used, then target size information may not be available and the method described in the previous section cannot be applied. In this section we describe an alternative method for computing t_c, where we assume a flat-earth model. Referring to figure 13.3, we have

$$\mathbb{L} = \frac{h}{\cos \varphi},$$

where $h = -p_d$ is the altitude. Differentiating in the inertial frame gives

$$\frac{\dot{\mathbb{L}}}{\mathbb{L}} = \frac{1}{\mathbb{L}} \left(\frac{\cos \varphi \dot{h} + h \dot{\varphi} \sin \varphi}{\cos^2 \varphi} \right)$$

$$= \frac{\cos \varphi}{h} \left(\frac{\cos \varphi \dot{h} + h \dot{\varphi} \sin \varphi}{\cos^2 \varphi} \right)$$

$$= \frac{\dot{h}}{h} + \dot{\varphi} \tan \varphi. \tag{13.32}$$

In the inertial frame, we have that

$$\cos \varphi = \check{\ell}^i \cdot \mathbf{k}^i, \tag{13.33}$$

where $\mathbf{k}^i = (0,\, 0,\, 1)^\top$, and therefore

$$\cos \varphi = \check{\ell}^i_z. \tag{13.34}$$

Differentiating equation (13.34) and solving for $\dot{\varphi}$ gives

$$\dot{\varphi} = -\frac{1}{\sin\varphi}\frac{d}{dt_i}\check{\ell}_z^i. \tag{13.35}$$

Therefore,

$$\dot{\varphi}\tan\varphi = -\frac{1}{\cos\varphi}\frac{d}{dt_i}\check{\ell}_z^i = -\frac{1}{\check{\ell}_z^i}\frac{d}{dt_i}\check{\ell}_z^i, \tag{13.36}$$

where $d\check{\ell}_z^i/dt_i$ can be determined by rotating the right-hand side of equation (13.28) into the inertial frame.

13.6 Precision Landing

Our objective in this section is to use the camera to guide the MAV to land precisely on a visually distinct target. The problem of guiding an aerial vehicle to intercept a moving target has been well studied. Proportional navigation (PN), in particular, has been an effective guidance strategy against maneuvering targets [98]. In this section, we present a method for implementing a 3-D pure PN guidance law using only vision information provided by a two-dimensional array of camera pixels.

PN generates acceleration commands that are proportional to the (pursuer-evader) line-of-sight (LOS) rates multiplied by the closing velocity. PN is often implemented as two 2-D algorithms implemented in the horizontal and vertical planes. The LOS rate is computed in the plane of interest and PN produces a commanded acceleration in that plane. While this approach works well for roll-stabilized skid-to-turn missiles, it is not appropriate for MAV dynamics. In this section we develop 3-D algorithms, and we show how to map the commanded body-frame accelerations to roll angle and pitch rate commands.

To derive the precision-landing algorithm, we will use the six-state navigation model given by

$$\dot{p}_n = V_g \cos\chi \cos\gamma \tag{13.37}$$

$$\dot{p}_e = V_g \sin\chi \cos\gamma \tag{13.38}$$

$$\dot{p}_d = -V_g \sin\gamma \tag{13.39}$$

$$\dot{\chi} = \frac{g}{V_g}\tan\phi\cos(\chi - \psi) \tag{13.40}$$

$$\dot{\phi} = u_1 \tag{13.41}$$

$$\dot{\gamma} = u_2, \tag{13.42}$$

where (p_n, p_e, p_d) are the north-east-down position of the MAV, V_g is the ground speed (assumed constant), χ is the course angle, γ is the flight path angle, ϕ is the roll angle, g is the gravitational constant, and u_1 and u_2 are control variables. Although equations (13.37)–(13.42) are valid for non-zero wind conditions, we will assume zero wind conditions in the foregoing discussion of precision landing.

The objective of this section is to design a vision-based guidance law that causes a MAV to intercept a ground-based target that may be moving. The position of the target is given by $\mathbf{p}_{\text{obj}}$. Similarly, let $\mathbf{p}_{\text{MAV}}$, $\mathbf{v}_{\text{MAV}}$, $\mathbf{a}_{\text{MAV}}$ denote the position, velocity, and acceleration of the MAV, respectively.

In the inertial frame, the position and velocity of the MAV can be expressed as

$$\mathbf{p}_{\text{MAV}}^i = (p_n, p_e, p_d)^\top$$

and

$$\mathbf{v}_{\text{MAV}}^i = (V_g \cos \chi \cos \gamma,\ V_g \sin \chi \cos \gamma,\ -V_g \sin \gamma)^\top.$$

In the vehicle-2 frame, however, the velocity vector of the MAV is given by

$$\mathbf{v}_{\text{MAV}}^{v_2} = (V_g, 0, 0)^\top.$$

Define $\boldsymbol{\ell} = \mathbf{p}_{\text{obj}} - \mathbf{p}_{\text{MAV}}$ and $\dot{\boldsymbol{\ell}} = \mathbf{v}_{\text{obj}} - \mathbf{v}_{\text{MAV}}$, and let $\mathbb{L} = \|\boldsymbol{\ell}\|$. The geometry associated with the precision landing problem is shown in figure 13.6. The proportional navigation strategy is to maneuver the MAV so that the line-of-sight rate $\dot{\boldsymbol{\ell}}$ is aligned with the negative line-of-sight vector $-\boldsymbol{\ell}$. Since $\boldsymbol{\ell} \times \dot{\boldsymbol{\ell}}$ is zero when $\dot{\boldsymbol{\ell}}$ and $\boldsymbol{\ell}$ are aligned, the acceleration will be proportional to the cross product. However, since $\boldsymbol{\ell}$ and $\dot{\boldsymbol{\ell}}$ cannot be directly computed from vision data, we normalize both quantities and define

$$\boldsymbol{\Omega}_\perp = \check{\boldsymbol{\ell}} \times \frac{\dot{\boldsymbol{\ell}}}{\mathbb{L}}. \tag{13.43}$$

With reference to figure 13.6, note that $\boldsymbol{\Omega}_\perp$ is directed into the page. Since the ground speed is not directly controllable, we will require that the commanded acceleration is perpendicular to the velocity vector of the MAV. Accordingly, let the commanded acceleration of the MAV be given by [99]

$$\mathbf{a}_{\text{MAV}} = N \boldsymbol{\Omega}_\perp \times \mathbf{v}_{\text{MAV}}, \tag{13.44}$$

where $N > 0$ is a tunable gain and is called the navigation constant.

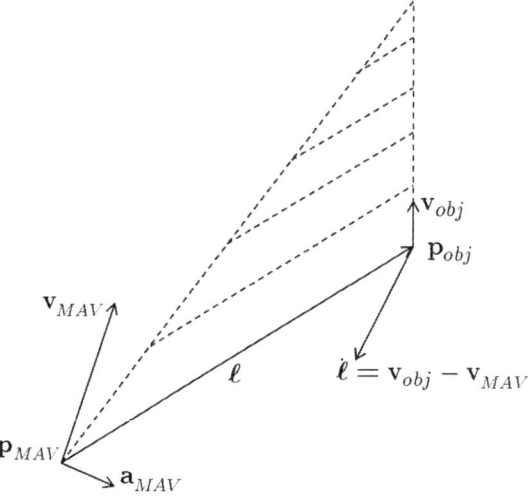

Figure 13.6 The geometry associated with precision landing.

The acceleration commands must be converted to control inputs u_1 and u_2, where the commanded acceleration is produced in the unrolled body frame, or the vehicle-2 frame. Therefore, the commanded acceleration $\mathbf{a}_{\text{MAV}}$ must be resolved in the vehicle-2 frame as

$$\mathbf{a}_{\text{MAV}}^{v2} = \mu N \mathbf{\Omega}_{\perp}^{v2} \times \mathbf{v}_{\text{MAV}}^{v2}$$

$$= \mu N \begin{pmatrix} \Omega_{\perp,x}^{v2} \\ \Omega_{\perp,y}^{v2} \\ \Omega_{\perp,z}^{v2} \end{pmatrix} \times \begin{pmatrix} V_g \\ 0 \\ 0 \end{pmatrix}$$

$$= \begin{pmatrix} 0 \\ \mu N V \Omega_{\perp,z}^{v2} \\ -\mu N V \Omega_{\perp,y}^{v2} \end{pmatrix}. \tag{13.45}$$

It is important to note that the commanded acceleration is perpendicular (by design) to the direction of motion, which is consistent with a constant-airspeed model.

The critical quantity $\mathbf{\Omega}_{\perp} = \check{\ell} \times \frac{\dot{\ell}}{\mathbb{L}}$ must be estimated from video camera data. Our basic approach will be to estimate $\mathbf{\Omega}_{\perp}$ in the camera frame, and then transform to the vehicle-2 frame using the expression

$$\mathbf{\Omega}_{\perp}^{v2} = \mathcal{R}_b^{v2} \mathcal{R}_g^b \mathcal{R}_c^g \mathbf{\Omega}_{\perp}^c. \tag{13.46}$$

Vision-guided Navigation

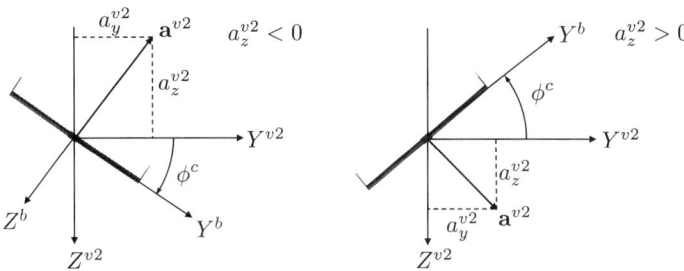

Figure 13.7 Polar converting logic that transforms an acceleration command $\mathbf{a}^{v2}$ to a commanded roll angle ϕ^c and a commanded normal acceleration $V_g \dot{\gamma}^c$.

The normalized line-of-sight vector $\check{\ell}^c$ can be estimated directly from camera data using equation (13.9). Differentiating $\ell^c/\mathbb{L}$ gives

$$\frac{d}{dt_i} \frac{\ell^c}{\mathbb{L}} = \frac{\mathbb{L}\dot{\ell}^c - \dot{\mathbb{L}}\ell^c}{\mathbb{L}^2} = \frac{\dot{\ell}^c}{\mathbb{L}} - \frac{\dot{\mathbb{L}}}{\mathbb{L}}\check{\ell}^c, \tag{13.47}$$

which, when combined with equation (13.28), results in the expression

$$\frac{\dot{\ell}^c}{\mathbb{L}} = \frac{\dot{\mathbb{L}}}{\mathbb{L}}\check{\ell}^c + Z(\epsilon)\dot{\epsilon} + \dot{\check{\ell}}^c_{\text{app}}, \tag{13.48}$$

where the inverse of time to collision $\dot{\mathbb{L}}/\mathbb{L}$ can be estimated using one of the techniques discussed in section 13.5.

Equation (13.45) gives an acceleration command in the vehicle-2 frame. In this section, we will describe how the acceleration command is converted into a roll command and a pitch rate command. The standard approach is to use polar control logic [100], which is shown in figure 13.7. From figure 13.7 it is clear that for $a_z^{v2} < 0$, we have

$$\phi^c = \tan^{-1}\left(\frac{a_y^{v2}}{-a_z^{v2}}\right)$$

$$V_g \dot{\gamma}^c = \sqrt{(a_y^{v2})^2 + (a_z^{v2})^2}.$$

Similarly, when $a_z^{v2} > 0$, we have

$$\phi^c = \tan^{-1}\left(\frac{a_y^{v2}}{a_z^{v2}}\right)$$

$$V_g \dot{\gamma}^c = -\sqrt{(a_y^{v2})^2 + (a_z^{v2})^2}.$$

Therefore, the general rule is

$$\phi^c = \tan^{-1}\left(\frac{a_y^{v2}}{|a_z^{v2}|}\right) \qquad (13.49)$$

$$\dot{\gamma}^c = -\text{sign}(a_z^{v2})\frac{1}{V_g}\sqrt{(a_y^{v2})^2 + (a_z^{v2})^2}. \qquad (13.50)$$

Unfortunately, equation (13.49) has a discontinuity at $(a_y^{v2}, a_z^{v2}) = (0, 0)$. For example, when $a_z^{v2} = 0$, the commanded roll angle is $\phi^c = \pi/2$ when $a_y^{v2} > 0$ and $\phi^c = -\pi/2$ when $a_y^{v2} < 0$. The discontinuity can be removed by multiplying equation (13.49) by the signed-sigmoidal function

$$\sigma(a_y^{v2}) = \text{sign}(a_y^{v2})\frac{1 - e^{-ka_y^{v2}}}{1 + e^{-ka_y^{v2}}}, \qquad (13.51)$$

where k is a positive control gain. The gain k adjusts the rate of the transition.

13.7 Chapter Summary

This chapter has provided a brief introduction to the rich topic of vision-based guidance of MAVs. We have focused on three basic scenarios: gimbal pointing, geolocation, and precision landing.

Notes and References

Vision-based guidance and control of MAVs is currently an active research topic (see, for example [51, 101, 102, 103, 104, 105, 106, 96, 107, 108, 109, 110]). The gimbal-pointing algorithm described in this chapter was presented in [111]. Geolocation algorithms using small MAV are described in [96, 112, 113, 114, 105, 110]. The removal of apparent motion, or egomotion, in the image plane is discussed in [115, 116, 104]. Time to collision can be estimated using structure from motion [117], ground-plane methods [118, 119], flow divergence [120], and insect-inspired methods [121]. Section 13.6 is taken primarily from [122]. Proportional navigation has been extensively analyzed in the literature. It has been shown to be optimal under certain conditions [123] and to produce zero-miss distances for a constant target acceleration [124]. If rich information regarding the time to go is available, augmented proportional navigation [125] improves the performance by adding terms that account for target and pursuer accelerations. A three-dimensional expression of PN can be found in [99, 126].

13.8 Design Project

13.1 Implement the gimbal pointing algorithm described in section 13.2. Download the files from the website that are associated with this chapter. Modify `param.m` so that the buildings have a maximum height of one meter, and modify the path planner so that the MAV moves between two fixed locations. Use the Simulink model `mavsim_chap13_gimbal.mdl` and modify the file `point_gimbal.m` to implement the gimbal pointing algorithm given by equations (13.13) and (13.14).

13.2 Implement the geolocation algorithms described in section 13.3. Use the gimbal-pointing routine developed in the previous problem to point the gimbal at the target. Use the Simulink model `mavsim_chap13_geolocation.mdl` and modify the file `geolocation.m` to implement the geolocation algorithm described in section 13.3.

APPENDIX A

Nomenclature and Notation

Nomenclature

- Unit vectors along the x, y, and z axes are denoted as $\mathbf{i}$, $\mathbf{j}$, and $\mathbf{k}$, respectively.

- A coordinate frame is denoted by $\mathcal{F}$, and a superscript denotes the label of the frame. For example, $\mathcal{F}^i$ is the inertial coordinate frame.

- Given a vector $\mathbf{p} \in \mathbb{R}^3$, the expression of $\mathbf{p}$ in the coordinates of frame $\mathcal{F}^a$ is denoted as $\mathbf{p}^a$.

- Given a vector $\mathbf{p} \in \mathbb{R}^3$, the first, second, and third components of $\mathbf{p}$ expressed with respect to the coordinates of $\mathcal{F}^a$ are denoted as p_x^a, p_y^a, and p_z^a, respectively.

- A rotation matrix from frame $\mathcal{F}^a$ to $\mathcal{F}^b$ is denoted as $\mathcal{R}_a^b$.

- The transpose of a matrix M is denoted as $M^\top$.

- Differentiation of a scalar with respect to time is denoted by a "dot" (e.g., $\dot{x}$). Differentiation of a vector with respect to frame $\mathcal{F}^a$ is denoted by $\frac{d}{dt_a}$.

- Trim conditions are denoted with a star superscript. For example, x^* is the trim state. Deviations from trim are denoted by an overbar (e.g., $\bar{x} = x - x^*$).

- Commanded signals will be denoted by a superscript 'c'. For example, the commanded course angle is χ^c and the commanded altitude is h^c.

- Zero-mean white Gaussian noise on the sensors is denoted by $\eta(t)$. The standard deviation is denoted by σ.

- A hat over a variable represents an estimate of the variable. For example, $\hat{x}$ could be the estimate of x from an extended Kalman filter.

- Tracking error signals are denoted by $\mathbf{e}_*$.

- In chapter 11 we use the notation $\overline{\mathbf{w}_a \mathbf{w}_b}$ to denote the line in $\mathbb{R}^3$ between waypoints $\mathbf{w}_a$ and $\mathbf{w}_b$.

Notation

The notation is listed in alphabetical order, where we have used the romanized names of Greek letters. For example, ω is listed as if it were spelled "omega." A "$*$" is used as a wild card when the same symbol is used for multiple quantities with different superscripts or subscripts. For example, a_{β_*} is used to denote a_{β_1} and a_{β_2}.

a_{β_*}	Constants for transfer function associated with side slip dynamics (chapter 5)
a_{ϕ_*}	Constants for transfer function associated with roll dynamics (chapter 5)
a_{θ_*}	Constants for transfer function associated with pitch dynamics (chapter 5)
a_{V_*}	Constants for transfer function associated with airspeed dynamics (chapter 5)
α	Angle of attack (chapter 2)
α_{az}	Gimbal azimuth angle (chapter 13)
α_{el}	Gimbal elevation angle (chapter 13)
b	Wing span (chapter 4)
b_*	Coefficients for reduced order model of the autopilot (chapter 9)
β	Side slip angle (chapter 2)
c	Mean aerodynamic chord of the wing (chapter 4)
C_D	Aerodynamic drag coefficient (chapter 4)
C_{ℓ_*}	Aerodynamic moment coefficient along the body frame x-axis (chapter 4)
C_L	Aerodynamic lift coefficient (chapter 4)
C_m	Aerodynamic pitching moment coefficient (chapter 4)
C_{n_*}	Aerodynamic moment coefficient along the body frame z-axis (chapter 4)
C_{p_*}	Aerodynamic moment coefficient along the body frame x-axis (chapter 5)
C_{prop}	Aerodynamic coefficient for the propeller (chapter 4)
C_{q_*}	Aerodynamic moment coefficient along the body frame y-axis (chapter 5)
C_{r_*}	Aerodynamic moment coefficient along the body frame z-axis (chapters 4, 5)
C_{X_*}	Aerodynamic force coefficient along the body frame x-axis (chapters 4, 5)
C_{Y_*}	Aerodynamic force coefficient along the body frame y-axis (chapter 4)
C_{Z_*}	Aerodynamic force coefficient along the body frame z-axis (chapters 4, 5)

Nomenclature and Notation

χ	Course angle (chapter 2)
χ_c	Crab angle: $\chi_c = \chi - \psi$ (chapter 2)
$\chi_d(e_{py})$	Desired course to track straight line path (chapter 10)
χ^∞	Desired approach angle for tracking a straight line path (chapter 10)
χ^o	Course angle of the orbit path $\mathcal{P}_{\text{orbit}}$ (chapter 10)
χ_q	Course angle of the straight line path $\mathcal{P}_{\text{line}}$ (chapter 10)
d	Distance between center of orbit and the MAV (chapter 10)
d_β	Disturbance signal associated with reduced side slip model (chapter 5)
d_χ	Disturbance signal associated with reduced course model (chapter 5)
d_h	Disturbance signal associated with reduced altitude model (chapter 5)
d_{ϕ_*}	Disturbance signals associated with reduced roll model (chapter 5)
d_{θ_*}	Disturbance signals associated with reduced pitch model (chapter 5)
d_{V_*}	Disturbance signals associated with reduced airspeed model (chapter 5)
δ_a	Control signal denoting the aileron deflection (chapter 4)
δ_e	Control signal denoting the elevator deflection (chapter 4)
δ_r	Control signal denoting the rudder deflection (chapter 4)
δ_t	Control signal denoting the throttle deflection (chapter 4)
$\mathbf{e}_p$	Path error for straight line path following (chapter 10)
ϵ_s	Pixel size (chapter 13)
ϵ_x	Pixel location along the camera x-axis (chapter 13)
ϵ_y	Pixel location along the camera y-axis (chapter 13)
η_*	Zero-mean Gaussian sensor noise (chapter 7)
f	Camera focal length (chapter 13)
$\mathbf{f}$	External force applied to the airframe, body frame components are denoted as f_x, f_y, and f_z (chapter 3, 4)
F	$= \sqrt{f^2 + \epsilon_x^2 + \epsilon_y^2}$, the distance to pixel location (ϵ_x, ϵ_y) in pixels (chapter 13)
F_{drag}	Force due to aerodynamic drag (chapter 4, 9)
F_{lift}	Force due to aerodynamic lift (chapter 4, 9)
F_{thrust}	Force due to thrust (chapter 9)
$\mathcal{F}^b$	Body coordinate frame (chapter 2)
$\mathcal{F}^i$	Inertial coordinate frame (chapter 2)
$\mathcal{F}^s$	Stability coordinate frame (chapter 2)
$\mathcal{F}^v$	Vehicle coordinate frame (chapter 2)
$\mathcal{F}^w$	Wind coordinate frame (chapter 2)

$\mathcal{F}^{v1}$	Vehicle-1 frame (chapter 2)
$\mathcal{F}^{v2}$	Vehicle-2 frame (chapter 2)
g	Gravitational acceleration (9.81 m/s²) (chapter 4)
γ	Inertial-referenced flight path angle (chapter 2)
γ_a	Air-mass-referenced flight path angle: $\gamma_a = \theta - \alpha$ (chapter 2)
Γ_*	Products of the inertia matrix, equation (3.13) (chapter 3)
h	Altitude: $h = -p_d$ (chapter 5)
h_{AGL}	Altitude above ground level (chapter 7)
$\mathcal{H}(\mathbf{r}, \mathbf{n})$	Half plane defined at position $\mathbf{w}$, with normal vector $\mathbf{n}$ (chapter 11)
$(\mathbf{i}^b, \mathbf{j}^b, \mathbf{k}^b)$	Unit vectors defining the body frame $\mathbf{i}^b$ points out the nose of the airframe, $\mathbf{j}^b$ points out the right wing, and $\mathbf{k}^b$ points through the bottom of the airframe (chapter 2)
$(\mathbf{i}^i, \mathbf{j}^i, \mathbf{k}^i)$	Unit vectors defining the inertial frame $\mathbf{i}^i$ points north, $\mathbf{j}^i$ points east, and $\mathbf{k}^i$ points down (chapter 2)
$(\mathbf{i}^v, \mathbf{j}^v, \mathbf{k}^v)$	Unit vectors defining the vehicle frame $\mathbf{i}^v$ points north, $\mathbf{j}^v$ points east, and $\mathbf{k}^v$ points down (chapter 2)
J	The inertia matrix Elements of the inertia matrix are denoted as J_x, J_y, J_z, and J_{xz} (chapter 3)
k_{d_*}	PID derivative gain (chapter 6)
k_{GPS}	Inverse of the time constant for GPS bias (chapter 7)
k_{i_*}	PID integral gain (chapter 6)
k_{motor}	Constant that specifies the efficiency of the motor (chapter 4)
k_{orbit}	Control gain for tracking orbital path (chapter 10)
k_{p_*}	PID proportional gain (chapter 6)
k_{path}	Control gain for tracking straight line path (chapter 10)
$K_{\theta_{DC}}$	DC gain of the transfer function from the elevator to the pitch angle (chapter 6)
ℓ	External moment applied to the airframe about the body frame x-axis (chapter 3)
$\boldsymbol{\ell}$	Line-of-sight vector from the MAV to a target location. $\boldsymbol{\ell} = (\ell_x, \ell_y, \ell_z)^\top$ (chapter 13)
$\check{\boldsymbol{\ell}}$	Unit vector in the direction of the line of sight: $\check{\boldsymbol{\ell}} = \boldsymbol{\ell}/\mathbb{L}$ (chapter 13)
L_*	State-space coefficients associated with lateral dynamics (chapter 5)

$\mathbb{L}$	Length of the line line of sight vector: $\mathbb{L} = \|\boldsymbol{\ell}\|$ (chapter 13)
$LPF(x)$	Low-pass filtered version of x (chapter 8)
λ	Direction of orbital path $\lambda = +1$ specifies a clockwise orbit; $\lambda = -1$ specifies a counter clockwise orbit (chapter 10)
$\lambda_{\text{dutch roll}}$	Poles of the Dutch roll mode (chapter 5)
λ_{phugoid}	Poles of the phugoid mode (chapter 5)
λ_{rolling}	Pole of the rolling mode (chapter 5)
λ_{short}	Pole of the short period mode (chapter 5)
λ_{spiral}	Pole of the spiral mode (chapter 5)
m	Mass of the airframe (chapter 3)
m	External moment applied to the airframe about the body frame y-axis (chapter 3)
m	External moments applied to the airframe. The body frame components are denoted as ℓ, m, and n (chapters 3, 4)
M	Width of the camera pixel array (chapter 13)
M_*	State-space coefficients associated with longitudinal dynamics (chapter 5)
n	External moment applied to the airframe about the body frame z-axis (chapter 3)
n_{lf}	Load factor (chapter 9)
N_*	State-space coefficients associated with lateral dynamics (chapter 5)
ν_*	Gauss-Markov process that models GPS bias (chapter 7)
ω_{n_*}	Natural frequency (chapter 6)
$\boldsymbol{\omega}_{b/i}$	Angular velocity of the body frame with respect to the inertial frame (chapter 2)
p	Roll rate of the MAV along the body frame x-axis (chapter 3)
p_d	Inertial down position of the MAV (chapter 3)
p_e	Inertial east position of the MAV (chapter 3)
$\mathcal{P}_{\text{line}}$	Set defining a straight line (chapter 10)
$\mathbf{p}_{\text{MAV}}$	Position of the MAV (chapter 13)
p_n	Inertial north position of the MAV (chapter 3)
$\mathbf{p}_{\text{obj}}$	Position of the object of interest (chapter 13)
P	Covariance of the estimation error associated with the Kalman filter (chapter 8)
$\mathcal{P}_{\text{orbit}}$	Set defining an orbit (Chapter 10)
ϕ	Roll angle (chapter 2, 3)
φ	Angle of the MAV relative to a desired orbit (chapter 10)
ψ	Heading angle (chapters 2, 3)

q	Pitch rate of the MAV along the body frame y-axis (chapter 3)
Q_*	Process covariance noise. Typically used to tune a Kalman filter (chapter 8)
r	Yaw rate of the MAV along the body frame z-axis (chapter 3)
ρ	Density of air (chapter 4)
ϱ	Angle between waypoint path segments (chapter 11)
R	Turning radius (chapter 5)
R_*	Covariance matrix for sensor measurement noise (chapter 8)
$\mathcal{R}_a^b$	Rotation matrix from frame a to frame b (chapters 2, 13)
S	Surface area of the wing (chapter 4)
S_{prop}	Area of the propeller (chapter 4)
σ_*	Standard deviation of zero-mean white Gaussian noise (chapter 7)
t_c	Time to collision: $t_c = \mathbb{L}/\dot{\mathbb{L}}$ (chapter 13)
T_s	Sample rate of the autopilot (chapters 6, 7, 8)
τ	Bandwidth of dirty differentiator (chapter 13)
$\mathcal{T}$	Terrain map (chapter 12)
θ	Pitch angle (chapter 2, 3)
u	Inertial velocity of the airframe projected onto $\mathbf{i}^b$, the body frame x-axis (chapter 2, 3)
u_{lat}	Input vector associated with lateral dynamics: $u_{\text{lat}} = (\delta_a, \delta_r)^\top$ (chapter 5)
u_{lon}	Input vector associated with longitudinal dynamics: $u_{\text{lon}} = (\delta_e, \delta_t)^\top$ (chapter 5)
u_r	Relative wind projected onto the body frame x-axis: $u_r = u - u_w$ (chapters 2, 4)
u_w	Inertial wind velocity projected onto $\mathbf{i}^b$, the body frame x-axis (chapters 2, 4)
υ	Camera field of view (chapter 13)
Υ	Return map used for path planning. The return map at position i is given by Υ_i (chapter 12)
v	Inertial velocity of the airframe projected onto $\mathbf{j}^b$, the body frame y-axis (chapters 2, 3)
v_r	Relative wind projected onto the body frame y-axis: $v_r = v - v_w$ (chapters 2, 4)
v_w	Inertial wind velocity projected onto $\mathbf{j}^b$, the body frame y-axis (chapters 2, 4)
$\mathbf{V}_a$	Airspeed vector defined as the velocity of the airframe with respect to the air mass (chapter 2)
V_a	Airspeed where $V_a = \|\mathbf{V}_a\|$ (chapter 2)

$\mathbf{V}_g$	Ground speed vector defined as the velocity of the airframe with respect to the inertial frame (chapter 2)
V_g	Ground speed where $V_g = \|\mathbf{V}_g\|$ (chapter 2)
$\mathbf{V}_w$	Wind speed vector defined as the velocity of the wind with respect to the inertial frame (chapter 2)
V_w	Wind speed where $V_w = \|\mathbf{V}_w\|$ (chapter 2)
w	Inertial velocity of the airframe projected onto $\mathbf{k}^b$, the body frame z-axis (chapters 2, 3)
w_d	Component of the wind in the down directions (chapter 2)
w_e	Component of the wind in the east directions (chapter 2)
w_n	Component of the wind in the north directions (chapter 2)
$\mathbf{w}_i$	Waypoint in $\mathbb{R}^3$ (chapter 11)
w_r	Relative wind projected onto the body frame z-axis: $w_r = w - w_w$ (chapters 2, 4)
w_w	Inertial wind velocity projected onto $\mathbf{k}^b$, the body frame z-axis (chapters 2, 4)
W_*	Bandwidth separation (chapter 6)
$\mathcal{W}$	Set of waypoints (chapter 11)
x	State variables (chapter 5)
x_{lat}	State variables associated with lateral dynamics: $x_{\text{lat}} = (v,\ p,\ r,\ \phi,\ \psi)^\top$ (chapter 5)
x_{lon}	State variables associated with longitudinal dynamics: $x_{\text{lon}} = (u,\ w,\ q,\ \theta,\ h)^\top$ (chapter 5)
X_*	State-space coefficients associated with longitudinal dynamics (chapter 5)
$y_{\text{abs pres}}$	Absolute pressure measurement signal (chapter 7)
$y_{\text{accel},*}$	Accelerometer measurement signal (chapter 7)
$y_{\text{diff pres}}$	Differential pressure measurement signal (chapter 7)
$y_{\text{GPS},*}$	GPS measurement signal GPS measurements are available for north, east, altitude, course, and groundspeed (chapter 7)
$y_{\text{gyro},*}$	Rate gyro measurement signal (chapter 7)
y_{mag}	Magnetometer measurement signal (chapter 7)
Y_*	State-space coefficients associated with lateral dynamics (chapter 5)
Z_*	State-space coefficients associated with longitudinal dynamics (chapter 5)
$Z(\epsilon)$	Transformation from pixel motion to motion of the line of sight vector in the camera frame (chapter 13)
ζ_*	Damping coefficient (chapter 6)

APPENDIX B

Quaternions

B.1 Quaternion Rotations

Quaternions provide an alternative way to represent the attitude of an aircraft. While it could be argued that it is more difficult to visualize the angular motion of a vehicle specified by quaternions instead of Euler angles, there are mathematical advantages to the quaternion representation that make it the method of choice for many aircraft simulations. Most significantly, the Euler angle representation has a singularity when the pitch angle θ is ± 90 deg. Physically, when the pitch angle is 90 deg, the roll and yaw angles are indistinguishable. Mathematically, the attitude kinematics specified by equation (3.3) are indeterminate since $\cos\theta = 0$ when $\theta = 90$ deg. The quaternion representation of attitude has no such singularity. While this singularity is not an issue for the vast majority of flight conditions, it is an issue for simulating aerobatic flight and other extreme maneuvers, some of which may not be intentional. The other advantage that the quaternion formulation provides is that it is more computationally efficient. The Euler angle formulation of the aircraft kinematics involves nonlinear trigonometric functions, whereas the quaternion formulation results in much simpler linear and algebraic equations. A thorough introduction to quaternions and rotation sequences is given by Kuipers [127]. An in-depth treatment to the use of quaternions specific to aircraft applications is given by Phillips [25].

In its most general form, a quaternion is an ordered list of four real numbers. We can represent the quaternion e as a vector in $\mathcal{R}^4$ as

$$e = \begin{pmatrix} e_0 \\ e_1 \\ e_2 \\ e_3 \end{pmatrix},$$

where e_0, e_1, e_2, and e_3 are scalars. When a quaternion is used to represent a rotation, we require that it be a *unit quaternion*, or in other words, $\|e\| = 1$.

It is common to refer e_0 as the scalar part of the unit quaternion and the vector defined by

$$\mathbf{e} = e_1 \mathbf{i}^i + e_2 \mathbf{j}^i + e_3 \mathbf{k}^i$$

Quaternions

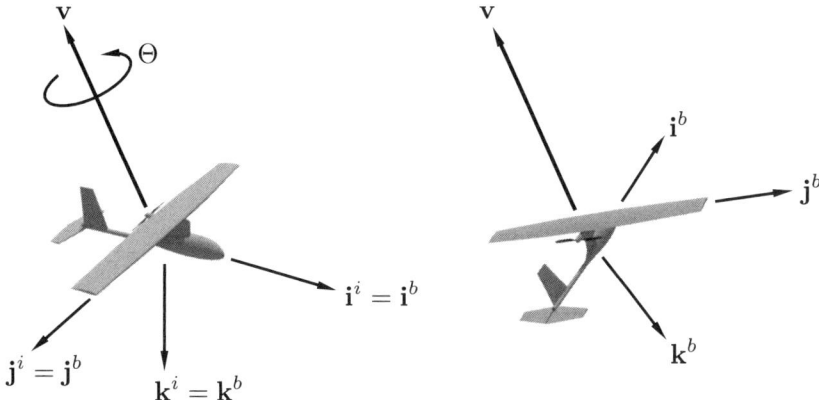

Figure B.1 Rotation represented by a unit quaternion. The aircraft on the left is shown with the body axes aligned with the inertial frame axes. The aircraft on the left has been rotated about the vector **v** by $\Theta = 86$ deg. This particular rotation corresponds to the Euler sequence $\psi = -90$ deg, $\theta = 15$ deg, $\phi = -30$ deg.

as the vector part. The unit quaternion can be interpreted as a single rotation about an axis in three-dimensional space. For a rotation through the angle Θ about the axis specified by the unit vector **v**, the scalar part of the unit quaternion is related to the magnitude of the rotation by

$$e_0 = \cos\left(\frac{\Theta}{2}\right).$$

The vector part of the unit quaternion is related to the axis of rotation by

$$\mathbf{v}\sin\left(\frac{\Theta}{2}\right) = \begin{pmatrix} e_1 \\ e_2 \\ e_3 \end{pmatrix}.$$

With this brief description of the quaternion, we can see how the attitude of a MAV can be represented with a unit quaternion. The rotation from the inertial frame to the body frame is simply specified as a single rotation about a specified axis, instead of a sequence of three rotations as required by the Euler angle representation.

B.2 Aircraft Kinematic and Dynamic Equations

Using a unit quaternion to represent the aircraft attitude, equations (3.14) through (3.17), which describe the MAV kinematics and dynamics, can be reformulated as

$$\begin{pmatrix} \dot{p}_n \\ \dot{p}_e \\ \dot{p}_d \end{pmatrix} = \begin{pmatrix} e_1^2 + e_0^2 - e_2^2 - e_3^2 & 2(e_1 e_2 - e_3 e_0) & 2(e_1 e_3 + e_2 e_0) \\ 2(e_1 e_2 + e_3 e_0) & e_2^2 + e_0^2 - e_1^2 - e_3^2 & 2(e_2 e_3 - e_1 e_0) \\ 2(e_1 e_3 - e_2 e_0) & 2(e_2 e_3 + e_1 e_0) & e_3^2 + e_0^2 - e_1^2 - e_2^2 \end{pmatrix} \begin{pmatrix} u \\ v \\ w \end{pmatrix}$$

(B.1)

$$\begin{pmatrix} \dot{u} \\ \dot{v} \\ \dot{w} \end{pmatrix} = \begin{pmatrix} rv - qw \\ pw - ru \\ qu - pv \end{pmatrix} + \frac{1}{m} \begin{pmatrix} f_x \\ f_y \\ f_z \end{pmatrix},$$

(B.2)

$$\begin{pmatrix} \dot{e}_0 \\ \dot{e}_1 \\ \dot{e}_2 \\ \dot{e}_3 \end{pmatrix} = \frac{1}{2} \begin{pmatrix} 0 & -p & -q & -r \\ p & 0 & r & -q \\ q & -r & 0 & p \\ r & q & -p & 0 \end{pmatrix} \begin{pmatrix} e_0 \\ e_1 \\ e_2 \\ e_3 \end{pmatrix}$$

(B.3)

$$\begin{pmatrix} \dot{p} \\ \dot{q} \\ \dot{r} \end{pmatrix} = \begin{pmatrix} \Gamma_1 pq - \Gamma_2 qr \\ \Gamma_5 pr - \Gamma_6(p^2 - r^2) \\ \Gamma_7 pq - \Gamma_1 qr \end{pmatrix} + \begin{pmatrix} \Gamma_3 l + \Gamma_4 n \\ \frac{1}{J_y} m \\ \Gamma_4 l + \Gamma_8 n \end{pmatrix}.$$

(B.4)

Note that the dynamic equations given by equations (B.2) and (B.4) are unchanged from equations (3.15) and (3.17) presented in the summary of chapter 3. However, care must be taken when propagating equation (B.3) to ensure that e remains a unit quaternion. If the dynamics are implemented using a Simulink s-function, then one possibility for maintaining $\|e\| = 1$ is to modify equation (B.3) so that in addition to the normal dynamics, there is also a term that seeks to minimize the cost function $J = \frac{1}{8}(1 - \|e\|^2)^2$. Since J is quadratic, we can use gradient descent to minimize J, and equation (B.3) becomes

$$\begin{pmatrix} \dot{e}_0 \\ \dot{e}_1 \\ \dot{e}_2 \\ \dot{e}_3 \end{pmatrix} = \frac{1}{2} \begin{pmatrix} 0 & -p & -q & -r \\ p & 0 & r & -q \\ q & -r & 0 & p \\ r & q & -p & 0 \end{pmatrix} \begin{pmatrix} e_0 \\ e_1 \\ e_2 \\ e_3 \end{pmatrix} - \lambda \frac{\partial J}{\partial e}$$

$$= \frac{1}{2} \begin{pmatrix} \lambda(1 - \|e\|^2) & -p & -q & -r \\ p & \lambda(1 - \|e\|^2) & r & -q \\ q & -r & \lambda(1 - \|e\|^2) & p \\ r & q & -p & \lambda(1 - \|e\|^2) \end{pmatrix} \begin{pmatrix} e_0 \\ e_1 \\ e_2 \\ e_3 \end{pmatrix},$$

where $\lambda > 0$ is a positive gain that specifies the strength of the gradient descent. In our experience, a value of $\lambda = 1000$ seems to work well, but in Simulink, a stiff solver like ODE15s must be used. This method for maintaining the orthogonality of the quaternion during integration is called Corbett-Wright orthogonality control and was first introduced in the 1950's for use with analog computers [25, 128].

With the exception of the gravity forces acting on the aircraft, all the external forces and moments act in the body frame of the aircraft and do not depend on the aircraft attitude relative to the inertial reference frame. The gravity force acts in the $\mathbf{k}^i$ direction, which can be expressed in the body frame using unit quaternions as

$$\mathbf{f}_g^b = mg \begin{pmatrix} 2(e_1 e_3 - e_2 e_0) \\ 2(e_2 e_3 + e_1 e_0) \\ e_3^2 + e_0^2 - e_1^2 - e_2^2 \end{pmatrix}.$$

B.2.1 12-state, 6-DOF Dynamic Model With Unit Quaternion Attitude Representation

The equations of motion for a MAV presented in section 5.1 utilized Euler angles to represent the attitude of the MAV. If we instead choose to take advantage of the superior numerical stability and efficiency of the unit quaternion attitude representation, the dynamic behavior of the MAV is described by the following equations:

$$\dot{p}_n = (e_1^2 + e_0^2 - e_2^2 - e_3^2)u + 2(e_1 e_2 - e_3 e_0)v + 2(e_1 e_3 + e_2 e_0)w \quad \text{(B.5)}$$

$$\dot{p}_e = 2(e_1 e_2 + e_3 e_0)u + (e_2^2 + e_0^2 - e_1^2 - e_3^2)v + 2(e_2 e_3 - e_1 e_0)w \quad \text{(B.6)}$$

$$\dot{h} = -2(e_1 e_3 - e_2 e_0)u - 2(e_2 e_3 + e_1 e_0)v - (e_3^2 + e_0^2 - e_1^2 - e_2^2)w \quad \text{(B.7)}$$

$$\dot{u} = rv - qw + 2g(e_1 e_3 - e_2 e_0)$$
$$+ \frac{\rho V_a^2 S}{2m} \left[C_X(\alpha) + C_{X_q}(\alpha) \frac{cq}{2V_a} + C_{X_{\delta_e}}(\alpha) \delta_e \right]$$
$$+ \frac{\rho S_{\text{prop}} C_{\text{prop}}}{2m} \left[(k_{\text{motor}} \delta_t)^2 - V_a^2 \right] \quad \text{(B.8)}$$

$$\dot{v} = pw - ru + 2g(e_2 e_3 + e_1 e_0)$$
$$+ \frac{\rho V_a^2 S}{2m} \left[C_{Y_0} + C_{Y_\beta} \beta + C_{Y_p} \frac{bp}{2V_a} + C_{Y_r} \frac{br}{2V_a} + C_{Y_{\delta_a}} \delta_a + C_{Y_{\delta_r}} \delta_r \right] \quad \text{(B.9)}$$

$$\dot{w} = qu - pv + g(e_3^2 + e_0^2 - e_1^2 - e_2^2)$$
$$+ \frac{\rho V_a^2 S}{2m} \left[C_Z(\alpha) + C_{Z_q}(\alpha) \frac{cq}{2V_a} + C_{Z_{\delta_e}}(\alpha) \delta_e \right] \quad \text{(B.10)}$$

$$\dot{e}_0 = -\frac{1}{2}(pe_1 + qe_2 + re_3) \quad \text{(B.11)}$$

$$\dot{e}_1 = \frac{1}{2}(pe_0 + re_2 - qe_3) \tag{B.12}$$

$$\dot{e}_2 = \frac{1}{2}(qe_0 - re_1 + pe_3) \tag{B.13}$$

$$\dot{e}_3 = \frac{1}{2}(re_0 + qe_1 - pe_2) \tag{B.14}$$

$$\dot{p} = \Gamma_1 pq - \Gamma_2 qr$$
$$+ \frac{1}{2}\rho V_a^2 Sb \left[C_{p_0} + C_{p_\beta}\beta + C_{p_p}\frac{bp}{2V_a} + C_{p_r}\frac{br}{2V_a} + C_{p_{\delta_a}}\delta_a + C_{p_{\delta_r}}\delta_r \right] \tag{B.15}$$

$$\dot{q} = \Gamma_5 pr - \Gamma_6(p^2 - r^2)$$
$$+ \frac{\rho V_a^2 Sc}{2J_y} \left[C_{m_0} + C_{m_\alpha}\alpha + C_{m_q}\frac{cq}{2V_a} + C_{m_{\delta_e}}\delta_e \right] \tag{B.16}$$

$$\dot{r} = \Gamma_7 pq - \Gamma_1 qr$$
$$+ \frac{1}{2}\rho V_a^2 Sb \left[C_{r_0} + C_{r_\beta}\beta + C_{r_p}\frac{bp}{2V_a} + C_{r_r}\frac{br}{2V_a} + C_{r_{\delta_a}}\delta_a + C_{r_{\delta_r}}\delta_r \right]. \tag{B.17}$$

The aerodynamic coefficients describing contributions of the roll and yaw moments are given by

$$C_{p_0} = \Gamma_3 C_{l_0} + \Gamma_4 C_{n_0}$$
$$C_{p_\beta} = \Gamma_3 C_{l_\beta} + \Gamma_4 C_{n_\beta}$$
$$C_{p_p} = \Gamma_3 C_{l_p} + \Gamma_4 C_{n_p}$$
$$C_{p_r} = \Gamma_3 C_{l_r} + \Gamma_4 C_{n_r}$$
$$C_{p_{\delta_a}} = \Gamma_3 C_{l_{\delta_a}} + \Gamma_4 C_{n_{\delta_a}}$$
$$C_{p_{\delta_r}} = \Gamma_3 C_{l_{\delta_r}} + \Gamma_4 C_{n_{\delta_r}}$$
$$C_{r_0} = \Gamma_4 C_{l_0} + \Gamma_8 C_{n_0}$$
$$C_{r_\beta} = \Gamma_4 C_{l_\beta} + \Gamma_8 C_{n_\beta}$$
$$C_{r_p} = \Gamma_4 C_{l_p} + \Gamma_8 C_{n_p}$$
$$C_{r_r} = \Gamma_4 C_{l_r} + \Gamma_8 C_{n_r}$$
$$C_{r_{\delta_a}} = \Gamma_4 C_{l_{\delta_a}} + \Gamma_8 C_{n_{\delta_a}}$$
$$C_{r_{\delta_r}} = \Gamma_4 C_{l_{\delta_r}} + \Gamma_8 C_{n_{\delta_r}}.$$

The inertia parameters specified by $\Gamma_1, \Gamma_2, \ldots, \Gamma_8$ are defined in equation (3.13). Angle of attack α, sideslip angle β, and airspeed V_a are calculated from the velocity components (u, v, w) and the wind velocity components (u_w, v_w, w_w) using the relations found in equation (2.8).

B.3 Conversion Between Euler Angles and Quaternions

Although not explicitly necessary for simulation purposes, we can calculate Euler angles from the attitude quaternion, and vice versa. For a quaternion representation of a rotation, the corresponding Euler angles can be calculated as

$$\phi = \text{atan2}(2(e_0 e_1 + e_2 e_3), (e_0^2 + e_3^2 - e_1^2 - e_2^2))$$
$$\theta = \text{asin}(2(e_0 e_2 - e_1 e_3))$$
$$\psi = \text{atan2}(2(e_0 e_3 + e_1 e_2), (e_0^2 + e_1^2 - e_2^2 - e_3^2)),$$

where $\text{atan2}(y, x)$ is the two-argument arctangent operator that returns the arctangent of y/x in the range $[-\pi, \pi]$ using the signs of both arguments to determine the quadrant of the return value. Only a single argument is required for the asin operator since the pitch angle is only defined in the range $[\pi/2, \pi/2]$.

From the yaw, pitch, and roll Euler angles (ψ, ϕ, θ), the corresponding quaternion elements are

$$e_0 = \cos\frac{\psi}{2}\cos\frac{\theta}{2}\cos\frac{\phi}{2} + \sin\frac{\psi}{2}\sin\frac{\theta}{2}\sin\frac{\phi}{2}$$
$$e_1 = \cos\frac{\psi}{2}\cos\frac{\theta}{2}\sin\frac{\phi}{2} - \sin\frac{\psi}{2}\sin\frac{\theta}{2}\cos\frac{\phi}{2}$$
$$e_2 = \cos\frac{\psi}{2}\sin\frac{\theta}{2}\cos\frac{\phi}{2} + \sin\frac{\psi}{2}\cos\frac{\theta}{2}\sin\frac{\phi}{2}$$
$$e_3 = \sin\frac{\psi}{2}\cos\frac{\theta}{2}\cos\frac{\phi}{2} - \cos\frac{\psi}{2}\sin\frac{\theta}{2}\sin\frac{\phi}{2}.$$

APPENDIX C

Animations in Simulink

In the study of aircraft dynamics and control, it is essential to be able to visualize the motion of the airframe. In this section we describe how to create animations in Matlab/Simulink.

C.1 Handle Graphics in Matlab

When a graphics function like `plot` is called in Matlab, the function returns a *handle* to the plot. A graphics handle is similar to a pointer in C/C++ in the sense that all of the properties of the plot can be accessed through the handle. For example, the Matlab command

```
>> plot_handle=plot(t,sin(t))
```

returns a pointer, or handle, to the plot of `sin(t)`. Properties of the plot can be changed by using the handle, rather than reissuing the `plot` command. For example, the Matlab command

```
>> set(plot_handle, `YData', cos(t))
```

changes the plot to `cos(t)` without redrawing the axes, title, label, or other objects that may be associated with the plot. If the plot contains drawings of several objects, a handle can be associated with each object. For example,

```
>> plot_handle1 = plot(t,sin(t))
>> hold on
>> plot_handle2 = plot(t,cos(t))
```

draws both `sin(t)` and `cos(t)` on the same plot, with a handle associated with each object. The objects can be manipulated separately without redrawing the other object. For example, to change `cos(t)` to `cos(2t)`, issue the command

```
>> set(plot_handle2, `YData', cos(2*t))
```

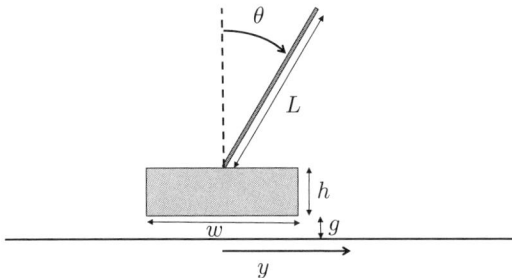

Figure C.1 Drawing for inverted pendulum. The first step in developing an animation is to draw a figure of the object to be animated and identify all of the physical parameters.

We can exploit this property to animate simulations in Simulink by redrawing only the parts of the animation that change in time, and thereby significantly reducing the simulation time. To show how handle graphics can be used to produce animations in Simulink, we will provide three detailed examples. In section C.2 we illustrate a 2-D animation of an inverted pendulum using the fill command. In section C.3 we illustrate a 3-D animation of a spacecraft using lines to produce a stick figure. In section C.4 we modify the spacecraft animation to use the vertices-faces data construction in Matlab.

C.2 Animation Example: Inverted Pendulum

Consider the image of the inverted pendulum shown in figure C.1, where the configuration is completely specified by the position of the cart y, and the angle of the rod from vertical θ. The physical parameters of the system are the rod length L, the base width w, the base height h, and the gap between the base and the track g. The first step in developing the animation is to determine the position of points that define the animation. For example, for the inverted pendulum in figure C.1, the four corners of the base are

$$(y + w/2, g), \ (y + w/2, g + h), \ (y - w/2, g + h), \quad \text{and} \quad (y - w/2, g),$$

and the two ends of the rod are given by

$$(y, g + h) \quad \text{and} \quad (y + L \sin \theta, g + h + L \cos \theta).$$

Since the base and the rod can move independently, each will need its own figure handle. The `drawBase` command can be implemented with the following Matlab code.

```
1    function handle
2      = drawBase(y, width, height, gap, handle, mode)
3        X = [y-width/2, y+width/2, y+width/2, y-width/2];
4        Y = [gap, gap, gap+height, gap+height];
5        if isempty(handle),
6           handle = fill(X,Y,`m',`EraseMode', mode);
7        else
8           set(handle,`XData',X,`YData',Y);
9        end
```

Lines 3 and 4 define the X and Y locations of the corners of the base. Note that in lines 1 and 2, handle is both an input and an output. If an empty array is passed into the function, then the fill command is used to plot the base in line 6. On the other hand, if a valid handle is passed into the function, then the base is redrawn using the set command in line 8.

The Matlab code for drawing the rod is similar and is listed below.

```
1    function handle
2      =drawRod(y, theta, L, gap, height, handle, mode)
3        X = [y, y+L*sin(theta)];
4        Y = [gap+height, gap + height + L*cos(theta)];
5        if isempty(handle),
6           handle = plot(X, Y, `g', `EraseMode', mode);
7        else
8           set(handle,`XData',X,`YData',Y);
9        end
```

The input mode is used to specify the EraseMode in Matlab. The EraseMode can be set to normal, none, xor, or background. A description of these different modes can be found by looking under Image Properties in the Matlab Helpdesk.

The main routine for the pendulum animation is listed below.

```
1    function drawPendulum(u)
2        % process inputs to function
3        y       = u(1);
4        theta   = u(2);
5        t       = u(3);
6
7        % drawing parameters
8        L = 1;
9        gap = 0.01;
10       width = 1.0;
11       height = 0.1;
12
13       % define persistent variables
```

```
14      persistent base_handle
15      persistent rod_handle
16
17      % first time function is called, initialize plot
18      %and persistent vars
19      if t==0,
20          figure(1), clf
21          track_width=3;
22          plot([-track_width,track_width],[0,0],`k');
23          hold on
24          base_handle
25              = drawBase(y, width, height, gap, [], `normal');
26          rod_handle
27              = drawRod(y, theta, L, gap, height, [], `normal');
28          axis([-track_width, track_width,
29              -L, 2*track_width-L]);
30
31      % at every other time step, redraw base and rod
32      else
33          drawBase(y, width, height, gap, base_handle);
34          drawRod(y, theta, L, gap, height, rod_handle);
35      end
```

The routine drawPendulum is called from the Simulink file shown in figure C.2, where there are three inputs: the position y, the angle θ, and the time t. Lines 3–5 rename the inputs to y, θ, and t. Lines 8–11 define the drawing parameters. We require that the handle graphics persist between function calls to drawPendulum. Since a handle is needed for both the base and the rod, we define two persistent variables in lines 14 and 15. The **if** statement in lines 19–34 is used to produce the animation. lines 20-28 are called once at the beginning of the simulation, and draw the initial animation. Line 20 brings the figure 1 window to the front and clears it. Lines 21 and 22 draw the ground along which the pendulum will move. Lines 24 and 25 calls the drawBase routine with an empty handle as input, and returns the handle base_handle to the base. The EraseMode is set to normal. Lines 26 and 27 calls the drawRod routine, and Lines 28 and 29 sets the axes of the figure. After the initial time step, all that needs to be changed are the locations of the base and rod. Therefore, in lines 32 and 33, the drawBase and drawRod routines are called with the figure handles as inputs.

C.3 Animation Example: Spacecraft Using Lines

The previous section described a simple 2-D animation. In this section we discuss a 3-D animation of a spacecraft with six degrees of freedom. Figure C.3 shows a simple line drawing of a spacecraft, where the bottom is meant to denote a solar panel that should be oriented toward the sun.

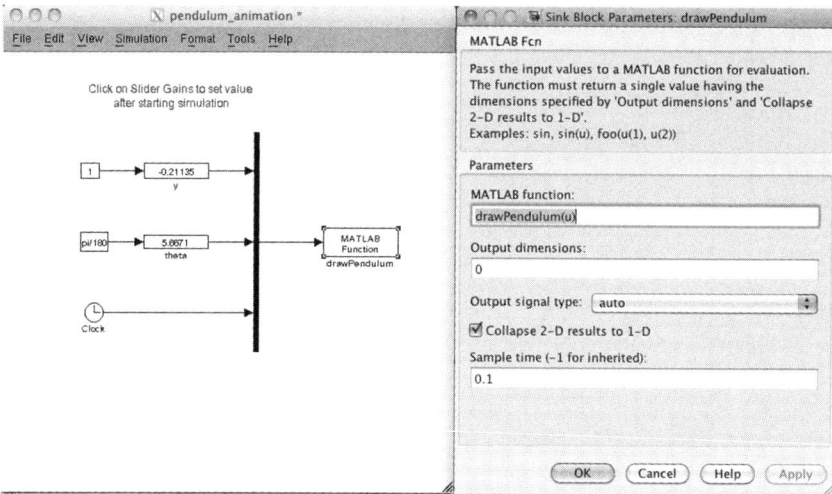

Figure C.2 Simulink file for debugging the pendulum simulation. There are three inputs to the Matlab m-file `drawPendulum`: the position y, the angle θ, and the time t. Slider gains for y and θ are used to verify the animation.

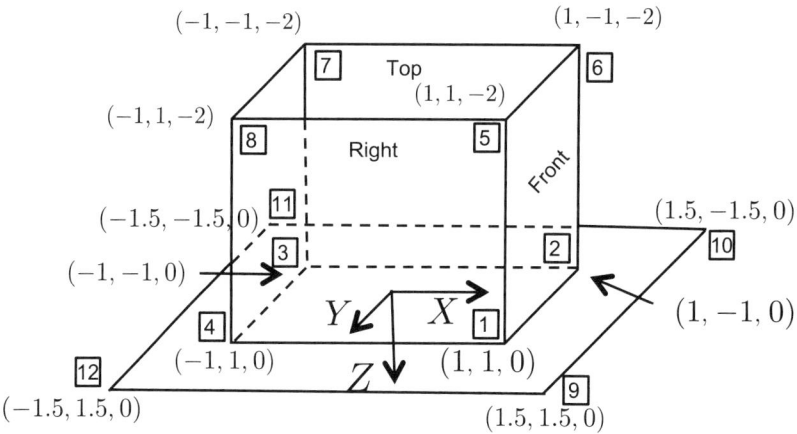

Figure C.3 Drawing used to create spacecraft animation. Standard aeronautics body axes are used, where the x-axis points out the front of the spacecraft, the y-axis points to the right, and the z-axis point out the bottom of the body.

The first step in the animation process is to label the points on the spacecraft and to determine their coordinates in a body-fixed coordinate system. We will use standard aeronautics axes with X pointing out the front of the spacecraft, Y pointing to the right, and Z pointing out the bottom. The points 1 through 12 are labeled in figure C.3 and specify how coordinates are assigned to each label. To create a line drawing, we need to connect the points in a way that draws each of the desired line segments. To do this as one continuous line, some of the segments will need to be repeated. To draw the spacecraft shown in figure C.3, we will transition through the following nodes: $1-2-3-4-1-5-6-2-6-7-3-7-8-4-8-5-1-9-10-2-10-11-3-11-12-4-12-9$. Matlab code that defines the local coordinates of the spacecraft is given below.

```
function XYZ=spacecraftPoints
% define points on the spacecraft in local NED
coordinates
XYZ = [...
    1     1     0;...  % point 1
    1    -1     0;...  % point 2
   -1    -1     0;...  % point 3
   -1     1     0;...  % point 4
    1     1     0;...  % point 1
    1     1    -2;...  % point 5
    1    -1    -2;...  % point 6
    1    -1     0;...  % point 2
    1    -1    -2;...  % point 6
   -1    -1    -2;...  % point 7
   -1    -1     0;...  % point 3
   -1    -1    -2;...  % point 7
   -1     1    -2;...  % point 8
   -1     1     0;...  % point 4
   -1     1    -2;...  % point 8
    1     1    -2;...  % point 5
    1     1     0;...  % point 1
    1.5   1.5   0;...  % point 9
    1.5  -1.5   0;...  % point 10
    1    -1     0;...  % point 2
    1.5  -1.5   0;...  % point 10
   -1.5  -1.5   0;...  % point 11
   -1    -1     0;...  % point 3
   -1.5  -1.5   0;...  % point 11
   -1.5   1.5   0;...  % point 12
   -1     1     0;...  % point 4
   -1.5   1.5   0;...  % point 12
    1.5   1.5   0;...  % point 9
]';
```

The configuration of the spacecraft is given by the Euler angles ϕ, θ, and ψ, which represent the roll, pitch, and yaw angles, respectively, and p_n, p_e, p_d, which represent the north, east, and down positions, respectively. The points on the spacecraft can be rotated and translated using the Matlab code listed below.

```matlab
function XYZ=rotate(XYZ,phi,theta,psi)
  % define rotation matrix
  R_roll = [...
          1, 0, 0;...
          0, cos(phi), -sin(phi);...
          0, sin(phi), cos(phi)];
  R_pitch = [...
          cos(theta), 0, sin(theta);...
          0, 1, 0;...
          -sin(theta), 0, cos(theta)];
  R_yaw = [...
          cos(psi), -sin(psi), 0;...
          sin(psi), cos(psi), 0;...
          0, 0, 1];
  R = R_roll*R_pitch*R_yaw;
  % rotate vertices
  XYZ = R*XYZ;
```

```matlab
function XYZ = translate(XYZ,pn,pe,pd)
  XYZ = XYZ + repmat([pn;pe;pd],1,size(XYZ,2));
```

Drawing the spacecraft at the desired location is accomplished using the following Matlab code:

```matlab
function handle
    = drawSpacecraftBody(pn,pe,pd,phi,theta,psi, handle, mode)
    % define points on spacecraft in local NED
    %coordinates
    NED = spacecraftPoints;
    % rotate spacecraft by phi, theta, psi
    NED = rotate(NED,phi,theta,psi);
    % translate spacecraft to [pn; pe; pd]
    NED = translate(NED,pn,pe,pd);
    % transform vertices from NED to XYZ
    R = [...
         0, 1, 0;...
         1, 0, 0;...
         0, 0, -1;...
         ];
    XYZ = R*NED;
    % plot spacecraft
    if isempty(handle),
      handle
      = plot3(XYZ(1,:),XYZ(2,:),XYZ(3,:), `EraseMode', mode);
    else
      set(handle,`XData',XYZ(1,:),`YData',XYZ(2,:),
      `ZData',XYZ(3,:));
      drawnow
    end
```

Animations in Simulink 267

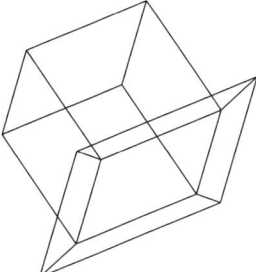

Figure C.4 Rendering of the spacecraft using lines and the `plot3` command.

Figure C.5 The mask function in Simulink allows the spacecraft points to be initialized at the beginning of the simulation.

Lines 11–16 are used to transform the coordinates from the north-east-down (NED) coordinate frame to the drawing frame used by Matlab which has the x-axis to the viewer's right, the y-axis into the screen, and the z-axis up. The `plot3` command is used in lines 19 and 20 to render the original drawing, and the `set` command is used to change the `XData`, `YData`, and `ZData` in lines 22 and 23. A Simulink file that can be used to debug the animation is on the book website. A rendering of the spacecraft is shown in figure C.4.

The disadvantage of implementing the animation using the function `spacecraftPoints` to define the spacecraft points is that this function is called each time the animation is updated. Since the points are static, they only need to be defined once. The Simulink mask function can be used to define the points at the beginning of the simulation. Masking the `drawSpacecraft` m-file in Simulink, and then clicking on `Edit Mask` brings up a window like the one shown in figure C.5. The spacecraft points can be defined in the initialization window, as shown in figure C.5, and passed to the `drawSpacecraft` m-file as a parameter.

C.4 Animation Example: Spacecraft Using Vertices and Faces

The stick-figure drawing shown in figure C.4 can be improved visually by using the vertex-face structure in Matlab. Instead of using the `plot3` command to draw a continuous line, we will use the `patch` command to draw faces defined by vertices and colors. The vertices, faces, and colors for the spacecraft are defined in the Matlab code listed below.

```matlab
function [V, F, patchcolors]=spacecraftVFC
% Define the vertices (physical location of vertices
  V = [...
    1    1    0;...  % point 1
    1   -1    0;...  % point 2
   -1   -1    0;...  % point 3
   -1    1    0;...  % point 4
    1    1   -2;...  % point 5
    1   -1   -2;...  % point 6
   -1   -1   -2;...  % point 7
   -1    1   -2;...  % point 8
    1.5  1.5  0;...  % point 9
    1.5 -1.5  0;...  % point 10
   -1.5 -1.5  0;...  % point 11
   -1.5  1.5  0;...  % point 12
  ];
% define faces as a list of vertices numbered above
  F = [...
        1, 2, 6, 5;...  % front
        4, 3, 7, 8;...  % back
        1, 5, 8, 4;...  % right
        2, 6, 7, 3;...  % left
        5, 6, 7, 8;...  % top
        9, 10, 11, 12;... % bottom
       ];
% define colors for each face
  myred    = [1, 0, 0];
  mygreen  = [0, 1, 0];
  myblue   = [0, 0, 1];
  myyellow = [1, 1, 0];
  mycyan   = [0, 1, 1];
  patchcolors = [...
    myred;...    % front
    mygreen;...  % back
    myblue;...   % right
    myyellow;... % left
    mycyan;...   % top
    mycyan;...   % bottom
    ];
```

Animations in Simulink

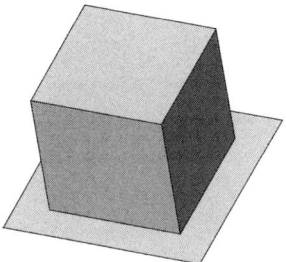

Figure C.6 Rendering of the spacecraft using vertices and faces.

The vertices are shown in figure C.3 and are defined in lines 3–16. The faces are defined by listing the indices of the points that define the face. For example, the front face, defined in line 19, consists of points $1-2-6-5$. Faces can be defined by N-points, where the matrix that defines the faces has N columns, and the number of rows is the number of faces. The color for each face is defined in lines 32–39. Matlab code that draws the spacecraft body is listed below.

```
 1    function handle
 2      = drawSpacecraftBody(pn,pe,pd,phi,theta,psi, handle, mode)
 3        [V, F, patchcolors] = spacecraftVFC;
 4        % define points on spacecraft
 5        V = rotate(V', phi, theta, psi)';}
 6        % rotate spacecraft
 7        V = translate(V', pn, pe, pd)';}
 8        % translate spacecraft
 9        R = [...
10              0, 1, 0;...
11              1, 0, 0;...
12              0, 0, -1;...
13            ];
14        V = V*R;   % transform vertices from NED to XYZ
15        if isempty(handle),
16        handle = patch('Vertices', V, 'Faces', F,...
17                       'FaceVertexCData',patchcolors,...
18                       'FaceColor','flat',...
19                       'EraseMode', mode);
20        else
21           set(handle,'Vertices',V,'Faces',F);
22        end
```

The transposes in lines 5–8 are used because the physical positions in the vertices matrix V are along the rows instead of the columns. A rendering of the spacecraft using vertices and faces is given in figure C.6. Additional examples using the vertex-face format can be found at the book website.

APPENDIX D

Modeling in Simulink Using S-functions

This chapter assumes basic familiarity with the Matlab/Simulink environment. For additional information, please consult the Matlab/Simulink documentation. Simulink is essentially a sophisticated tool for solving interconnected hybrid ordinary differential and difference equations. Each block in Simulink is assumed to have the structure

$$\dot{x}_c = f(t, x_c, x_d, u); \qquad x_c(0) = x_{c0} \tag{D.1}$$

$$x_d[k+1] = g(t, x_c, x_d, u); \qquad x_d[0] = x_{d0} \tag{D.2}$$

$$y = h(t, x_c, x_d, u), \tag{D.3}$$

where $x_c \in \mathbb{R}^{n_c}$ is a continuous state with initial condition x_{c0}, $x_d \in \mathbb{R}^{n_d}$ is a discrete state with initial condition x_{d0}, $u \in \mathbb{R}^m$ is the input to the block, $y \in \mathbb{R}^p$ is the output of the block, and t is the elapsed simulation time. An *s-function* is a Simulink tool for explicitly defining the functions f, g, and h and the initial conditions x_{c0} and x_{d0}. As explained in the Matlab/Simulink documentation, there are a number of methods for specifying an s-function. In this appendix, we will overview two different methods: a level-1 m-file s-function and a C-file s-function. The C-file s-function is compiled into C-code and executes much faster than m-file s-functions.

D.1 Example: Second-order Differential Equation

In this section we will show how to implement a system specified by the standard second-order transfer function

$$Y(s) = \frac{\omega_n^2}{s^2 + 2\zeta\omega_n s + \omega_n^2} U(s) \tag{D.4}$$

using both a level-1 m-file s-function and a C-file s-function. The first step in either case is to represent equation (D.4) in state space form.

Using control canonical form [30], we have

$$\begin{pmatrix} \dot{x}_1 \\ \dot{x}_2 \end{pmatrix} = \begin{pmatrix} -2\zeta\omega_n & -\omega_n^2 \\ 1 & 0 \end{pmatrix} \begin{pmatrix} x_1 \\ x_2 \end{pmatrix} + \begin{pmatrix} 1 \\ 0 \end{pmatrix} u \qquad (D.5)$$

$$y = \begin{pmatrix} 0 & \omega_n^2 \end{pmatrix} \begin{pmatrix} x_1 \\ x_2 \end{pmatrix}. \qquad (D.6)$$

D.1.1 Level-1 M-file S-function

The code listing for an m-file s-function that implements the system described by equations (D.5) and (D.6) is shown below. Line 1 defines the main m-file function. The inputs to this function are always the elapsed time t; the state x, which is a concatenation of the continuous state and discrete state; the input u; and a `flag`, followed by user defined input parameters, which in this case are ζ and ω_n. The Simulink engine calls the s-function and passes the parameters t, x, u, and `flag`. When `flag==0`, the Simulink engine expects the s-function to return the structure `sys`, which defines the block; initial conditions `x0`; an empty string `str`; and an array `ts` that defines the sample times of the block. When `flag==1`, the Simulink engine expects the s-function to return the function $f(t, x, u)$; when `flag==2`, the Simulink engine expects the s-function to return $g(t, x, u)$; and when `flag==3` the Simulink engine expects the s-function to return $h(t, x, u)$. The switch statement that calls the proper functions based on the value of `flag` is shown in lines 2–11. The block setup and the definition of the initial conditions are shown in lines 13–27. The number of continuous states, discrete states, outputs, and inputs are defined in lines 16 19, respectively. The direct feedthrough term on line 20 is set to one if the output depends explicitly on the input u. For example, if $D \neq 0$ in the linear state-space output equation $y = Cx + Du$. The initial conditions are defined on line 24. The sample times are defined on line 27. The format for this line is `ts = [period offset]`, where `period` defines the sample period and is 0 for continuous time or -1 for inherited, and where `offset` is the sample time offset, which is typically 0. The function $f(t, x, u)$ is defined in lines 30–32, and the output function $h(t, x, u)$ is defined in lines 35–36. A Simulink file that calls this m-file s-function is included on the book website.

```
1   function [sys,x0,str,ts] = second_order_m(t,x,u,flag,
2                               zeta,wn)
3       switch flag,
4       case 0,
5           [sys,x0,str,ts]=mdlInitializeSizes;
6           % initialize block
```

```
 7      case 1,
 8          sys=mdlDerivatives(t,x,u,zeta,wn);
 9          % define xdot = f(t,x,u)
10      case 3,
11          sys=mdlOutputs(t,x,u,wn);
12          % define xup = g(t,x,u)
13      otherwise,
14          sys = [];
15      end
16
17  %==============================================================%
18  function [sys,x0,str,ts]=mdlInitializeSizes
19      sizes = simsizes;
20      sizes.NumContStates  = 2;
21      sizes.NumDiscStates  = 0;
22      sizes.NumOutputs     = 1;
23      sizes.NumInputs      = 1;
24      sizes.DirFeedthrough = 0;
25      sizes.NumSampleTimes = 1;
26      sys = simsizes(sizes);
27
28      x0  = [0; 0];    % define initial conditions
29      str = [];        % str is always an empty matrix
30      % initialize the array of sample times
31      ts  = [0 0];     % continuous sample time
32
33  %==============================================================%
34  function xdot=mdlDerivatives(t,x,u,zeta,wn)
35      xdot(1) = -2*zeta*wn*x(1) - wn^2*x(2) + u;
36      xdot(2) = x(1);
37
38  %==============================================================%
39  function y=mdlOutputs(t,x,u,wn)
40      y = wn^2*x(2);
```

D.1.2 C-file S-function

The code listing for a C-file s-function that implements the system defined by equations (D.5) and (D.6) is shown below. The function name must be specified as in line 3. The number of parameters that are passed to the s-function is specified in line 17, and macros that access the parameters are defined in lines 6 and 7. Line 8 defines a macro that allows easy access to the input of the block. The block structure is defined using `mdlInitializeSizes` in lines 15–36. The number of continuous states, discrete states, inputs, and outputs is defined in lines 21–27. The sample time and offset are specified in lines 41–46. The initial conditions for the states are specified in lines 52–57. The function $f(t, x, u)$ is defined in lines 76–85, and the function $h(t, x, u)$ is defined in lines 62–69. The C-file s-function is compiled using the Matlab command >> `mex secondOrder_c.c`. A Simulink file that calls this C-file s-function is included on the book website.

Modeling in Simulink

```c
/*  File    : secondOrder_c.c
 */
#define S_FUNCTION_NAME secondOrder_c
#define S_FUNCTION_LEVEL 2
#include "simstruc.h"
#define zeta_PARAM(S) mxGetPr(ssGetSFcnParam(S,0))
#define wn_PARAM(S)   mxGetPr(ssGetSFcnParam(S,1))
#define U(element) (*uPtrs[element])
/* Pointer to Input Port0 */

/* Function: mdlInitializeSizes
 * Abstract:
 *    The sizes information is used by Simulink to
 *    determine the S-function blocks characteristics
 *    (number of inputs, outputs, states, etc.).
 */
static void mdlInitializeSizes(SimStruct *S)
{
    ssSetNumSFcnParams(S, 2);
    /* Number of expected parameters */
    if (ssGetNumSFcnParams(S)
        != ssGetSFcnParamsCount(S)) { return;
    /* Parameter mismatch will be reported by Simulink */
    }
    ssSetNumContStates(S, 2);
    ssSetNumDiscStates(S, 0);
    if (!ssSetNumInputPorts(S, 1)) return;
    ssSetInputPortWidth(S, 0, 1);
    ssSetInputPortDirectFeedThrough(S, 0, 1);
    if (!ssSetNumOutputPorts(S, 1)) return;
    ssSetOutputPortWidth(S, 0, 1);
    ssSetNumSampleTimes(S, 1);
    ssSetNumRWork(S, 0);
    ssSetNumIWork(S, 0);
    ssSetNumPWork(S, 0);
    ssSetNumModes(S, 0);
    ssSetNumNonsampledZCs(S, 0);
    ssSetOptions(S, SS_OPTION_EXCEPTION_FREE_CODE);
}

/* Function: mdlInitializeSampleTimes */
static void mdlInitializeSampleTimes(SimStruct *S)
{
    ssSetSampleTime(S, 0, CONTINUOUS_SAMPLE_TIME);
    ssSetOffsetTime(S, 0, 0.0);
    ssSetModelReferenceSampleTimeDefaultInheritance(S);
}

#define MDL_INITIALIZE_CONDITIONS
/* Function: mdlInitializeConditions
 *    Set initial conditions
 */
static void mdlInitializeConditions(SimStruct *S)
{
```

```c
    real_T *x0 = ssGetContStates(S);
    x0[0] = 0.0;
    x0[1] = 0.0;
}

/* Function: mdlOutputs
 *          output function
 */
static void mdlOutputs(SimStruct *S, int_T tid)
{
    real_T *y    = ssGetOutputPortRealSignal(S,0);
    real_T *x    = ssGetContStates(S);
    InputRealPtrsType uPtrs
     = ssGetInputPortRealSignalPtrs(S,0);

    UNUSED_ARG(tid); /* not used */
    const real_T *wn   = wn_PARAM(S);
    y[0] = wn[0]*wn[0]*x[1];
}

#define MDL_DERIVATIVES
/* Function: mdlDerivatives
 *          Calculate state-space derivatives
 */
static void mdlDerivatives(SimStruct *S)
{
    real_T *dx   = ssGetdX(S);
    real_T *x    = ssGetContStates(S);
    InputRealPtrsType uPtrs
     = ssGetInputPortRealSignalPtrs(S,0);

    const real_T *zeta = zeta_PARAM(S);
    const real_T *wn   = wn_PARAM(S);
    dx[0] = -2*zeta[0]*wn[0]*x[0] - wn[0]*wn[0]*x[1] + U(0);
    dx[1] = x[0];
}

/* Function: mdlTerminate
 *    No termination needed.
 */
static void mdlTerminate(SimStruct *S)
{
    UNUSED_ARG(S); /* unused input argument */
}

#ifdef  MATLAB_MEX_FILE
#include "simulink.c"
#else
#include "cg_sfun.h"
#endif
```

APPENDIX E

Airframe Parameters

This appendix gives the physical parameters for two small unmanned aircraft: a Zagi flying wing, shown in figure E.1(a), and the Aerosonde UAV, shown in figure E.1(b). Mass, geometry, propulsion, and aerodynamic parameters for the Zagi flying wing are given in table E.1. Mass, geometry, propulsion, and aerodynamic parameters for the Aerosonde are given in table E.2 [129].

Figure E.1 (a) The Zagi airframe. (b) The Aerosonde UAV.

E.1 Zagi Flying Wing

TABLE E.1
Parameters for a Zagi flying wing

Parameter	Value	Longitudinal Coef.	Value	Lateral Coef.	Value
m	1.56 kg	C_{L_0}	0.09167	C_{Y_0}	0
J_x	0.1147 kg m^2	C_{D_0}	0.01631	C_{l_0}	0
J_y	0.0576 kg m^2	C_{m_0}	−0.02338	C_{n_0}	0
J_z	0.1712 kg m^2	C_{L_α}	3.5016	C_{Y_β}	−0.07359
J_{xz}	0.0015 kg m^2	C_{D_α}	0.2108	C_{l_β}	−0.02854
S	0.2589 m^2	C_{m_α}	−0.5675	C_{n_β}	−0.00040
b	1.4224 m	C_{L_q}	2.8932	C_{Y_p}	0

TABLE E.1
Continued.

Parameter	Value	Longitudinal Coef.	Value	Lateral Coef.	Value
c	0.3302 m	C_{D_q}	0	C_{l_p}	−0.3209
S_{prop}	0.0314 m^2	C_{m_q}	−1.3990	C_{n_p}	−0.01297
ρ	1.2682 kg/m^3	$C_{L_{\delta_e}}$	0.2724	C_{Y_r}	0
k_{motor}	20	$C_{D_{\delta_e}}$	0.3045	C_{l_r}	0.03066
k_{T_p}	0	$C_{m_{\delta_e}}$	−0.3254	C_{n_r}	−0.00434
k_Ω	0	C_{prop}	1.0	$C_{Y_{\delta_a}}$	0
e	0.9	M	50	$C_{l_{\delta_a}}$	0.1682
		α_0	0.4712	$C_{n_{\delta_a}}$	−0.00328
		ϵ	0.1592		
		C_{D_p}	0.0254		

E.2 Aerosonde UAV

TABLE E.2
Aerodynamic coefficients for the Aerosonde UAV

Parameter	Value	Longitudinal Coef.	Value	Lateral Coef.	Value
m	13.5 kg	C_{L_0}	0.28	C_{Y_0}	0
J_x	0.8244 kg-m^2	C_{D_0}	0.03	C_{l_0}	0
J_y	1.135 kg-m^2	C_{m_0}	−0.02338	C_{n_0}	0
J_z	1.759 kg-m^2	C_{L_α}	3.45	C_{Y_β}	−0.98
J_{xz}	0.1204 kg-m^2	C_{D_α}	0.30	C_{l_β}	−0.12
S	0.55 m^2	C_{m_α}	−0.38	C_{n_β}	0.25
b	2.8956 m	C_{L_q}	0	C_{Y_p}	0
c	0.18994 m	C_{D_q}	0	C_{l_p}	−0.26
S_{prop}	0.2027 m^2	C_{m_q}	−3.6	C_{n_p}	0.022
ρ	1.2682 kg/m^3	$C_{L_{\delta_e}}$	−0.36	C_{Y_r}	0
k_{motor}	80	$C_{D_{\delta_e}}$	0	C_{l_r}	0.14
k_{T_p}	0	$C_{m_{\delta_e}}$	−0.5	C_{n_r}	−0.35
k_Ω	0	C_{prop}	1.0	$C_{Y_{\delta_a}}$	0
e	0.9	M	50	$C_{l_{\delta_a}}$	0.08
		α_0	0.4712	$C_{n_{\delta_a}}$	0.06
		ϵ	0.1592	$C_{Y_{\delta_r}}$	−0.17
		C_{D_p}	0.0437	$C_{l_{\delta_r}}$	0.105
		$C_{n_{\delta_r}}$	−0.032		

APPENDIX F

Trim and Linearization in Simulink

F.1 Using the Simulink `trim` Command

Simulink provides a built-in routine for computing trim conditions for general Simulink diagrams. Useful instructions for using this command can be obtained by typing `help trim` at the Matlab prompt. As described in section 5.3, given the parameters V_a^*, γ^*, and R^*, the objective is to find x^* and u^* such that $\dot{x}^* = f(x^*, u^*)$ where x and u are defined in equations (5.17) and (5.18), $\dot{x}^*$ is given by equation (5.21), and where $f(x, u)$ is defined by the right-hand side of equations (5.1)–(5.12).

The format for the Simulink `trim` command is

```
[X,U,Y,DX]=TRIM('SYS',X0,U0,Y0,IX,IU,IY,DX0,IDX),
```

where X is the computed trim state x^*, U is the computed trim input u^*, Y is the computed trim output y^*, and DX is the computed derivative of the state $\dot{x}^*$. The system is specified by the Simulink model SYS.mdl, where the state of the model is defined by the union of all of the states in the subsystems of SYS.mdl and the inputs and outputs are defined by Simulink Inports and Outports respectively. Figure F.1 shows a Simulink model that could be used to compute aircraft trim. The inputs to the system as specified by the four Inports are the servo commands delta_e, delta_a, delta_r, and delta_t. The states of this block are the states of the Simulink model, which in our case are $\xi = (p_n, p_e, p_d, u, v, w, \phi, \theta, \psi, p, q, r)^\top$, and the outputs are specified by the three Outports as the airspeed V_a, the angle of attack α, and the sideslip angle β. Our purpose in specifying V_a, α, and β as outputs is that we wish to force the Simulink `trim` command to maintain $V_a = V_a^*$ and α^* is often a quantity of interest. If we have access to a rudder, then we can enforce a coordinated turn by forcing the trim command to maintain $\beta^* = 0$. If a rudder is not available, then β will not necessarily be zero in a turn.

Since the trim calculation problem reduces to solving a system of nonlinear algebraic equations, which may have many solutions, the Simulink `trim` command requires that initial guesses for the state X0, input U0, output Y0, and derivative of the state DX0 be specified. If we know from the outset, that some of the states, inputs, outputs, or derivatives of states are fixed and specified by their initial conditions, then those constraints are indicated by the index vectors IX, IU, IY, and IDX.

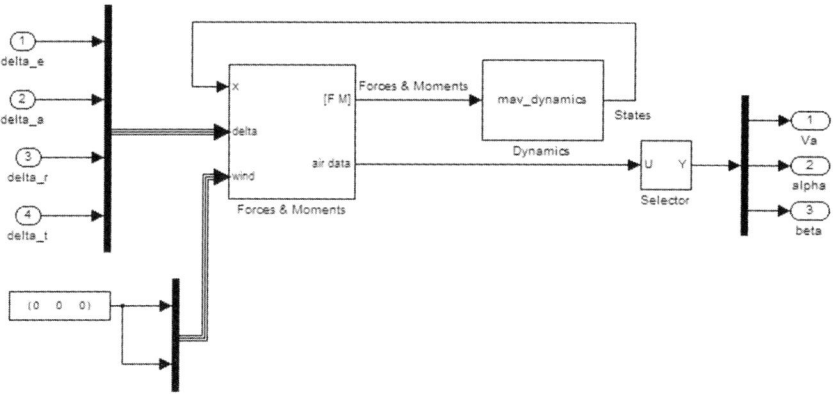

Figure F.1 Simulink diagram used to to compute trim and linear state space models.

For our situation we know that

$$\dot{x}^* = ([\text{don't care}], [\text{don't care}], -V_a^* \sin \gamma^*, 0, 0, 0, 0, 0, V_a^*/R^*, 0, 0, 0)^\top,$$

therefore we let
```
DX = [0; 0; -Va*sin(gamma); 0; 0; 0; 0; 0;
      Va/R; 0; 0; 0]
IDX = [3; 4; 5; 6; 7; 8; 9; 10; 11; 12].
```

Similarly, the initial state, inputs, and outputs can be specified as
```
X0 = [0; 0; 0; Va; 0; 0; 0; gamma; 0; 0; 0; 0]
IX0 = []
U0 = [0; 0; 0; 1]
IU0 = []
Y0 = [Va; gamma; 0]
IY0 = [1,3].
```

F.2 Numerical Computation of Trim

If simulation is developed in an environment other than Simulink, it may be necessary to write a stand-alone trim routine. This section briefly describes how this might be done. The parameters V_a^*, γ^*, and R^* fully describe the climbing-turn trim maneuver and therefore will be inputs to the trim-finding algorithm. In the calculations that follow, we will show that the variables α, β, and ϕ, along with the input parameters V_a^*, γ^*, and R^*, fully define the trim states and inputs. Thus, if we can find trim values α^*, β^*, and ϕ^* for the specified V_a^*, γ^*, and R^*, we will be able to solve analytically for the trim states and trim inputs. The first step is to show that the state variables and the input commands can be expressed in terms of V_a^*, γ^*, R^*, α^*, β^*, and ϕ^*. Since V_a^*, γ^*, and R^* are user-specified inputs to the algorithm, computing the trim states will then consist of an optimization algorithm over α, β, and ϕ to find α^*, β^*, and ϕ^*. These values will then be used to find the trim states x^* and u^*.

Body frame velocities u^*, v^*, w^*
From equation (2.7), the body frame velocities can be expressed in terms of V_a^*, α^*, and β^* as

$$\begin{pmatrix} u^* \\ v^* \\ w^* \end{pmatrix} = V_a^* \begin{pmatrix} \cos\alpha^* \cos\beta^* \\ \sin\beta^* \\ \sin\alpha^* \cos\beta^* \end{pmatrix}.$$

Pitch angle θ^*
By definition of the flight path angle (with $V_w = 0$), we have

$$\theta^* = \alpha^* + \gamma^*.$$

Angular rates p, q, r
The angular rates can be expressed in terms of the Euler angles by using equation (3.2). Therefore,

$$\begin{pmatrix} p^* \\ q^* \\ r^* \end{pmatrix} = \begin{pmatrix} 1 & 0 & -\sin\theta^* \\ 0 & \cos\phi^* & \sin\phi^*\cos\theta^* \\ 0 & -\sin\phi^* & \cos\phi^*\cos\theta^* \end{pmatrix} \begin{pmatrix} \dot\phi^* = 0 \\ \dot\theta^* = 0 \\ \dot\psi^* = \dfrac{V_a^*}{R^*} \end{pmatrix}$$

$$= \frac{V_a^*}{R^*} \begin{pmatrix} -\sin\theta^* \\ \sin\phi^*\cos\theta^* \\ \cos\phi^*\cos\theta^* \end{pmatrix},$$

where θ^* has already been expressed in terms of γ^* and α^*.

Elevator δ_e
Given p^*, q^*, and r^*, we can solve equation (5.11) for δ_e^*, giving

$$\delta_e^* = \frac{\left[\dfrac{J_{xz}(p^{*2}-r^{*2})+(J_x-J_z)p^*r^*}{\frac{1}{2}\rho(V_a^*)^2 cS}\right] - C_{m_0} - C_{m_\alpha}\alpha^* - C_{m_q}\dfrac{cq^*}{2V_a^*}}{C_{m_{\delta_e}}}. \tag{F.1}$$

Throttle δ_t

Equation (5.4) can be solved for δ_t^*, giving

$$\delta_t^* = \sqrt{\frac{2m\left(-r^*v^* + q^*w^* + g\sin\theta^*\right) - \rho(V_a^*)^2 S\left[C_X(\alpha^*) + C_{X_q}(\alpha^*)\frac{cq^*}{2V_a^*} + C_{X_{\delta_e}}(\alpha^*)\delta_e^*\right]}{\rho S_{\text{prop}} C_{\text{prop}} k_{\text{motor}}^2} + \frac{(V_a^*)^2}{k_{\text{motor}}^2}}. \quad \text{(F.2)}$$

Aileron δ_a ***and Rudder*** δ_r

The aileron and rudder commands are found by solving equations (5.10) and (5.12):

$$\begin{pmatrix} \delta_a^* \\ \delta_r^* \end{pmatrix} = \begin{pmatrix} C_{p_{\delta_a}} & C_{p_{\delta_r}} \\ C_{r_{\delta_a}} & C_{r_{\delta_r}} \end{pmatrix}^{-1}$$

$$\times \begin{pmatrix} \dfrac{-\Gamma_1 p^* q^* + \Gamma_2 q^* r^*}{\frac{1}{2}\rho(V_a^*)^2 Sb} - C_{p_0} - C_{p_\beta}\beta^* - C_{p_p}\dfrac{bp^*}{2V_a^*} - C_{p_r}\dfrac{br^*}{2V_a^*} \\ \dfrac{-\Gamma_7 p^* q^* + \Gamma_1 q^* r^*}{\frac{1}{2}\rho(V_a^*)^2 Sb} - C_{r_0} - C_{r_\beta}\beta^* - C_{r_p}\dfrac{bp^*}{2V_a^*} - C_{r_r}\dfrac{br^*}{2V_a^*} \end{pmatrix}.$$

(F.3)

F.2.1 Trim Algorithm

All of the state variables of interest and the control inputs have been expressed in terms of V_a^*, γ^*, R^*, α^*, β^* and ϕ^*. The inputs to the trim algorithm are V^*, γ^*, and R^*. To find α^*, β^*, and ϕ^*, we need to solve the following optimization problem:

$$(\alpha^*, \phi^*, \beta^*) = \arg\min \left\|\dot{x}^* - f(x^*, u^*)\right\|^2.$$

This can be performed numerically using a gradient descent algorithm that will be described in the next section. The trim algorithm is summarized in algorithm 13.

F.2.2 Numerical Implementation of Gradient Descent

The objective of this section is to describe a simple gradient descent algorithm that solves the optimization problem

$$\min_{\xi} J(\xi),$$

Trim and Linearization in Simulink

Algorithm 13 Trim

1: Input: Desired airspeed V_a^*, desired flight path angle γ^*, and desired turn radius R^*
2: Compute: $(\alpha^*, \beta^*, \phi^*) = \arg\min \|\dot{x}^* - f(x^*, u^*)\|^2$
3: Compute trimmed states:

$$\begin{pmatrix} u^* = V_a^* \cos\alpha^* \cos\beta^* \\ v^* = V_a^* \sin\beta^* \\ w^* = V_a^* \sin\alpha^* \cos\beta^* \\ \theta^* = \alpha^* + \gamma^* \\ p^* = -\dfrac{V_a^*}{R^*} \sin\theta^* \\ q^* = \dfrac{V_a^*}{R^*} \sin\phi^* \cos\theta^* \\ r^* = \dfrac{V_a^*}{R^*} \cos\phi^* \cos\theta^* \end{pmatrix}$$

4: Compute trimmed input:

$$\begin{pmatrix} \delta_e^* = [\text{Equation (F.1)}] \\ \delta_t^* = [\text{Equation (F.2)}] \\ \begin{pmatrix} \delta_a^* \\ \delta_r^* \end{pmatrix} = [\text{Equation (F.3)}] \end{pmatrix}$$

---

where $J : \mathbb{R}^m \to \mathbb{R}$ is assumed to be continuously differentiable with well-defined local minima. The basic idea is to follow the negative gradient of the function given an initial starting location $\xi^{(0)}$. In other words, we let

$$\dot{\xi} = -\kappa \frac{\partial J}{\partial \xi}(\xi), \tag{F.4}$$

where κ is a positive constant that defines the descent rate. A discrete approximation of equation (F.4) is given by

$$\xi^{(k+1)} = \xi^{(k)} - \kappa_d \frac{\partial J}{\partial \xi}(\xi^{(k)}),$$

where κ_d is κ divided by the discrete step size.

For the trim calculation, the partial derivative $\frac{\partial J}{\partial \xi}$ is difficult to determine analytically. However, it can be efficiently computed numerically. By definition, we have

$$\frac{\partial J}{\partial \xi} = \begin{pmatrix} \frac{\partial J}{\partial \xi_1} \\ \vdots \\ \frac{\partial J}{\partial \xi_m} \end{pmatrix},$$

where

$$\frac{\partial J}{\partial \xi_i} = \lim_{\epsilon \to 0} \frac{J(\xi_1, \cdots, \xi_i + \epsilon, \cdots, \xi_m) - J(\xi_1, \cdots, \xi_i, \cdots, \xi_m)}{\epsilon},$$

which can be numerically approximated as

$$\frac{\partial J}{\partial \xi_i} \approx \frac{J(\xi_1, \cdots, \xi_i + \epsilon, \cdots, \xi_m) - J(\xi_1, \cdots, \xi_i, \cdots, \xi_m)}{\epsilon},$$

where ϵ is a small constant.

For the trim algorithm, the objective $J(\alpha, \beta, \phi)$ is equal to $\|\dot{x}^* - f(x^*, u^*)\|^2$, which is computed using algorithm 14. The gradient descent optimization algorithm is summarized in algorithm 15.

F.3 Using the Simulink `linmod` Command to Generate a State-space Model

Simulink also provides a built-in routine for computing a linear state-space model for a general Simulink diagram. Helpful instruction can be obtained by typing `help linmod` at the Matlab prompt. The format for the `linmod` command is

`[A,B,C,D]=LINMOD('SYS',X,U),`

where X and U are the state and input about which the Simulink diagram is to be linearized, and [A,B,C,D] is the resulting state-space model. If the `linmod` command is used on the Simulink diagram shown in figure F.1, where there are twelve states and four inputs, the resulting state space equations will include the models given in equations (5.43)

Trim and Linearization in Simulink

Algorithm 14 Computation of $J = \left\| \dot{x}^* - f(x^*, u^*) \right\|^2$

1: Input: $\alpha^*, \beta^*, \phi^*, V_a^*, R^*, \gamma^*$
2: Compute $\dot{x}^*$:

$$\dot{x}^* = [\text{equation (5.21)}]$$

3: Compute trimmed states:

$$\begin{pmatrix} u^* = V_a^* \cos \alpha^* \cos \beta^* \\ v^* = V_a^* \sin \beta^* \\ w^* = V_a^* \sin \alpha^* \cos \beta^* \\ \theta^* = \alpha^* + \gamma^* \\ p^* = -\dfrac{V_a^*}{R^*} \sin \theta^* \\ q^* = \dfrac{V_a^*}{R^*} \sin \phi^* \cos \theta^* \\ r^* = \dfrac{V_a^*}{R^*} \cos \phi^* \cos \theta^* \end{pmatrix}$$

4: Compute trimmed input:

$$\begin{pmatrix} \delta_e^* = [\text{equation (F.1)}] \\ \delta_t^* = [\text{equation (F.2)}] \\ \begin{pmatrix} \delta_a^* \\ \delta_r^* \end{pmatrix} = [\text{equation (F.3)}] \end{pmatrix}$$

5: Compute $f(x^*, u^*)$:

$$f(x^*, u^*) = [\text{Equation (5.3)–(5.12)}].$$

6: Compute J:

$$J = \left\| \dot{x}^* - f(x^*, u^*) \right\|^2$$

---

and (5.50). To obtain equation (5.43) for example, you could use the following steps:

```
[A,B,C,D]=linmod(filename,x_trim,u_trim)
E1 = [...
0, 0, 0, 0, 1, 0, 0, 0, 0, 0, 0, 0;...
0, 0, 0, 0, 0, 0, 0, 0, 1, 0, 0;...
0, 0, 0, 0, 0, 0, 0, 0, 0, 0, 1;...
```

Algorithm 15 Minimize $J(\xi)$

1: Input: $\alpha^{(0)}, \beta^{(0)}, \phi^{(0)}, V_a, R, \gamma$
2: **for** $k = 1$ to N **do**
3: $\quad \alpha^+ = \alpha^{(k-1)} + \epsilon$
4: $\quad \beta^+ = \beta^{(k-1)} + \epsilon$
5: $\quad \phi^+ = \phi^{(k-1)} + \epsilon$
6: $\quad \frac{\partial J}{\partial \alpha} = \frac{J(\alpha^+, \beta^{(k-1)}, \phi^{(k-1)}) - J(\alpha^{(k-1)}, \beta^{(k-1)}, phi^{(k-1)})}{\epsilon}$
7: $\quad \frac{\partial J}{\partial \beta} = \frac{J(\alpha^{(k-1)}, \beta^+, \phi^{(k-1)}) - J(\alpha^{(k-1)}, \beta^{(k-1)}, phi^{(k-1)})}{\epsilon}$
8: $\quad \frac{\partial J}{\partial \phi} = \frac{J(\alpha^{(k-1)}, \beta^{(k-1)}, \phi^+) - J(\alpha^{(k-1)}, \beta^{(k-1)}, phi^{(k-1)})}{\epsilon}$
9: $\quad \alpha^{(k)} = \alpha^{(k-1)} - \kappa \frac{\partial J}{\partial \alpha}$
10: $\quad \beta^{(k)} = \beta^{(k-1)} - \kappa \frac{\partial J}{\partial \beta}$
11: $\quad \phi^{(k)} = \phi^{(k-1)} - \kappa \frac{\partial J}{\partial \phi}$
12: **end for**

```
0, 0, 0, 0, 0, 0, 1, 0, 0, 0, 0, 0;...
0, 0, 0, 0, 0, 0, 0, 0, 1, 0, 0, 0;...
]
 E2 = [...
0, 1, 0, 0;...
0, 0, 1, 0;...
]
 A_lat = E1 * A * E1'
 B_lat = E1 * B * E2'
```

F.4 Numerical Computation of State-space Model

Another way to find A and B is to approximate $\frac{\partial f}{\partial x}$ and $\frac{\partial f}{\partial u}$ numerically. The ith column of $\frac{\partial f}{\partial x}$ can be approximated as

$$\begin{pmatrix} \frac{\partial f_1}{\partial x_i} \\ \frac{\partial f_2}{\partial x_i} \\ \vdots \\ \frac{\partial f_n}{\partial x_i} \end{pmatrix} (x^*, u^*) \approx \frac{f(x^* + \epsilon e_i, u^*) - f(x^*, u^*)}{\epsilon},$$

where e_i has a one in the ith element and zeros elsewhere. Similarly, the ith column of $\frac{\partial f}{\partial u}$ can be approximated as

$$\begin{pmatrix} \frac{\partial f_1}{\partial u_i} \\ \frac{\partial f_2}{\partial u_i} \\ \vdots \\ \frac{\partial f_n}{\partial u_i} \end{pmatrix} (x^*, u^*) \approx \frac{f(x^*, u^* + \epsilon e_i) - f(x^*, u^*)}{\epsilon}.$$

These calculations can be conveniently done by taking advantage of software functions created to calculate $f(x, u)$, that were originally developed for solving the aircraft nonlinear equations of motion or for calculating trim states.

APPENDIX G

Essentials from Probability Theory

Let $X = (x_1, \ldots, x_n)^\top$ be a random vector whose elements are random variables. The mean, or expected value of X, is denoted by

$$\mu = \begin{pmatrix} \mu_1 \\ \vdots \\ \mu_n \end{pmatrix} = \begin{pmatrix} E\{x_1\} \\ \vdots \\ E\{x_n\} \end{pmatrix} = E\{X\},$$

where

$$E\{x_i\} = \int \xi f_i(\xi)\, d\xi$$

and $f(\cdot)$ is the probability density function for x_i. Given any pair of components x_i and x_j of X, we denote their covariance as

$$\text{cov}(x_i, x_j) = \Sigma_{ij} = E\{(x_i - \mu_i)(x_j - \mu_j)\}.$$

The covariance of any component with itself is the variance, that is,

$$\text{var}(x_i) = \text{cov}(x_i, x_i) = \Sigma_{ii} = E\{(x_i - \mu_i)(\xi - \mu_i)\}.$$

The standard deviation of x_i is the square root of the variance:

$$\text{stdev}(x_i) = \sigma_i = \sqrt{\Sigma_{ii}}.$$

The covariances associated with a random vector X can be grouped into a matrix known as the covariance matrix:

$$\Sigma = \begin{pmatrix} \Sigma_{11} & \Sigma_{12} & \cdots & \Sigma_{1n} \\ \Sigma_{21} & \Sigma_{22} & \cdots & \Sigma_{2n} \\ \vdots & & \ddots & \vdots \\ \Sigma_{n1} & \Sigma_{n2} & \cdots & \Sigma_{nn} \end{pmatrix} = E\{(X - \mu)(X - \mu)^\top\} = E\{XX^\top\} - \mu\mu^\top.$$

Note that $\Sigma = \Sigma^\top$ so that Σ is both symmetric and positive semi-definite, which implies that its eigenvalues are real and nonnegative.

Essentials from Probability Theory

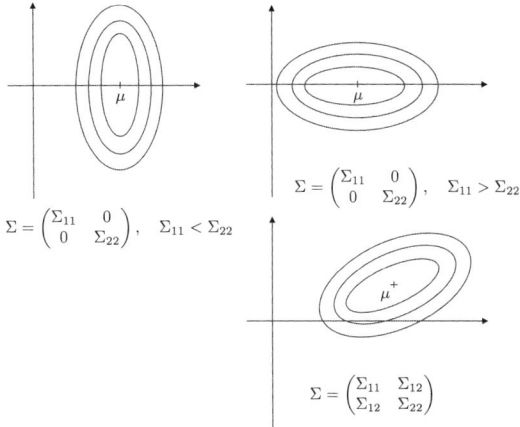

Figure G.1 Level curves for the pdf of a 2-D Gaussian random variable.

The probability density function for a Gaussian random variable is given by

$$f_x(x) = \frac{1}{\sqrt{2\pi}\sigma_x} e^{-\frac{(x-\mu_x)^2}{\sigma_x^2}},$$

where μ_x is the mean of x and σ_x is the standard deviation. The vector equivalent is given by

$$f_X(X) = \frac{1}{\sqrt{2\pi \det \Sigma}} \exp\left[-\frac{1}{2}(X-\boldsymbol{\mu})^\top \Sigma^{-1}(X-\boldsymbol{\mu})\right],$$

in which case we write

$$X \sim \mathcal{N}(\boldsymbol{\mu}, \Sigma)$$

and say that X is normally distributed with mean $\boldsymbol{\mu}$ and covariance Σ.

Figure G.1 shows the level curves for a 2-D Gaussian random variable with different covariance matrices.

APPENDIX H

Sensor Parameters

This appendix gives the noise and error parameters for several commercially available sensors typical of what would be used in a MAV autopilot. There are numerous error sources for each of the sensors. Some are random, like electrical noise. Others, such as nonlinearity, temperature sensitivity, and cross-axis sensitivity, are more deterministic in nature. We assume that these deterministic error sources can be mitigated through careful calibration and compensation, as is done by most autopilot manufacturers, with the majority of the remaining error having random characteristics.

Rate gyros, accelerometers, and pressure sensors are sampled each time through the autopilot control loop. A common sample rate is in the range of 50 to 100 Hz. We denote the control loop sample period as T_s. Digital compasses and GPS receivers are both digital devices with slower update rates that depend on the specific model of the sensor. We denote these sample periods as $T_{s,\text{compass}}$ and $T_{s,\text{GPS}}$ and specify values for typical sensors below.

H.1 Rate Gyros

An example of a MEMS rate gyro suitable for a MAV autopilot is the Analog Devices ADXRS450. It has a range of ± 350 deg/s, a bandwidth of 80 Hz, and a noise density of 0.015 deg/s/$\sqrt{\text{Hz}}$. The standard deviation of the measurement error due to sensor noise is

$$\sigma_{\text{gyro},*} = N\sqrt{B},$$

where B is the bandwidth and N is the noise density. For the ADXRS540 this results in $\sigma_{\text{gyro},*} = 0.13$ deg/s. Sources of deterministic error include cross-axis sensitivity (± 3 percent), nonlinearity (0.05 percent full-scale range RMS), and acceleration sensitivity (0.03 deg/s/g).

H.2 Accelerometers

An example of a MEMS accelerometer that could be used in an autopilot is the Analog Devices ADXL325. It has a range of $\pm 6\ g$ and a variable bandwidth of 0.5 to 550 Hz. For an autopilot, a typical bandwidth would be 100 Hz. It has a noise density of 250 $\mu g/\sqrt{Hz}$. The standard deviation of the measurement error due to sensor noise is

$$\sigma_{\text{accel},*} = N\sqrt{B},$$

which for the ADXL325 results in $\sigma_{\text{accel},*} = 0.0025$ g. Sources of deterministic error include cross-axis sensitivity (± 1 percent) and nonlinearity (± 0.2 percent of full-scale range).

H.3 Pressure Sensors

An example of an absolute pressure sensor that can be used for altitude measurement is the Freescale Semiconductor MP3H6115A. It has a range of 15 to 115 kPa, and its maximum error is characterized as having a bound of 1.5 percent of the full scale, or ± 1.5 kPa. The accuracy of the sensor, as indicated by the maximum error, is limited due to linearity errors, temperature sensitivity, and pressure hysteresis. Practical experience with MEMS absolute pressure sensors has shown that careful calibration can reduce these errors can be reduced to about 0.125 kPa of temperature-related bias drift and about 0.01 kPa of sensor noise. Therefore, we have $\beta_{\text{abs pres}} = 0.125$ kPa and $\sigma_{\text{abs pres}} = 0.01$ kPa.

The Freescale Semiconductor MPXV5004G is an example of a differential pressure sensor appropriate for use as an airspeed sensor. It has a range of 0 to 4 kPa, and its maximum error has a bound of 2.5 percent of the full-scale range, or 0.1 kPa. The accuracy of the sensor is limited by linearity errors, temperature sensitivity, and pressure hysteresis. Practical experience with this type of differential pressure sensor has shown that careful calibration can reduce these errors to about 0.02 kPa of temperature-related bias drift and about 0.002 kPa of sensor noise. Therefore, we have $\beta_{\text{diff pres}} = 0.020$ kPa and $\sigma_{\text{diff pres}} = 0.002$ kPa.

H.4 Digital Compass/Magnetometer

An example of a digital compass suitable for a small unmanned aircraft is the Honeywell HMR3300. It is a three-axis, tilt-compensated device with a built-in microcontroller for signal conditioning. When level it has an accuracy of ± 1 degree and ± 3 degrees of accuracy up to ± 30 degrees of tilt. It has a 0.5 degree repeatability and an 8 Hz update rate ($T_{s,\text{compass}} = 0.125$ s). Assuming that a portion of the accuracy and repeatability errors are due to uncertainty in the declination angle and that a portion is due to electromagnetic interference, reasonable parameters for the sensor noise standard deviation and bias error are $\sigma_{\text{mag}} = 0.3$ degrees and $\beta_{\text{mag}} = 1$ degree.

H.5 GPS

Sources of GPS measurement error and modeling of GPS measurement error are discussed in detail in section 7.5. Sample periods for GPS can vary between 0.2 and 2 seconds. For our purposes we will assume that the sample period for GPS is given by $T_{s,\text{GPS}} = 1.0$ s.

Bibliography

[1] R. C. Nelson, *Flight Stability and Automatic Control*. Boston, MA: McGraw-Hill, 2nd ed., 1998.

[2] J. Roskam, *Airplane Flight Dynamics and Automatic Flight Controls, Parts I & II*. Lawrence, KS: DARcorporation, 1998.

[3] J. H. Blakelock, *Automatic Control of Aircraft and Missiles*. New York: John Wiley & Sons, 1965.

[4] J. H. Blakelock, *Automatic Control of Aircraft and Missiles*. New York: John Wiley & Sons, 2nd ed., 1991.

[5] M. V. Cook, *Flight Dynamics Principles*. New York: John Wiley & Sons, 1997.

[6] B. Etkin and L. D. Reid, *Dynamics of Flight: Stability and Control*. New York: John Wiley & Sons, 1996.

[7] B. L. Stevens and F. L. Lewis, *Aircraft Control and Simulation*. Hoboken, NJ: John Wiley & Sons, Inc., 2nd ed., 2003.

[8] D. T. Greenwood, *Principles of Dynamics*. Englewood Cliffs, NJ: Prentice Hall, 2nd ed., 1988.

[9] T. R. Kane and D. A. Levinson, *Dynamics: Theory and Applications*. New York: McGraw Hill, 1985.

[10] M. W. Spong and M. Vidyasagar, *Robot Dynamics and Control*. New York: John Wiley & Sons, Inc., 1989.

[11] M. D. Shuster, "A survey of attitude representations," *The Journal of the Astronautical Sciences*, vol. 41, pp. 439–517, October–December 1993.

[12] T. R. Yechout, S. L. Morris, D. E. Bossert, and W. F. Hallgren, *Introduction to Aircraft Flight Mechanics*. AIAA Education Series, American Institute of Aeronautics and Astronautics, 2003.

[13] M. Rauw, *FDC 1.2 - A SIMULINK Toolbox for Flight Dynamics and Control Analysis*, February 1998. Available at `http://dutchroll.sourceforge.net/fdc.html`.

[14] H. Goldstein, *Classical Mechanics*. Cambridge, MA: Addison-Wesley, 1951.

[15] A. V. Rao, *Dynamics of Particles and Rigid Bodies: A Systematic Approach*. Cambridge: Cambridge University Press, 2006.

[16] M. J. Sidi, *Spacecraft Dynamics and Control*. Cambridge Aerospace Series. New York: Cambridge University Press, 1997.

[17] S. S. Patankar, D. E. Schinstock, and R. M. Caplinger, "Application of pendulum method to UAV momental ellipsoid estimation," in *6th AIAA Aviation Technology, Integration and Operations Conference (ATIO)*, AIAA 2006-7820, September 2006.

[18] M. R. Jardin and E. R. Mueller, "Optimized measurements of UAV mass moment of inertia with a bifilar pendulum," in *Proceedings of the AIAA Guidance, Navigation, and Control Conference and Exhibit*, AIAA 2007-6822, August 2007.

[19] J. B. Marion, *Classical Dynamics of Particles and Systems*. New York: Academic Press, 2nd ed., 1970.

[20] W. E. Wiesel, *Spaceflight Dynamics*. New York: McGraw Hill, 2nd ed., 1997.

[21] J. R. Wertz, ed., *Spacecraft Attitude Determination and Control*. Dordrecht, Neth.: Kluwer Academic Publishers, 1978.

[22] R. F. Stengel, *Flight Dynamics*. Princeton, NJ: Princeton University Press, 2004.

[23] K. S. Fu, R. C. Gonzalez, and C.S.G. Lee, *Robotics: Control, Sensing, Vision, and Intelligence*. New York: McGraw-Hill, 1987.

[24] J. W. Langelaan, N. Alley, and J. Niedhoefer, "Wind field estimation for small unmanned aerial vehicles," in *AIAA Guidance, Navigation, and Control Conference*, AIAA 2010-8177, August 2010.

[25] W. F. Phillips, *Mechanics of Flight*. New Jersey: Wiley, 2nd ed., 2010.

[26] R. Rysdyk, "UAV path following for constant line-of-sight observation," in *AIAA Journal of Guidance, Control, and Dynamics*, vol. 29, no. 5, pp. 1092–1100, 2006.

[27] J. Osborne and R. Rysdyk, "Waypoint guidance for small UAVs in wind," in *Proceedings of the AIAA Infotech@Aerospace Conference*, September 2005.

[28] G. F. Franklin, J. D. Powell, and M. Workman, *Digital Control of Dynamic Systems*. Menlo Park, CA: Addison Wesley, 3rd ed., 1998.

[29] R. W. Beard, "Embedded UAS autopilot and sensor systems," in *Encyclopedia of Aerospace Engineering* (R. Blockley and W. Shyy, eds.), pp. 4799–4814. Chichester, UK: John Wiley & Sons, Ltd, 2010.

[30] G. F. Franklin, J. D. Powell, and A. Emami-Naeini, *Feedback Control of Dynamic Systems*. Menlo Park, CA: Addison Wesley, 4th ed., 2002.

[31] "U.S. standard atmosphere, 1976." U.S. Government Printing Office, Washington, D.C., 1976.

[32] National Oceanic and Atmospheric Administration, "The world magnetic model." http://www.ngdc.noaa.gov/geomag/WMM/, 2011.

[33] J.-M. Zogg, *GPS: Essentials of Satellite Navigation*. http://zogg-jm.ch/Dateien/GPS_Compendium(GPS-x-02007).pdf, u-blox AG, 2009.

[34] B. W. Parkinson, J. J. Spilker, P. Axelrad, and P. Enge, eds., *Global Positioning System: Theory and Applications*. Reston, VA: American Institute for Aeronautics and Astronautics, 1996.

[35] E. D. Kaplan, ed., *Understanding GPS: Principles and Applications*. Norwood, MA: Artech House, 1996.

[36] M. S. Grewal, L. R. Weill, and A. P. Andrews, *Global Positioning Systems, Inertial Navigation, and Integration*. New Jersey: John Wiley & Sons, 2nd ed., 2007.

[37] J. Rankin, "GPS and differential GPS: An error model for sensor simulation," in *Position, Location, and Navigation Symposium*, pp. 260–266, 1994.

[38] R. Figliola and D. Beasley, *Theory and Design for Mechanical Measurements*. New York: John Wiley & Sons, Inc., 2006.

[39] S. D. Senturia, *Microsystem Design*. Dordrecht, Neth.: Kluwer Academic Publishers, 2001.

[40] J. W. Gardner, V. K. Varadan, and O. O. Awadelkarim, *Microsensors, MEMS, and Smart Devices*. New York: John Wiley & Sons, 2001.

[41] V. Kaajakari, *Practical MEMS: Design of Microsystems, Accelerometers, Gyroscopes, RF MEMS, Optical MEMS, and Microfluidic Systems*. Small Gear Publishing, 2009.

[42] R. E. Kalman, "A new approach to linear filtering and prediction problems," *Transactions of the ASME, Journal of Basic Engineering*, vol. 82, pp. 35–45, 1960.

[43] F. L. Lewis, *Optimal Estimation: With an Introduction to Stochastic Control Theory*. New York: John Wiley & Sons, 1986.

[44] A. Gelb, ed., *Applied Optimal Estimation*. Cambridge, MA: MIT Press, 1974.

[45] B.D.O. Anderson and J. B. Moore, *Linear Optimal Control*. Englewood Cliffs, NJ: Prentice Hall, 1971.

[46] R. G. Brown, *Introduction to Random Signal Analysis and Kalman Filtering*. New York: John Wiley & Sons, Inc., 1983.

[47] A. M. Eldredge, "Improved state estimation for miniature air vehicles," Master's thesis, Brigham Young University, 2006.

[48] R. W. Beard, "State estimation for micro air vehicles," in *Innovations in Intelligent Machines I*, J. S. Chahl, L. C. Jain, A. Mizutani, and M. Sato-Ilic, eds., pp. 173–199, Berlin Heidelberg: Springer Verlag, 2007.

[49] A. D. Wu, E. N. Johnson, and A. A. Proctor, "Vision-aided inertial navigation for flight control," *Journal of Aerospace Computing, Information, and Communication*, vol. 2, pp. 348–360, September 2005.

[50] T. P. Webb, R. J. Prazenica, A. J. Kurdila, and R. Lind, "Vision-based state estimation for autonomous micro air vehicles," *AIAA Journal of Guidance, Control, and Dynamics*, vol. 30, May–June 2007.

[51] S. Ettinger, M. Nechyba, P. Ifju, and M. Waszak, "Vision-guided flight stability and control for micro air vehicles," *Advanced Robotics*, vol. 17, no. 3, pp. 617–640, 2003.

[52] J. D. Anderson, *Introduction to Flight*. McGraw Hill, 1989.

[53] D. R. Nelson, D. B. Barber, T. W. McLain, and R. W. Beard, "Vector field path following for miniature air vehicles," *IEEE Transactions on Robotics*, vol. 37, pp. 519–529, June 2007.

[54] D. R. Nelson, D. B. Barber, T. W. McLain, and R. W. Beard, "Vector field path following for small unmanned air vehicles," in *American Control Conference*, (Minneapolis, MN), pp. 5788–5794, June 2006.

[55] D. A. Lawrence, E. W. Frew, and W. J. Pisano, "Lyapunov vector fields for autonomous unmanned aircraft flight control," *AIAA Journal of Guidance, Control, and Dynamics*, vol. 31, pp. 1220–1229, September–October 2008.

[56] O. Khatib, "Real-time obstacle avoidance for manipulators and mobile robots," in *Proceedings of the IEEE International Conference on Robotics and Automation*, vol. 2, pp. 500–505, April 1985.

[57] K. Sigurd and J. P. How, "UAV trajectory design using total field collision avoidance," in *Proceedings of the AIAA Guidance, Navigation and Control Conference*, August 2003.

[58] S. Park, J. Deyst, and J. How, "A new nonlinear guidance logic for trajectory tracking," in *Proceedings of the AIAA Guidance, Navigation and Control Conference*, AIAA 2004-4900, August 2004.

[59] I. Kaminer, A. Pascoal, E. Hallberg, and C. Silvestre, "Trajectory tracking for autonomous vehicles: An integrated approach to guidance and control," *AIAA Journal of Guidance, Control and Dynamics*, vol. 21, no. 1, pp. 29–38, January–February 1998.

[60] T. W. McLain and R. W. Beard, "Coordination variables, coordination functions, and cooperative timing missions," *AIAA Journal of Guidance, Control and Dynamics*, vol. 28, no. 1, pp. 150–161, January 2005.

[61] L. E. Dubins, "On curves of minimal length with a constraint on average curvature, and with prescribed initial and terminal positions and tangents," *American Journal of Mathematics*, vol. 79, no. 3, pp. 497–516, July 1957.

[62] E. P. Anderson, R. W. Beard, and T. W. McLain, "Real time dynamic trajectory smoothing for uninhabited aerial vehicles," *IEEE Transactions on Control Systems Technology*, vol. 13, pp. 471–477, May 2005.

[63] G. Yang and V. Kapila, "Optimal path planning for unmanned air vehicles with kinematic and tactical constraints," in *Proceedings of the IEEE Conference on Decision and Control*, (Las Vegas, NV), pp. 1301–1306, December 2002.

[64] P. Chandler, S. Rasumussen, and M. Pachter, "UAV cooperative path planning," in *Proceedings of the AIAA Guidance, Navigation, and Control Conference*, (Denver, CO), AIAA 2000-4370, August 2000.

[65] D. Hsu, R. Kindel, J.-C. Latombe, and S. Rock, "Randomized kinodynamic motion planning with moving obstacles," in *Algorithmic and Computational Robotics: New Directions*, pp. 247–264f. Natick, MA: A. K. Peters, 2001.

[66] F. Lamiraux, S. Sekhavat, and J.-P. Laumond, "Motion planning and control for Hilare pulling a trailer," *IEEE Transactions on Robotics and Automation*, vol. 15, pp. 640–652, August 1999.

[67] R. M. Murray and S. S. Sastry, "Nonholonomic motion planning: Steering using sinusoids," *IEEE Transactions on Automatic Control*, vol. 38, pp. 700–716, May 1993.

[68] T. Balch and R. C. Arkin, "Behavior-based formation control for multirobot teams," *IEEE Transactions on Robotics and Automation*, vol. 14, pp. 926–939, December 1998.

[69] R. C. Arkin, *Behavior-based Robotics*. Cambridge, MA: MIT Press, 1998.

[70] R. Sedgewick, *Algorithms*. Addison-Wesley, 2nd ed., 1988.

[71] F. Aurenhammer, "Voronoi diagrams – a survey of fundamental geometric data struct," *ACM Computing Surveys*, vol. 23, pp. 345–405, September 1991.

[72] T. H. Cormen, C. E. Leiserson, and R. L. Rivest, *Introduction to Algorithms*. New York: McGraw-Hill, 2002.

[73] T. K. Moon and W. C. Stirling, *Mathematical Methods and Algorithms*. Englewood Cliffs, NJ: Prentice Hall, 2000.

[74] S. M. LaValle and J. J. Kuffner, "Randomized kinodynamic planning," *International Journal of Robotic Research*, vol. 20, pp. 378–400, May 2001.

[75] J.-C. Latombe, *Robot Motion Planning*. Dordrecht, Neth.: Kluwer Academic Publishers, 1991.

[76] H. Choset, K. M. Lynch, S. Hutchinson, G. Kantor, W. B. and Lydia E. Kavraki, and S. Thrun, *Principles of Robot Motion: Theory, Algorithms, and Implementation*. Cambridge, MA: MIT Press, 2005.

[77] S. M. LaValle, *Planning Algorithms*. Cambridge University Press, 2006.

[78] T. McLain and R. Beard, "Cooperative rendezvous of multiple unmanned air vehicles," in *Proceedings of the AIAA Guidance, Navigation and Control Conference*, (Denver, CO), AIAA 2000-4369, August 2000.

[79] T. W. McLain, P. R. Chandler, S. Rasmussen, and M. Pachter, "Cooperative control of UAV rendezvous," in *Proceedings of the American Control Conference*, (Arlington, VA), pp. 2309–2314, June 2001.

[80] R. W. Beard, T. W. McLain, M. Goodrich, and E. P. Anderson, "Coordinated target assignment and intercept for unmanned air vehicles," *IEEE Transactions on Robotics and Automation*, vol. 18, pp. 911–922, December 2002.

[81] H. Choset and J. Burdick, "Sensor-based exploration: The hierarchical generalized Voronoi graph," *The International Journal of Robotic Research*, vol. 19, pp. 96–125, February 2000.

[82] H. Choset, S. Walker, K. Eiamsa-Ard, and J. Burdick, "Sensor-based exploration: Incremental construction of the hierarchical generalized Voronoi graph," *The International Journal of Robotics Research*, vol. 19, pp. 126–148, February 2000.

[83] D. Eppstein, "Finding the k shortest paths," *SIAM Journal of Computing*, vol. 28, no. 2, pp. 652–673, 1999.

[84] S. M. LaValle, "Rapidly-exploring random trees: A new tool for path planning." TR 98-11, Computer Science Dept., Iowa State University, October 1998.

[85] J. J. Kuffner and S. M. LaValle, "RRT-connect: An efficient approach to single-query path planning," in *Proceedings of the IEEE International Conference on Robotics and Automation*, (San Francisco, CA), pp. 995–1001, April 2000.

[86] M. Zucker, J. Kuffner, and M. Branicky, "Multipartite RRTs for rapid replanning in dynamic environments," in *Proceedings of the IEEE International Conference on Robotics and Automation*, (Rome, Italy), April 2007.

[87] S. Karaman and E. Frazzoli, "Incremental sampling-based algorithms for optimal motion planning," *International Journal of Robotic Research*, (in review).

[88] A. Ladd and L. E. Kavraki, "Generalizing the analysis of PRM," in *Proceedings of the IEEE International Conference on Robotics and Automation*, (Washington, DC), pp. 2120–2125, May 2002.

[89] E. Frazzoli, M. A. Dahleh, and E. Feron, "Real-time motion planning for agile autonomous vehicles," *Journal of Guidance, Control, and Dynamics*, vol. 25, pp. 116–129, January–February 2002.

[90] E. U. Acar, H. Choset, and J. Y. Lee, "Sensor-based coverage with extended range detectors," *IEEE Transactions on Robotics*, vol. 22, pp. 189–198, February 2006.

[91] C. Luo, S. X. Yang, D. A. Stacey, and J. C. Jofriet, "A solution to vicinity problem of obstacles in complete coverage path planning," in *Proceedings of the IEEE International Conference on Robotics and Automation*, (Washington DC), pp. 612–617, May 2002.

[92] Z. J. Butler, A. A. Rizzi, and R. L. Hollis, "Cooperative coverage of rectilinear environments," in *Proceedings of the IEEE International Conference on Robotics and Automation*, (San Francisco, CA), pp. 2722–2727, April 2000.

[93] J. Cortes, S. Martinez, T. Karatas, and F. Bullo, "Coverage control for mobile sensing networks," in *Proceedings of the IEEE International Conference on Robotics and Automation*, (Washington, DC), pp. 1327–1332, May 2002.

[94] M. Schwager, J.-J. Slotine, and J. J. Daniela Russell, "Consensus learning for distributed coverage control," in *Proceedings of the International Conference on Robotics and Automation*, (Pasadena, CA), pp. 1042–1048, May 2008.

[95] J. H. Evers, "Biological inspiration for agile autonomous air vehicles," in *Symposium on Platform Innovations and System Integration for Unmanned Air, Land, and Sea Vehicles*, (Florence, Italy), NATO Research and Technology Organization AVT-146, paper no. 15, May 2007.

[96] D. B. Barber, J. D. Redding, T. W. McLain, R. W. Beard, and C. N. Taylor, "Vision-based target geo-location using a fixed-wing miniature air vehicle," *Journal of Intelligent and Robotic Systems*, vol. 47, pp. 361–382, December 2006.

[97] Y. Ma, S. Soatto, J. Kosecka, and S. Sastry, *An Invitation to 3-D Vision: From Images to Geometric Models*. New York: Springer-Verlag, 2003.

[98] P. Zarchan, *Tactical and Strategic Missile Guidance*, vol. 124 of *Progress in Astronautics and Aeronautics*. Washington, DC: American Institute of Aeronautics and Astronautics, 1990.

[99] M. Guelman, M. Idan, and O. M. Golan, "Three-dimensional minimum energy guidance," *IEEE Transactions on Aerospace and Electronic Systems*, vol. 31, pp. 835–841, April 1995.

[100] J. G. Lee, H. S. Han, and Y. J. Kim, "Guidance performance analysis of bank-to-turn (BTT) missiles," in *Proceedings of the IEEE International Conference on Control Applications*, (Kohala, HI), pp. 991–996, August 1999.

[101] E. Frew and S. Rock, "Trajectory generation for monocular-vision based tracking of a constant-velocity target," in *Proceedings of the 2003 IEEE International Conference on Robotics and Automation*, (Taipei, Taiwan), September 2003.

[102] R. Kumar, S. Samarasekera, S. Hsu, and K. Hanna, "Registration of highly-oblique and zoomed in aerial video to reference imagery," in *Proceedings of the IEEE Computer Society Computer Vision and Pattern Recognition Conference*, (Barcelona, Spain), June 2000.

[103] D. Lee, K. Lillywhite, S. Fowers, B. Nelson, and J. Archibald, "An embedded vision system for an unmanned four-rotor helicopter," in *SPIE Optics East, Intelligent Robots and Computer Vision XXIV: Algorithms, Techniques, and Active Vision*, vol. 6382-24, 63840G, (Boston, MA), October 2006.

[104] J. Lopez, M. Markel, N. Siddiqi, G. Gebert, and J. Evers, "Performance of passive ranging from image flow," in *Proceedings of the IEEE International Conference on Image Processing*, vol. 1, pp. 929–932, September 2003.

[105] M. Pachter, N. Ceccarelli, and P. R. Chandler, "Vision-based target geolocation using camera equipped MAVs," in *Proceedings of the IEEE Conference on Decision and Control*, (New Orleans, LA), December 2007.

[106] R. J. Prazenica, A. J. Kurdila, R. C. Sharpley, P. Binev, M. H. Hielsberg, J. Lane, and J. Evers, "Vision-based receding horizon control for micro air vehicles in urban environments," *AIAA Journal of Guidance, Dynamics, and Control*, (in review).

[107] I. Wang, V. Dobrokhodov, I. Kaminer, and K. Jones, "On vision-based target tracking and range estimation for small UAVs," in *AIAA Guidance, Navigation, and Control Conference and Exhibit*, pp. 1–11, August 2005.

[108] Y. Watanabe, A. J. Calise, E. N. Johnson, and J. H. Evers, "Minimum-effort guidance for vision-based collision avoidance," in *Proceedings of the AIAA Atmospheric Flight Mechanics Conference and Exhibit*, (Keystone, Co), American Institute of Aeronautics and Astronautics, AIAA 2006-6608, August 2006.

[109] Y. Watanabe, E. N. Johnson, and A. J. Calise, "Optimal 3-D guidance from a 2-D vision sensor," in *Proceedings of the AIAA Guidance, Navigation, and Control Conference*, (Providence, RI), American Institute of Aeronautics and Astronautics, AIAA 2004-4779, August 2004.

[110] I. H. Whang, V. N. Dobrokhodov, I. I. Kaminer, and K. D. Jones, "On vision-based tracking and range estimation for small UAVs," in *Proceedings of the AIAA Guidance, Navigation, and Control Conference and Exhibit*, (San Francisco, CA), August 2005.

[111] R. W. Beard, D. Lee, M. Quigley, S. Thakoor, and S. Zornetzer, "A new approach to observation of descent and landing of future Mars mission using bioinspired technology innovations," *AIAA Journal of Aerospace Computing, Information, and Communication*, vol. 2, no. 1, pp. 65–91, January 2005.

[112] M. E. Campbell and M. Wheeler, "A vision-based geolocation tracking system for UAVs," in *Proceedings of the AIAA Guidance, Navigation, and Control Conference and Exhibit*, (Keystone, Co), AIAA 2006-6246, August 2006.

[113] V. N. Dobrokhodov, I. I. Kaminer, and K. D. Jones, "Vision-based tracking and motion estimation for moving targets using small UAVs," in *Proceedings of the AIAA Guidance, Navigation, and Control Conference and Exhibit*, (Keystone, Co), AIAA 2006-6606, August 2006.

[114] E. W. Frew, "Sensitivity of cooperative target geolocation to orbit coordination," *Journal of Guidance, Control, and Dynamics*, vol. 31, pp. 1028–1040, July–August 2008.

[115] D. Murray and A. Basu, "Motion tracking with an active camera," *IEEE Transactions on Pattern Analysis and Machine Intelligence*, vol. 16, pp. 449–459, May 1994.

[116] S. Hutchinson, G. D. Hager, and P. I. Corke, "A tutorial on visual servo control," *IEEE Transactions on Robotics and Automation*, vol. 12, pp. 651–670, October 1996.

[117] J. Oliensis, "A critique of structure-from-motion algorithms," *Computer Vision and Image Understanding (CVIU)*, vol. 80, no. 2, pp. 172–214, 2000.

[118] J. Santos-Victor and G. Sandini, "Uncalibrated obstacle detection using normal flow," *Machine Vision and Applications*, vol. 9, no. 3, pp. 130–137, 1996.

[119] L. Lorigo, R. Brooks, and W. Grimson, "Visually guided obstacle avoidance in unstructured environments," in *Proceedings of IROS '97*, (Grenoble, Fr.), September 1997.

[120] R. Nelson and Y. Aloimonos, "Obstacle avoidance using flow field divergence," *IEEE Transactions on Pattern Analysis and Machine Intelligence*, vol. 11, pp. 1102–1106, October 1989.

[121] F. Gabbiani, H. Krapp, and G. Laurent, "Computation of object approach by a wide field visual neuron," *Journal of Neuroscience*, vol. 19, no. 3, pp. 1122–1141, February 1999.

[122] R. W. Beard, J. W. Curtis, M. Eilders, J. Evers, and J. R. Cloutier, "Vision-aided proportional navigation for micro air vehicles," in *Proceedings of the AIAA Guidance, Navigation and Control Conference*, (Hilton Head, NC), American Institute of Aeronautics and Astronautics, AIAA 2007-6609, August 2007.

[123] A. E. Bryson and Y. C. Ho, *Applied Optimal Control*. Waltham, MA: Blaisdell Publishing Company, 1969.

[124] M. Guelman, "Proportional navigation with a maneuvering target," *IEEE Transactions on Aerospace and Electronic Systems*, vol. 8(3), pp. 364–371, May 1972.

[125] C. F. Lin, *Modern Navigation, Guidance, and Control Processing*. Englewood Cliffs, NJ: Prentice Hall, 1991.

[126] J. Waldmann, "Line-of-sight rate estimation and linearizing control by an imaging seeker in a tactical missile guided by proportional navigation," *IEEE Transactions on Control Systems Technology*, vol. 10, pp. 556–567, July 2002.

[127] J. B. Kuipers, *Quaternions and Rotation Sequences: A Primer with Applications to Orbits, Aerospace, and Virtual Reality*. Princeton, NJ: Princeton University Press, 1999.

[128] J. P. Corbett and F. B. Wright, "Stabilization of computer circuits," in *WADC TR 57-25* (E. Hochfeld, ed.), (Wright-Patterson Air Force Base, OH), 1957.

[129] G. Platanitis and S. Shkarayev, "Integration of an autopilot for a micro air vehicle," in *Infotech@Aerospace*, AIAA 2005-7066, September 2005.

[130] R. Rysdyk, "Course and heading changes in significant wind," in *AIAA Journal of Guidance, Control, and Dynamics*, vol. 33, no. 4, pp. 1311–12, July-August 2010.

Index

Absolute Pressure Sensor, 145
Absolution Pressure Sensor, 126
Accelerometers, 120, 147, 156
Aerodynamic Coefficients, 51
Aileron, 42
Airspeed, 18, 20, 22, 54, 57
Altitude, 62
Angle of Attack, 16, 20, 41, 57
Autopilot; Airspeed Hold using Pitch, 110; Airspeed Hold using Throttle, 111; Altitude Hold Using Pitch, 108; Course Hold, 102; Longitudional, 106; Pitch Attitude Hold, 106; Roll Attitude Hold, 99; Side Slip Hold, 104

Bandwidth Separation, 104

Coordinate Frames, 8, 226; Body Frame, 14, 227; Camera Frame, 227; Gimbal Frame, 227; Gimbal-1 Frame, 226; Inertial Frame, 12; Stability Frame, 16; Vehicle Frame, 12; Vehicle-1 Frame, 13; Vehicle-2 Frame, 14; Wind Frame, 17
Coordinated Turn, 64, 69, 166, 167
Course Angle, 20, 21, 140
Crab Angle, 20, 22

Design Models, 4, 60, 164
Differential Pressure Sensor, 129, 146
Dubins Path; Computation, 195; Definition, 194; RRT Coverage, 222; RRT Paths through Obstacle Field, 217; Tracking, 200
Dutch-roll Mode, 91
Dynamics; Guidance Model, 171; Rotational Motion, 33; Translational Motion, 32

Ego Motion, 234
Elevator, 42
Estiamtion; Heading, 158
Estimation; Airspeed, 146; Altitude, 145; Angular Rates, 145; Course, 148, 158; Groundspeed, 148; Position, 148, 158; Roll and Pitch Angles, 147, 156; Wind, 158
Euler Angles, 12, 29; Heading (Yaw)), 13; Pitch, 14; Roll, 14

Fillet, 189

Flat Earth Model, 231, 239
Flight Path Angle; Inertial Referenced, 20, 165, 167
Flight-path Angle; Air Mass Referenced, 22
Focal Length, 228
Forces; Drag, 41, 44, 45, 47; Gravitation, 40; Lift, 41, 44, 47; Propulsion, 53

Gimbal; Azimuth Angle, 227, 231; Dynamics, 229; Elevation Angle, 227, 231
GPS, 134, 158
Ground Speed, 18, 22, 54, 140

Half Plane, 188, 191, 200
Heading Angle, 13, 20, 23

Inertial Matrix, 35

Kalman Filter; Basic Explanation, 149; Continuous-Discrete to Estimate Roll and Pitch, 156; Continuous-Discrete to Position, Course, Wind, Heading, 158; Derivation, 151; Extended Kalman Filter, 156; Geolocation, 232
Kinematics; Guidance Model, 168; Position, 30, 61, 165, 166; Rotation, 31, 61

Lateral Motion, 68, 78, 99
Linearization, 78
Load Factor, 165
Longitudinal Motion, 43, 50, 71, 82, 105
Low Pass Filter, 234
Low-pass Filter, 144

Minimum Turning Radius, 65

Normalized Line of Sight Vector, 229

$\mathcal{P}_{\text{line}}$, 175
$\mathcal{P}_{\text{orbit}}$, 181
Path Following, 174
Phugoid Mode, 89
PID: Digital Implementation, 114
Pitch Angle, 14
Pitching Moment, 44, 45
Pitot Tube, 130
Precision Landing, 240
Proportional Navigation, 240, 242

Rapidly Exploring Random Trees (RRT), 212; 3-D Terrain, 216; Point to Point Algorithm, 215; Smoothing Algorithm, 217; Using Dubins Paths, 217
Rate Gyros, 124, 145, 156
Right Handed Rotation, 10
Roll Angle, 14
Roll Mode, 90
Rotation Matrices, 9; Body to Gimbal, 227; Body to Gimbal-1, 227; Body to Stability, 17; Body to Wind, 18; Gimbal to Camera, 227; Gimbal-1 to Gimbal, 227; Stability to Wind, 18; Vehicle to Body, 15; Vehicle to Vehicle-1, 14; Vehicle-1 to Vehicle-2, 14; Vehicle-2 to Body, 15
Rudder, 42

Saturation Constraints, 97
Short-period Mode, 87
Side Slip Angle, 17, 20, 41, 57, 71
Simulation Model, 4, 61, 277
Spiral-divergence Mode, 90
Stall, 46
State-space Models; Lateral, 78; Longitudinal, 82
State Variables, 28
Successive Loop Closure, 95

Time to Collision, 238
Transfer Functions; Aileron to Roll, 69, 99; Airspeed to Altitude, 74; Elevator to Pitch, 73, 106; Pitch to Airspeed, 110; Pitch to Altitude, 74, 108; Roll to Course, 70, 102; Roll to Heading, 70; Rudder to Side Slip, 71, 104; Throttle to Airspeed, 77, 111
Trim, 65, 78

Vector Field; Orbit, 183; Straight Line, 179
Voronoi Path Planning, 207

Waypoint Configuration, 200
Waypoint Path, 187
Wind Gusts, 54; Dryden Model, 55
Wind Speed, 18, 22, 54
Wind Triangle, 20, 22, 160